Selected Titles in This Series

165 **V. V. Ishkhanov, B. B. Lur′e, and D. K. Faddeev,** The embedding problem in Galois theory, 1997

164 **E. I. Gordon,** Nonstandard methods in commutative harmonic analysis, 1997

163 **A. Ya. Dorogovtsev, D. S. Silvestrov, A. V. Skorokhod, and M. I. Yadrenko,** Probability theory: Collection of problems, 1997

162 **M. V. Boldin, G. I. Simonova, and Yu. N. Tyurin,** Sign-based methods in linear statistical models, 1997

161 **Michael Blank,** Discreteness and continuity in problems of chaotic dynamics, 1997

160 **V. G. Osmolovskiĭ,** Linear and nonlinear perturbations of the operator div, 1997

159 **S. Ya. Khavinson,** Best approximation by linear superpositions (approximate nomography), 1997

158 **Hideki Omori,** Infinite-dimensional Lie groups, 1997

157 **V. B. Kolmanovskiĭ and L. E. Shaĭkhet,** Control of systems with aftereffect, 1996

156 **V. N. Shevchenko,** Qualitative topics in integer linear programming, 1997

155 **Yu. Safarov and D. Vassiliev,** The asymptotic distribution of eigenvalues of partial differential operators, 1997

154 **V. V. Prasolov and A. B. Sossinsky,** Knots, links, braids and 3-manifolds. An introduction to the new invariants in low-dimensional topology, 1997

153 **S. Kh. Aranson, G. R. Belitsky, and E. V. Zhuzhoma,** Introduction to the qualitative theory of dynamical systems on surfaces, 1996

152 **R. S. Ismagilov,** Representations of infinite-dimensional groups, 1996

151 **S. Yu. Slavyanov,** Asymptotic solutions of the one-dimensional Schrödinger equation, 1996

150 **B. Ya. Levin,** Lectures on entire functions, 1996

149 **Takashi Sakai,** Riemannian geometry, 1996

148 **Vladimir I. Piterbarg,** Asymptotic methods in the theory of Gaussian processes and fields, 1996

147 **S. G. Gindikin and L. R. Volevich,** Mixed problem for partial differential equations with quasihomogeneous principal part, 1996

146 **L. Ya. Adrianova,** Introduction to linear systems of differential equations, 1995

145 **A. N. Andrianov and V. G. Zhuravlev,** Modular forms and Hecke operators, 1995

144 **O. V. Troshkin,** Nontraditional methods in mathematical hydrodynamics, 1995

143 **V. A. Malyshev and R. A. Minlos,** Linear infinite-particle operators, 1995

142 **N. V. Krylov,** Introduction to the theory of diffusion processes, 1995

141 **A. A. Davydov,** Qualitative theory of control systems, 1994

140 **Aizik I. Volpert, Vitaly A. Volpert, and Vladimir A. Volpert,** Traveling wave solutions of parabolic systems, 1994

139 **I. V. Skrypnik,** Methods for analysis of nonlinear elliptic boundary value problems, 1994

138 **Yu. P. Razmyslov,** Identities of algebras and their representations, 1994

137 **F. I. Karpelevich and A. Ya. Kreinin,** Heavy traffic limits for multiphase queues, 1994

136 **Masayoshi Miyanishi,** Algebraic geometry, 1994

135 **Masaru Takeuchi,** Modern spherical functions, 1994

134 **V. V. Prasolov,** Problems and theorems in linear algebra, 1994

133 **P. I. Naumkin and I. A. Shishmarev,** Nonlinear nonlocal equations in the theory of waves, 1994

132 **Hajime Urakawa,** Calculus of variations and harmonic maps, 1993

131 **V. V. Sharko,** Functions on manifolds: Algebraic and topological aspects, 1993

130 **V. V. Vershinin,** Cobordisms and spectral sequences, 1993

129 **Mitsuo Morimoto,** An introduction to Sato's hyperfunctions, 1993

128 **V. P. Orevkov,** Complexity of proofs and their transformations in axiomatic theories, 1993

(*Continued in the back of this publication*)

The Embedding Problem in Galois Theory

Translations of

MATHEMATICAL MONOGRAPHS

Volume 165

The Embedding Problem in Galois Theory

V. V. Ishkhanov

B. B. Lur'e

D. K. Faddeev

American Mathematical Society

Providence, Rhode Island

В. В. Ишханов
Б. Б. Лурье
Д. К. Фаддеев

ЗАДАЧА ПОГРУЖЕНИЯ В ТЕОРИИ ГАЛУА

«Наука», Москва, 1990

Translated from the Russian by N. B. Lebedinskaya

1991 *Mathematics Subject Classification.* Primary 12F12; Secondary 11R32, 11S20.

ABSTRACT. The "embedding problem" is a part of modern algebra that is related to the arithmetics of algebraic number fields, theory of algebras, group homology, and the inverse Galois problem. In the book the authors give formulations of the embedding problem and present certain necessary and sufficient conditions for its solvability. The cases of local and global number fields are treated in more detail.

The book can be used by researchers and graduate students working in algebra and number theory.

Library of Congress Cataloging-in-Publication Data

Ishkhanov, V. V. (Vladimir Vaganovich)
[Zadacha pogruzheniia v teorii Galua. English]
The embedding problem in Galois theory / V. V. Ishkhanov, B. B. Lur′e, D. K. Faddeev ; [translated from the Russian by N. B. Lebedinskaya].
p. cm. — (Translations of mathematical monographs, ISSN 0065-9282 ; v. 165)
Includes bibliographical references and index.
ISBN 0-8218-4592-6 (hc : alk. paper)
1. Galois theory. I. Lur′e, B. B. (Boris Beniaminovich) II. Faddeev, D. K. (Dmitriĭ Konstantinovich) III. Title. IV. Series.
QA214.I8413 1997
512′.3—dc21 97-12187
CIP

---

10 9 8 7 6 5 4 3 2 1 02 01 00 99 98 97

Contents

Foreword

Parascientific folklore contains the following scheme parodying a certain kind of scientific research. Consider a difficult problem A. Rather than solve it as such, consider a more general and more difficult problem B. Next, weaken the requirements in the problem B and try to solve the problem C obtained in such a way. We shall call this activity an *approach to the problem* A, even though "the intersection" of A and C is empty.

From the formal point of view the embedding problem fits in the above scheme. The central problem of modern Galois theory involves the inverse problem that is as follows: given a field k and a group G, construct an extension L/k with Galois group G. The embedding problem for fields generalizes the inverse problem and consists in finding the conditions under which one can construct a field L normal over k, with group G, such that L extends a given normal extension K/k with Galois group G/A. Moreover, the requirements applied to the object L to be found are usually weakened: it is not necessary for L to be a field, but L must be a Galois algebra over the field k, with group G. In this setting the embedding problem is rich in content. But the inverse problem in terms of Galois algebras is poor in content because a Galois algebra providing a solution of the inverse problem always exists and may be easily constructed.

Nevertheless, the embedding problem is by no means a whim but a fruitful approach to the solution of the inverse problem in Galois theory. We hope that the reader of the book will gain the evidence of this. Here we indicate only a few arguments that confirm this thesis.

1. If the group G contains a normal subgroup A, then a natural approach to solving the inverse problem for the field k and the group G lies in choosing a field K that extends the field k with Galois group G/A such that it can in turn be embedded in an appropriate extension L. We see that the embedding problem naturally arises while solving the inverse problem (historically, it originated in just this manner).

2. In trying to choose a field among the Galois algebras that solve the embedding problem under consideration, we usually must overcome difficulties that are of a technical but not a fundamental nature, because the request that the solution be a field imposes inequality-type conditions.

3. Finally, several basic results in Galois theory are obtained by way of solving the embedding problem. The solution of the inverse problem for solvable groups over algebraic number fields that is obtained in papers by Shafarevich is primarily such a result.

The embedding problem has a rather long history. As early as the 1920s, Scholz established the solvability of the embedding problem for the algebraic number fields in the case where the group G is a semidirect product of an Abelian group A and the complementary subgroup. In 1932, under rather severe restrictions on the field K and the group G, Brauer found a necessary and sufficient condition for the solvability of the embedding problem in the case where A is Abelian (A is called the *kernel* of the embedding problem).

In 1944, in a joint paper by Delauné and Faddeev, a necessary condition of embeddability called the *compatibility condition* was formulated. Under the assumptions of Brauer, this condition turns out to coincide with Brauer's necessary and sufficient condition. In 1947, the compatibility condition was rediscovered by Hasse in somewhat different form. Hasse also conjectured that this condition is not only necessary but also sufficient for embeddability. Hasse's conjecture was disproved in two different ways by Faddeev and Shafarevich in 1954.

We have mentioned above the achievements of Shafarevich in the solution of the inverse problem during the 1950s.

In homological terms a complete solution of the embedding problem with Abelian kernel was given by Yakovlev in 1963.

Embedding theory was a common topic of Faddeev's lecture courses at St. Petersburg University. This book is based on these lectures and contains the main results on the embedding problem, which belong for the most part to the schools of Faddeev and Shafarevich.

In embedding theory research tools from the theory of algebras and homological algebra are used successfully. In our book both of these approaches are introduced.

In Chapter 1, the reader is acquainted with the formulation of the embedding problem (more precisely, with various formulations) and with the most general techniques applied to the embedding problem.

In Chapter 2, the compatibility condition is stated and studied; it is an important necessary embedding condition. The first sections, like Chapter 1, are general in nature. In the last sections it is shown that in the most interesting cases the compatibility condition is reduced to rather simply verifiable conditions (by Lur′e).

In Chapter 3, the embedding problem with Abelian kernel is studied. Here, the case investigated by Brauer is considered; reduction theorems (by Kochendörffer and Faddeev) that enable one to reduce the embedding problem to p-groups are proved. It is proved that the embedding problem for local fields has a positive solution (by Demushkin and Shafarevich). Central to the chapter is the theory devised by Yakovlev.

Chapter 4 is devoted to the study of the embedding problem with noncommutative p-kernel for local fields. Under certain additional assumptions it can be reduced to the problem with commutative kernel. The embedding problem that has a solution in the proper sense (the solution is a field) is also studied.

Chapter 5 is concerned with the embedding problem for algebraic number fields (the kernel is not assumed to be Abelian). In the first two sections the problem with noncommutative kernel of order p^3 is solved (by Lur′e). Next, a theorem by Neukirch (see §4) is proved that yields a complete solution of the embedding problem with solvable kernel of odd order in the absence, in the field to be embedded, of primitive roots of unity of degree dividing the period of the kernel. The concluding sections of the chapter are devoted to studying the results on the inverse problem of Galois theory in the case of solvable groups of arbitrary order obtained by Shafarevich. A simpler proof of the Shafarevich theorem based on the techniques of Galois cohomology is presented. In addition, in the last section the embedding problem in which the order and the index of the kernel are relatively prime is solved in the proper sense (by Ishkhanov).

The exposition of the material in the book assumes the knowledge of the bases of algebra, representation theory, algebraic number theory, and technique of Galois cohomology. References [**1, 4, 6, 29**] provide the necessary background. Other information not related immediately to the embedding problem is included in the Appendix.

Generally accepted notation (for example, $\mathbb{Z}$ is the ring of integers and so on) is not specifically stipulated.

The authors are indebted to Professor Yakovlev for useful discussions and also to Professor Oskolkov for his help in preparing the manuscript for publication.

* * *

While this book was in press, our teacher, Associated Member of the Russian Academy of Sciences Dmitrii Konstantinovich Faddeev, passed away.

Faddeev was one of the founders of embedding theory and an initiator in the writing of this book. He wrote a major part of the text and coordinated the efforts of the authors.

Since this foreword was prepared with the participation of Faddeev, his basic contribution to the embedding problem was not reflected adequately.

Dmitrii Konstantinovich Faddeev, eminent scientist and wonderful person, will always remain in the memory of those who lived and worked with him.

V. Ishkhanov, B. Lur′e

CHAPTER 1

Preliminary Information about the Embedding Problem

§1. Statement of the problem

Let A be a normal extension of a field k with Galois group G. Let N be a normal subgroup of G. The field K of elements of A invariant with respect to automorphisms from N is a normal extension of k with group F isomorphic to G/N in a natural way; namely, under restriction to K the automorphisms of A from the cosets of G by N merge. The question arises: What conditions must the field K with group F isomorphic to G/N satisfy in order that it can be embedded in the field A with group G in the sense indicated above? The study of this question forms the subject of the embedding theory in Galois theory.

The embedding problem can also be stated as follows. Let a normal extension K of the field k with Galois group F and a surjective homomorphism φ of a group G onto F be given. Under what conditions can one construct an extension A of the field k containing K in such a way that G is the Galois group of the extension A/k and the homomorphism of restriction to K of the automorphisms from G coincides with φ?

Thus, the embedding problem is related to the exact sequence of groups

$$1 \to N \to G \xrightarrow{\varphi} F \to 1.$$

The kernel N of the mapping φ is called *the kernel of the embedding problem.* We denote such a problem by $(K/k, G, \varphi)$ or $(K/k, G, \varphi, N)$ if the indication to N is essential in the context. Sometimes it is expedient to omit the mapping φ, using the notation $(K/k, G, N)$.

Hereinafter we assume that G is finite, unless otherwise stated.

The conditions under which the embedding problem $(K/k, G, \varphi)$ is solvable are called *embedding conditions.* The notion of a solution of the embedding problem admits various formulations. In the least precise form this is merely an extension A/k without indicating how the elements of G act as automorphisms of A. On the other hand, a solution in the strict sense assumes an indication of not only a field A but also an isomorphism $G \to \mathrm{Gal}(A/k)$. The intermediate statement consists of presenting a field A and an isomorphism $N \to \mathrm{Gal}(A/k)$. A tantamount question can be formulated as follows. What solutions of the problem $(K/k, G, \varphi)$ may be treated as equivalent? For more details, see §§1.14 and 3.15.*

If $F = 1$, i.e., if the fields K and k coincide, then the embedding problem becomes the inverse problem in Galois theory, i.e., to the problem of the existence of a normal extension A/k with given Galois group G.

*The notation §1.14 means §14 of Chapter 1, and so on.—*Editor's remark.*

Consider the simplest example of the embedding problem. Let G be a cyclic group of fourth order: $G = \{g\}$, $g^4 = 1$. Let k be a field of characteristic different from 2 and $K = k(\sqrt{d})$ a quadratic extension of k. The epimorphism φ of G onto the Galois group of K/k has the group $N = \{g^2\}$ as kernel, and the elements g and g^3 induce the automorphism $\sqrt{d} \to -\sqrt{d}$ on K. We also assume that k does not contain $\sqrt{-1}$, otherwise the embedding problem can be solved trivially:

$$A = k(\sqrt[4]{d}).$$

Assume that the problem is solvable and A/k is a solution. Consider a primitive element θ of A and put $m = (\theta - \theta^{g^2})/(\theta^g - \theta^{g^3})$. By the primitivity of θ, the element m is well defined and is different from zero. Furthermore, $m^g = (\theta^g - \theta^{g^3})/(\theta^{g^2} - \theta) = -m^{-1}$, $m^{g^2} = -(m^g)^{-1} = m$, so that $m \in K$. Clearly, $m \cdot m^g = -1$, i.e., the norm of m is equal to -1.

It is almost obvious that in a quadratic extension $K = k(\sqrt{d})$ of k, an element with norm -1 exists if and only if d is representable as a sum of two squares of elements from k. Indeed, if $N(m) = -1$ and $m = x + y\sqrt{d}$, $x, y \in k$, then $y \neq 0$ and $d = (x/y)^2 + (1/y)^2$ in the relation $-1 = N(m) = x^2 - dy^2$. Thus, for the solvability of the problem it is necessary that d be representable as a sum of two squares. We show that this condition is also sufficient. Let $d = u^2 + v^2$, $u, v \in k$, $v \neq 0$. Then, putting $m = (u + \sqrt{d})/v$, we obtain $m \cdot m^g = -1$, where g is the automorphism $\sqrt{d} \to -\sqrt{d}$ of $K = k(\sqrt{d})$. We put $\lambda = 1 + m^2$. Then $\lambda \neq 0$ and $\lambda^g = 1 + 1/m^2 = \lambda/m^2$. Let $\theta = \sqrt{\lambda}$ and $A = K(\theta)$. Let $\bar{g}$ be an isomorphism of A extending the automorphism g of K. Then $(\theta^{\bar{g}})^2 = \lambda^g = \lambda/m^2 = (\theta/m)^2$, whence, up to sign, $\theta^{\bar{g}} = \theta/m \in A$. Thus, $\bar{g}$ is an automorphism of A/k. Further, $\theta^{\bar{g}^2} = \theta^{\bar{g}}/m^g = \theta/(m \cdot m^g) = -\theta$; $\theta^{\bar{g}^3} = -\theta/m$; $\theta^{\bar{g}^4} = \theta$, i.e., $\bar{g}^4 = 1$. Consequently, $A = K(\theta)$ is normal over k and has over k a cyclic group of order 4 as Galois group, which proves sufficiency.

Thus, even for the simplest problem under consideration the embedding condition is nontrivial.

§2. A module of the regular representation of a finite group

As is known (see [**6**]), in a normal extension K of k with Galois group G there exists a normal basis consisting of all elements a^g conjugate to a certain element $a \in K$. Thus, the field K as a vector space over k with operators from G is a module of the regular representation of the group G. In the sequel we shall deal with other objects which, as G-modules, also are modules of the regular representation. We present some properties of such modules as a theorem.

THEOREM 1.2. *Let G be a finite group, T a subgroup, A a module (over k) of the regular representation of G, B the subspace of T-invariant elements of A. Then*

(1) the dimension of B over k is equal to $(G:T)$;

(2) if $g \in G$ and $b^g = b$ for all $b \in B$, then $g \in T$;

(3) the subspace B is invariant with respect to arbitrary $g \in G$ if and only if T is an invariant subgroup of G. In this case, B is isomorphic to a module of the regular representation of the group G/T.

PROOF. Consider G as the union of cosets with respect to T; namely, $G = \bigcup_\Psi \Psi = \bigcup_\Psi \overline{\Psi}T$. Here we denote cosets by Ψ and their representatives by $\overline{\Psi}$.

Let $\{a^g\}$ be a normal basis for A, and let $x = \sum_{g\in G} c_g a^g$, $c_g \in k$. Then $x^h = \sum_{g\in G} c_g a^{gh} = \sum_{g\in G} c_{gh^{-1}} a^g$, and $x^h = x$ if and only if $c_g = c_{gh^{-1}}$. If this occurs for arbitrary $h \in T$, then $c_{g_1} = c_{g_2}$ provided that g_1 and g_2 belong to the same coset $\Psi = \overline{\Psi}T$. Conversely, if this condition is fulfilled, then the element x is T-invariant. It is clear from what we have said that the elements $b_\Psi = \sum_{g\in\Psi} a^g$ constitute a basis for the subspace B of T-invariant elements. Hence, the validity of (1) follows.

Now we assume that $h \in G$ leaves invariant all elements of B; for that it is necessary that all $b_\Psi = \sum_{g\in\Psi} a^g$ be h-invariant. This is the case if and only if $\Psi h = \Psi$ for any Ψ, i.e., $\overline{\Psi}Th = \overline{\Psi}T$, whence $Th = T$, i.e., $h \in T$. Thus, the proof of (2) is complete.

Finally, let B be invariant for any $g \in G$. This is equivalent to the fact that $b_\Psi^g \in B$ for any $g \in G$ and for any Ψ and, therefore, to the fact that the Ψg are right cosets of G with respect to T. Let $\Psi g = \overline{\Psi}Tg = lT$ for some $l \in G$. Then $\overline{\Psi}g \in lT$ and $lT = \overline{\Psi}gT$. Thus, $\overline{\Psi}Tg = \overline{\Psi}gT$ and $Tg = gT$. This is valid for all $g \in G$, whence T is a normal subgroup of G. The cosets Ψ are elements of the quotient group, and, clearly, if $g \in \Psi_1$ then $\Psi g = \Psi\Psi_1$. Thus, the elements $b_\Psi = \sum_{g\in\Psi} a^g$ move under the action of the elements from some coset Ψ_1 in the same way as the elements of the quotient group under multiplication by Ψ_1 from the right. Therefore, B is isomorphic to a module of the regular representation of G/T. The proof of (3) is complete. □

§3. S-Algebras

An associative and commutative finite-dimensional separable algebra A over a field k is called an *S-algebra.* As is known, such an algebra is isomorphic to a direct sum of separable extensions of the base field k. The components of the unit in the direct summands are units in these summands and constitute a system of minimal idempotents of the algebra such that the product of any two different idempotents vanishes. The sum of these idempotents is equal to the unit of the algebra. It is naturally identified with the unit of the field k and defines a standard inclusion of the field k in the algebra. Each of the fields isomorphic to direct summands of the algebra is isomorphic to the quotient ring of the polynomial ring $k[t]$ by an irreducible separable polynomial for which a primitive element of the corresponding field is a root. If the field k is infinite, then in any finite separable extension there are infinitely many primitive elements, so that the irreducible polynomials associated with the direct summands of the S-algebra may be chosen to be distinct even if the corresponding direct summands are isomorphic. Thus, the S-algebra A is also isomorphic to the quotient ring of the ring $k[t]$ by a certain, possibly reducible, polynomial without multiple roots in any extension of the field k.

The simplest S-algebra is a completely splitting algebra $A = \sum^\oplus ke_i$, where k is the base field and e_i are orthogonal idempotents. After tensoring an S-algebra A with a field L containing the splitting fields of all direct summands, A becomes the algebra $A \otimes L$ over L that splits completely.

§4. Galois algebras

An S-algebra A over a field k is called a *Galois algebra with group* G if there exists a homomorphism of G to the group of automorphisms of A over k and A has a G-normal basis over k. Thus, a Galois algebra is simultaneously an S-algebra and a k-module of the regular representation of G.

We observe that the indication of the group G in the definition of a Galois algebra is necessary: the same S-algebra can sometimes be provided with the structure of a Galois algebra for different groups. For example, if A is a completely splitting S-algebra, then its full group of automorphisms over k is the entire symmetric group of permutations of direct summands, so that A can be provided with the structure of a Galois algebra for any group of order equal to the number of direct summands.

A normal extension A of the field k with Galois group G is a Galois algebra by virtue of the existence of a normal basis. More general Galois algebras arise, in particular, from normal extensions when one extends the base field. More precisely, let A be a normal extension of k with group G and let L be an extension of k. Then $\tilde{A} = A \otimes L$ is an algebra over L (under the natural embedding of L in $\tilde{A}$), and the operators from G can be extended naturally to $\tilde{A}$ by the rule $(x \otimes c)^g = x^g \otimes c$ for $x \in A$, $c \in L$.

When we study the embedding problem it turns out to be reasonable to weaken somewhat the restrictions on the object we are looking for by requiring that it be not a field but only a Galois algebra. More precisely, it is worthwhile to formulate the problem of the embedding of a given field K/k with group F in a Galois algebra with group G for which F is the homomorphic image under φ in such a way that K coincides with the algebra of elements of A invariant under all automorphisms from the kernel of the homomorphism $\varphi\colon G \to F$.

When a solution of the embedding problem is a field, such a solution is said to be *proper*, and the embedding problem is called *solvable in the proper sense.*

The solution of the problem of the embedding in a Galois algebra admits a broad and a narrow interpretation, similar to the solution of the problem of the embedding in a field.

As a theorem, we state a property that completely characterizes Galois algebras, so that it can be taken for a definition.

THEOREM 1.4. *Let A be an S-algebra over a field k and assume that a homomorphism of the group G to the group of k-automorphisms of A is given. In order that A be a Galois algebra with group G it is necessary and sufficient that the dimension of A be equal to the order of G and the algebra of G-invariant elements of A coincide with k.*

PROOF. If A is a Galois algebra, then it is obvious that its dimension is equal to the order of G. Now let $a \in A$ be a generator of a normal basis, i.e., $\{a^g\}_{g \in G}$ is a basis of A over k. Let $x \in A$ be an invariant element, i.e., such an element that $x^g = x$ for arbitrary $g \in G$. We express it in terms of the basis: $x = \sum_{g \in G} a^g x_g$ for $x_g \in k$. Then $x^h = \sum_g a^{gh} x_g = \sum_g a^g x_{gh^{-1}}$. The relation $x^h = x$ is equivalent to the system of relations $x_g = x_{gh^{-1}}$. Since these relations must hold for all $g, h \in G$, we have $x = x_1 \sum_g a^g$, i.e., the algebra of G-invariant elements is one-dimensional and, therefore, it coincides with the field k if k is naturally embedded in A. Thus, the necessity is proved.

Now we prove the sufficiency. Let an S-algebra A satisfy the conditions $[A\!:\!k] = (G:1)$ and $A^G = k$.

First we assume that A splits completely, i.e., $A = \sum_{i=1}^{n}{}^{\oplus} e_i k$. Here, by the first condition, $n = (G:1)$ and the e_i are the components of unit in the direct decomposition. Let $e_1 = e_1'$ be one of them and let $e_1', e_2', \dots, e_m'$, be distinct elements that can be obtained from e_1 by the automorphisms from G. All e_i' are minimal idempotents, so that they occur in the set $e_1, \dots, e_n$. By orthogonality, the sum $e_1' + \cdots + e_m'$ is an idempotent different from 0, and it is invariant under the automorphisms from G. By the second condition, $e_1' + \cdots + e_m' \in k$ and, by idempotency, $e_1' + \cdots + e_m' = 1$. Thus, $m = n$ and all of the idempotents obtained from e_1 by the automorphisms from G are distinct and the set of these idempotents coincides with the set $e_1, \dots, e_n$ of basis idempotents. Thus, a normal basis $\{e_1^g\}$, $g \in G$, exists in A.

Now we proceed with consideration of the general case. Let an S-algebra A satisfy the assumptions of the theorem. We tensor it with the extension L of k containing the splitting fields of all components of the algebra. Then the algebra $\tilde{A} = A \otimes L$ splits completely over L (under the natural inclusion of L in $A \otimes L$). The automorphisms from G can be extended naturally to $\tilde{A}$ by the rule $(a \otimes y)^g = a^g \otimes y$ for $a \in A$, $y \in L$. The algebra of G-invariant elements of $\tilde{A}$ is equal to L which is the base field for $\tilde{A}$ because the invariance condition can be formulated as a system of homogeneous linear equations with respect to the coordinates in an L-basis of $\tilde{A}$; for such a basis one can take any basis of A over k, and the dimension of the solution space for such a system is not changed under extension of a field from k to L, i.e., it remains equal to 1. The dimension of $\tilde{A}$ over L is obviously equal to $[A\!:\!k] = (G\!:\!1)$. Therefore, the algebra $\tilde{A}$, splitting over L, satisfies the assumptions of the theorem and, by the above results, it admits a G-normal L-basis. Thus, the representation of G over L in the G-module $\tilde{A}$ is equivalent to the regular representation. A theorem of representation theory (see [**29**]) states if two representations realized in the field k are equivalent over the extension L of the field k, then they are equivalent in k as well. Therefore, the representation of G in A as in a G-module is equivalent to the regular one, i.e., a normal basis in A for the group G exists. The proof of the theorem is complete. □

In proving the theorem we essentially used the multiplicative structure of the algebra. It is easily seen that a G-module whose dimension is equal to the order of G and whose submodule of invariants is one-dimensional need not be a regular representation module.

We observe that we have proved in passing a theorem on the existence of a normal basis in a normal extension of a field because the assumptions of the theorem are fulfilled for a normal extension.

§5. The standard representation of a Galois algebra

Let G be a finite group, H a subgroup of G, and M a space over k that is a right H-module. Let $k[G]$ be the group ring of G regarded as a right G-module and as a left H-module. Then the tensor product $M \otimes k[G]$ over $k[H]$ is a right G-module $\widetilde{M}$ over k which is called the *induced module for M* and is denoted by

$\widetilde{M} = \operatorname{ind}_H^G M$. As is known (see [29]), this notion plays a fundamental role in representation theory.

We recall in more detail the structure of $\widetilde{M}$. We set $\mathbb{Z}[G] = \sum_{g\in G}^{\oplus} \mathbb{Z}u_g$, assuming that $u_g^{g_1} = u_{gg_1}$ for $g, g_1 \in G$ and $hu_g = u_{hg}$ for $h \in H$, $g \in G$. Then $\widetilde{M} = \sum_{g\in G} M \otimes u_g$. Let $G = \bigcup_\rho \rho = \bigcup_\rho H\overline{\rho}$ be the decomposition of G with respect to H. We assume that $g \in \rho$, i.e., $g = h\overline{\rho}$ for $h \in H$. Then $x \otimes u_g = x \otimes hu_{\overline{\rho}} = x^h \otimes u_{\overline{\rho}}$ for $x \in M$. Therefore, $\widetilde{M} = \sum_\rho M \otimes u_{\overline{\rho}}$. This sum is direct. The right operator $g \in G$ acts on the element $\sum_\rho x_\rho \otimes u_{\overline{\rho}}$, $x_\rho \in M$, by the formula $(\sum_\rho x_\rho \otimes u_{\overline{\rho}})^g = \sum_\rho x_\rho \otimes u_{\overline{\rho}g}$. The element $u_{\overline{\rho}g}$ belongs to the coset ρg with representative $\overline{\rho g}$. Thus, $u_{\overline{\rho}g} = u_{\overline{\rho}g\overline{\rho g}^{-1}\overline{\rho g}} = (\overline{\rho}g\overline{\rho g}^{-1})u_{\overline{\rho g}}$ and $(\sum_\rho x_\rho \otimes u_{\overline{\rho}})^g = \sum_\rho x_\rho^{\overline{\rho}g\overline{\rho g}^{-1}} \otimes u_{\overline{\rho g}}$. In such a way one may describe the action of right operators $g \in G$ in terms of "coordinates" of the set $\widetilde{M}$.

If the module M is not only an H-module but also a G-module, then the description of $\widetilde{M} = \operatorname{ind}_H^G M$ can be given in a simpler way. With the element $\mu = \sum_\rho x_\rho \otimes u_{\overline{\rho}} \in M$ we associate a function f defined on the cosets ρ with values in M, namely, $f(\rho) = x_\rho^{\overline{\rho}}$. Then, to the element $\mu^g = \sum_\rho x_\rho^{\overline{\rho}g\overline{\rho g}^{-1}} \otimes u_{\overline{\rho g}}$ there corresponds the function f^g with values $f^g(\rho g) = x_\rho^{\overline{\rho}g} = (f(\rho))^g$, so that $f^g(\rho) = (f(\rho g^{-1}))^g$. This correspondence is a G-isomorphism. Thus, in this case $\widetilde{M}$ is isomorphic as a G-module to the group of functions on the cosets ρ with the natural action of operators: $f^g(\rho) = (f(\rho g^{-1}))^g$.

It is of interest to note that in this representation $\widetilde{M}$ seems to depend on the action of the elements of G on M, although in fact $\widetilde{M}$ depends only on the action of the elements from H.

Now we return to the consideration of a Galois algebra. We prove the following theorem.

THEOREM 1.5. *For a Galois algebra A with group G there exists a field K with Galois group H that is a subgroup of G, and the Galois algebra A is isomorphic, as a G-module, to $\operatorname{ind}_H^G K$, where the field K is regarded as an H-module.*

PROOF. As an S-algebra, the algebra A with group G decomposes into a direct sum of fields. Let K be one of the components, E the unit of K, so that $K = AE$. We denote by H the subgroup of G whose elements do not change E. Let $G = \bigcup \rho = \bigcup_\rho H\overline{\rho}$ be the decomposition of G with respect to H. Clearly, $E^{g_1} = E^{g_2}$ if and only if g_1 and g_2 belong to the same coset, so that the $E^{\overline{\rho}}$ are distinct among the E^g. These elements, together with E, are minimal idempotents; the sum of them is different from zero, it is idempotent and invariant under the action of all $g \in G$. So $\sum_\rho E^{\overline{\rho}} = 1$ and $A = \sum_\rho^{\oplus} AE^{\overline{\rho}}$. The summands $AE^{\overline{\rho}}$ are fields isomorphic to K because $AE^{\overline{\rho}} = (AE)^{\overline{\rho}}$. For each ρ we fix an isomorphism of K to $AE^{\overline{\rho}}$ assuming that the image of $x \in K$ in $AE^{\overline{\rho}}$ is $(xE)^{\overline{\rho}}$. We denote this image by $x \cdot E^{\overline{\rho}}$ (this symbol should not be regarded as a product of $x \in K = AE$ by $E^{\overline{\rho}}$ in the algebra A). We want to see how the element $g \in G$ acts in this notation. For any $g \in G$ we have

$$(x \cdot E^{\overline{\rho}})^g = (xE)^{\overline{\rho}g} = ((xE)^{\overline{\rho}g\overline{\rho g}^{-1}})^{\overline{\rho g}} = x^{\overline{\rho}g\overline{\rho g}^{-1}} \cdot E^{\overline{\rho g}}.$$

Thus, the elements $g \in G$ act on the elements of the Galois algebra in the same way as in the tensor product $\Upsilon K \bigotimes_H \mathbb{Z}[G]$, if we regard K as an H-module. It remains to establish that K is a normal extension of the field k with group H. It is obvious that the elements from H give automorphisms of K. It remains to prove that K is normal over $k \approx kE$ with group H. The dimension of the field K over k is $(G:H)$ times less than the dimension of A over k and, therefore, is equal to the order of H. Next, if $x \in K$ is invariant for arbitrary $h \in H$, then the element $y = \sum_\rho x \cdot E^{\overline{\rho}} \in A$ is invariant for arbitrary $g \in G$ because $y^g = \sum_\rho x^{\overline{\rho} g \overline{\rho g}^{-1}} \cdot E^{\overline{\rho g}} = \sum_\rho x \cdot E^{\overline{\rho g}} = y$. Therefore, $y \in k$ and $x = yE \in kE$, i.e., x belongs to the image of the field k under its natural inclusion in the field $K = AE$. Thus, the field K is normal over k with group H. □

The field K is called the *field-kernel* of A. A Galois algebra is fully determined by the field-kernel normal over the base field with Galois group H and by the inclusion $i\colon H \to G$ of this group in the group G.

Conversely, we construct the Galois algebra $A = \operatorname{ind}_H^G K$, starting from the field K with group H and the inclusion $i\colon H \to G$. Indeed, the dimension of A over k is equal to $(H:1)(G:H) = (G:1)$, A is naturally an algebra over k in which the operators from G act. It remains to verify that a G-normal basis exists in A. But this follows from the fact that the module A obtained from a module of the regular representation of the subgroup H by inducing is a module of the regular representation for G. We recall why this is the case.

Let M be a module of the regular representation for H and $\widetilde{M} = \operatorname{ind}_H^G M = \sum_\rho^\oplus M \otimes E^{\overline{\rho}}$. We denote by the letter x a generator of a normal basis of the module M and show that $y = x \otimes E$ generates a normal basis of $\widetilde{M}$. In fact, let $g = \overline{h\rho}$ for $h \in H$. Then $y^g = x^h \otimes E^{\overline{\rho}}$, so that $\{y^g\}$ is a basis of $M \otimes E^{\overline{\rho}}$ for $g \in \rho$, and all of y^g comprise a union of bases of all of $M \otimes E^{\overline{\rho}}$, i.e., a basis of $\widetilde{M}$.

The representation of a Galois algebra in the form $A = \operatorname{ind}_H^G K = \sum_\rho^\oplus K \cdot E^{\overline{\rho}}$ is called the *standard representation* and denoted by $(K; i\colon H \to G)$. If we assume that K is a Galois algebra, then, by the above arguments, $A = \operatorname{ind}_H^G K$ is a Galois algebra.

§6. The algebra of T-invariant elements of a Galois algebra

Assume that a Galois algebra $A = (K; i\colon H \to G)$ and a subgroup T of G are given. We want to know the structure of the subalgebra $B = A^T$ consisting of all elements of A left invariant under all automorphisms $t \in T$. In the course of studying it will be necessary to consider, along with the decomposition $G = \bigcup \rho = \bigcup_\rho H\overline{\rho}$ of G with respect to the subgroup H, the decomposition of G with respect to the double module $G = \bigcup \pi = \bigcup_\pi H\overline{\pi}T$. Each double coset π is equal to ρT, where ρ is a coset in the decomposition with respect to H that is contained in the coset π.

THEOREM 1.6. *The algebra $B = A^T$ splits into a direct sum of fields. The direct summands are in one-to-one correspondence with the cosets π of the decomposition of G with respect to the double module. The component BE_π of the algebra B which corresponds to the coset π is isomorphic to the field K^{H_π}, where $H_\pi = H \cap \overline{\rho} T \overline{\rho}^{-1}$, $\overline{\rho}$ being a representative from any coset $\rho = H\overline{\rho}$ that is contained in π. The corresponding component AE_π of the algebra A is a Galois algebra over the field BE_π with group T. Thus, the algebra A as a whole has a T-normal basis over B.*

PROOF. We write the algebra A in the standard form $A = \sum_{\rho}^{\oplus} K \cdot E^{\overline{\rho}}$. Let $y \in A$, $y = \sum_{\rho} x_{\rho} \cdot E^{\overline{\rho}}$ for $x_{\rho} \in K$. Then $y^t = \sum_{\rho} x_{\rho}^{\overline{\rho t}\,\overline{\rho t}^{-1}} \cdot E^{\overline{\rho t}}$ and the relation $y^t = y$ is equivalent to the system of relations $x_{\rho t} = x_{\rho}^{\overline{\rho t}\,\overline{\rho t}^{-1}}$ for all ρ. These relations for the elements from B must hold for all $t \in T$ as well. Thus, the "coefficients" of $E^{\overline{\rho}}$ are conjugate, provided that the cosets ρ occur in the same coset $\pi = \rho T$ with respect to the double module. The coefficients that correspond to cosets ρ occurring in different cosets π are unrelated. In particular, the minimal idempotent element of the algebra B for which the coefficient of $E^{\overline{\rho}}$ is equal to 1 has coefficient 1 at all $E^{\overline{\rho}_1}$ for $\rho_1 \subset \rho T$ and coefficient 0 at all other $E^{\overline{\rho}_2}$. Thus, $E_{\pi} = \sum_{\rho \in \pi} E^{\overline{\rho}}$ are minimal idempotents of B. The component BE_{π} of B that corresponds to the coset π is isomorphic to the field of "admissible" coefficients x_{ρ} of E_{ρ}, where $\rho \in \pi$. Now we find out what restrictions the system of equalities established above imposes on x_{ρ}. For the consistency of this system it is necessary and sufficient that $x_{\rho}^{\overline{\rho t_1}\,\overline{\rho t_1}^{-1}} = x_{\rho}^{\overline{\rho t_2}\,\overline{\rho t_2}^{-1}}$ whenever $\rho t_1 = \rho t_2$. In turn, this is equivalent to the fact that x_{ρ} is not changed by the automorphisms $\overline{\rho} t_1 \overline{\rho t_1}^{-1} (\overline{\rho} t_2 \overline{\rho t_2}^{-1})^{-1}$ if $\rho t_1 = \rho t_2$, i.e., if $H\overline{\rho} t_1 = H\overline{\rho} t_2$, $\overline{\rho} t_1 t_2^{-1} \overline{\rho}^{-1} \in H$; at the same time, $\overline{\rho} t_1 t_2^{-1} \overline{\rho}^{-1} \in \overline{\rho} T \overline{\rho}^{-1}$. If this requirement is fulfilled, then $\overline{\rho} t_1 \overline{\rho t_1}^{-1} (\overline{\rho} t_2 \overline{\rho t_2}^{-1})^{-1} = \overline{\rho} t_1 t_2^{-1} \overline{\rho}^{-1}$. Thus, for the consistency of the system of relations $x_{\rho t} = x_{\rho}^{\overline{\rho t}\,\overline{\rho t}^{-1}}$ it is necessary and sufficient that x_{ρ} belong to the subfield K_{π} of the field K of elements invariant with respect to the group $H \cap \overline{\rho} T \overline{\rho}^{-1}$. The component $B_{\pi} = BE_{\pi}$ of the algebra B is isomorphic to the same field K_{π}. The degree over k of this field is equal to $(H : (H \cap \overline{\rho} T \overline{\rho}^{-1}))$.

As was observed above, $\rho t_1 = \rho t_2$ if and only if $t_1 t_2^{-1} \in \overline{\rho}^{-1} H \overline{\rho} \cap T$, i.e., if and only if t_1 and t_2 belong to the same coset in the decomposition of the group T with respect to $\overline{\rho}^{-1} H \overline{\rho} \cap T$. The number of these cosets is equal to the number of cosets in the decomposition of the group $\overline{\rho} T \overline{\rho}^{-1}$ with respect to $H \cap \overline{\rho} T \overline{\rho}^{-1}$. We denote the order of the group $H \cap \overline{\rho} T \overline{\rho}^{-1}$ for $\rho \in \pi$ by the symbol c_{π}. In this notation the number of those $E^{\overline{\rho}}$ that occur in E_{π} is equal to $|T| : c_{\pi}$ and the dimension of the algebra AE^{π} is equal to $(K : k) \cdot |T| : c_{\pi} = |H|\,|T| : c_{\pi}$. The degree of the field $K_{\pi} = BE_{\pi}$ is equal to $|H| : c_{\pi}$. Therefore, the dimension of AE_{π} over K_{π} is equal to the order $|T|$ of the group T. By Theorem 1.4, the algebra AE_{π} is a Galois algebra over the field K_{π} with group T.

All the statements of the theorem are proved. □

COROLLARY 1. *In order that the algebra $B = A^T$ be a field it is necessary and sufficient that $HT = G$.*

Indeed, this is the case if and only if the decomposition of G with respect to the double module consists of a single coset.

COROLLARY 2. *If T is a normal subgroup of G, then $B = A^T$ is a Galois algebra with group G/T.*

These corollaries follow from the theorem just proved and part 3 of Theorem 1.2. From Theorem 1.6 it also follows that the field-kernel of this algebra is the subfield of K invariant with respect to the normal subgroup $H \cap T$ of H.

COROLLARY 3. *If T is a normal subgroup of G, then the algebra $B = A^T$ is a field (normal over k) if and only if $HT = G$.*

COROLLARY 4. *If the algebra A is a solution of the embedding problem $(K/k, G, \varphi, N)$, then the condition $HN = G$ is fulfilled and the field-kernel of the solution is a solution of the embedding problem $(K/k, H, \varphi, H \cap N)$.*

In fact, the field K/k must be the algebra A^N of invariants of the algebra A with respect to the normal subgroup N of G. The last statement of Theorem 1.6 enables us to draw a conclusion concerning the field-kernel.

COROLLARY 5. *If the embedding problem $(K/k, G, \varphi, N)$ is solvable and N is contained in the Frattini group of G, then all of the solutions of the problem are fields.*

In fact, the Frattini group is, by definition, the intersection of all maximal subgroups of G, so that the relation $HN = G$ is not possible for any proper subgroup of G.

The problem of embedding a field in a Galois algebra is a special case of the formally more general problem of *embedding a Galois algebra in a Galois algebra.* The precise statement of this problem is the same as for a field. Namely, for a Galois algebra B with group F and an exact sequence $1 \longrightarrow N \longrightarrow G \xrightarrow{\varphi} F \longrightarrow 1$, it is required to find an algebra A with group G containing the algebra B in such a way that the restriction epimorphism of G to B coincides with the epimorphism φ.

However, this problem in essence contains nothing new. Namely, the following corollary is valid.

COROLLARY 6. *For the solvability of the problem of embedding an algebra B with group F, having the field-kernel K with group $H \subset F$, in an algebra A with group G it is necessary that the embedding problem $(K/k, \widetilde{H}, \widetilde{\varphi}, N)$ be solvable, where $\widetilde{H}$ is the preimage of H in G, $\widetilde{\varphi}$ is the restriction of φ to $\widetilde{H}$.*

In fact, if A is a solution of the embedding problem, then $B = A^N$. By Theorem 1.6, the component AE of A located over the component $K = BE$ of B is a Galois algebra with group $\varphi^{-1}(H) = \widetilde{H}$.

The condition stated is not only necessary but also sufficient. It is easily seen that if $\tilde{A}$ is a solution of the embedding problem $(K/k, \widetilde{H}, \widetilde{\varphi}, N)$, then a solution of the problem $(B/k, G, \varphi, N)$ is given by the module $\operatorname{ind}_{\widetilde{H}}^{G} \tilde{A}$ regarded as an algebra in the standard representation $(\tilde{A}; i \colon \widetilde{H} \to G)$.

§7. Generalized Galois algebras

Let B be an S-algebra over the field k, and let G be a given subgroup of the group of automorphisms (over k) of the algebra B. We denote by A the subalgebra of G-invariant elements of B. We say that B is a *generalized Galois algebra* with group G over the algebra A if the ratio $\dim B / \dim A$ of the dimensions (over k) is equal to the order n of G.

LEMMA. *Let B be a finite-dimensional algebra with unit over the field k, and let G be a finite subgroup of order n of the group of automorphisms of B such that the subalgebra of G-invariant elements coincides with k. Then $\dim B \leq n$.*

PROOF. We denote by $u_1, u_2, \ldots, u_m$ a basis of B over k. We regard the automorphisms from G as right operators and write them as superscripts. Then

$$\left\| \begin{matrix} u_1^g \\ \vdots \\ u_m^g \end{matrix} \right\| = A_g \left\| \begin{matrix} u_1 \\ \vdots \\ u_m \end{matrix} \right\|,$$

where A_g is an $(m \times m)$ matrix over k. Assume that

$$z = c_1 u_1 + \cdots + c_m u_m = \|c_1, \ldots, c_m\| \left\| \begin{matrix} u_1 \\ \vdots \\ u_m \end{matrix} \right\|, \qquad c_i \in k,$$

is invariant under all $g \in G$. We exclude the trivial case $g = 1$. Then

$$z^g = \|c_1, \ldots, c_m\| \left\| \begin{matrix} u_1^g \\ \vdots \\ u_m^g \end{matrix} \right\| = \|c_1, \ldots, c_m\| A_g \left\| \begin{matrix} u_1 \\ \vdots \\ u_m \end{matrix} \right\|,$$

and because of the relation $z^g = z$ we conclude that*

$$(c_1, \ldots, c_m) A_g = (c_1, \ldots, c_m), \qquad g \in G, \quad g \neq 1.$$

Thus, the rows $(c_1, \ldots, c_m)$ are solutions of a system of $n - 1$ homogeneous linear equations. By assumption, the space of solutions of this system is one-dimensional. Therefore, the number of equations cannot be less than $m-1$, so that $n-1 \geq m-1$ and $m = \dim B \leq n$, as required. □

We note that if B is an S-algebra and $\dim B = n$, then, by Theorem 1.4, B is a Galois algebra with group G over k.

THEOREM 1.7. *A generalized Galois algebra B with group G over the S-algebra $A = B^G$ is a direct sum of Galois algebras with group G over fields which are direct summands of A.*

PROOF. Let $A = Ae_1 \oplus \cdots \oplus Ae_m = k_1 \oplus \cdots \oplus k_m$ be a decomposition of the S-algebra A into the direct sum of fields $k_i = Ae_i$, where $\{e_i\}$ is a system of minimal idempotents. Then $B = Be_1 \oplus \cdots \oplus Be_m$. Clearly, $(Be_i)^g = Be_i$ for all $g \in G$ because $e_i \in A$, so that $e_i^g = e_i$. Let n be the order of G. By the lemma, $\dim Be_i/k_i \leq n$. At the same time, by the assumption of the theorem, $\dim B / \dim A = (\sum \dim Be_i)/(\sum \dim k_i) = n$ (the dimensions are considered over the base field k). Consequently, for every i, $\dim Be_i / \dim k_i = \dim Be_i/k_i = n$, and by Theorem 1.4, Be_i is a Galois algebra with group G over the field k_i, as required. □

---

*Hereinafter, we shall write a row matrix $\|c_1 \ \ldots \ c_m\|$ in the form $(c_1, \ldots, c_m)$.—*Editor's remark.*

§8. Speiser's theorem for a Galois algebra

In the theory of algebraic extensions the so-called "Hilbert Theorem 90" plays an important role. The statement of this theorem is as follows: if K is an extension of the field k with cyclic group and if the norm (from K in k) of an element $x \in K$ is equal to 1, then $x = y^g y^{-1}$, where y is a certain element from K, g is a generator of the Galois group of K over k. This theorem can be generalized to any normal extension K of k in the following form. Let G be the Galois group of K over k. If x_g, $g \in G$, is a system of elements from K satisfying the relations $x_{g_1}^{g_2} x_{g_2} = x_{g_1 g_2}$ for arbitrary $g_1, g_2 \in G$, then there exists an element $y \in K$ such that $x_g = y^g y^{-1}$ for arbitrary $g \in G$. In the language of homological algebra this theorem means that the one-dimensional cohomology group $H^1(G, K^*)$ of the group G with the coefficients in the multiplicative group K^* of the field K is trivial. This generalization is a special case of a theorem proved by Speiser [**102**] and is usually called *Speiser's theorem.*

Our aim in this section is to extend Speiser's theorem to Galois algebras.

THEOREM 1.8. *Let x_g be a cocycle of the group G in the multiplicative group of the Galois algebra B with group G, i.e., a system of elements x_g, $g \in G$, invertible in B, satisfying the relations $x_{g_1}^{g_2} x_{g_2} = x_{g_1 g_2}$. Then there exists an invertible element y such that $x_g = y^g y^{-1}$.*

PROOF. Consider the k-linear function $l(z) = \sum_{g \in G} z^g x_g$ on the algebra B. Clearly,

$$(l(z))^{g_1} = \sum_{g \in G} z^{gg_1} x_g^{g_1} = \sum_{g \in G} z^{gg_1} x_{gg_1} x_{g_1}^{-1} = l(z) \cdot x_{g_1}^{-1}.$$

We prove that $l(z)$ does not vanish identically. Let e be a minimal idempotent of B, $K = Be$ the field-kernel of B, H the subgroup of G that is the Galois group of the field K over k. We have $l(ze)e = \sum_{g \in G} z^g e_g x_g e = \sum_{h \in H} z^h (ex_h)$, since $e^g e = 0$ for $g \notin H$. If $u_1, \dots, u_m$ is a basis of K over k, then the rows $(u_1^h, \dots, u_m^h)$, $h \in H$, are linearly independent, and the simultaneous fulfillment of the relations $\sum_{h \in H} u_i^h ex_h = 0$ is impossible because otherwise the elements $ex_h \in K$ yield a solution of a system of m homogeneous linear equations with m unknowns which has linearly independent rows of coefficients. Thus, there exists $z \in B$ (moreover, $z \in K = BE$) such that $l(z) \neq 0$. We show that all the components $l(z)e^{g_1}$ of $l(z)$ are different from zero; therefore, $l(z)$ is invertible in B. In fact,

$$l(z)e^{g_1} = x_{g_1}(l(z))^{g_1} e^{g_1} = x_{g_1}(l(z)e)^{g_1},$$

and all of the components of $l(z)$ do or do not vanish simultaneously. Thus, there exists an element z such that $l(z)e \neq 0$ and, therefore, $l(z)$ is invertible in B. Let $l(z) = y^{-1}$. Then $x_{g_1} = y^{-1} y^{g_1}$, as is required to prove. □

This theorem can be extended to generalized Galois algebras as well. Namely, if B is a generalized Galois algebra with group G over an S-algebra A and for $g \in G$, $x_g \in B$ is a cocycle, i.e., a system of invertible elements satisfying the relations $x_{g_1}^{g_2} x_{g_2} = x_{g_1 g_2}$, then there exists an invertible element $y \in B$ such that $x_g = y^g y^{-1}$. Indeed, a generalized Galois algebra is a direct sum of Galois algebras with the same group over the fields.

§9. The semidirect embedding problem

Let G be a semidirect extension of the kernel N of the homomorphism φ, so that in G there exists a subgroup F_1 such that $F_1N = G$ and $F_1 \cap N = 1$. In this case F_1 is isomorphic to G/N.

THEOREM 1.9. *The semidirect problem of the embedding of a field in a Galois algebra is always solvable, and there exists a solution whose field-kernel is isomorphic to the extendable field. Conversely, if the problem of the embedding of a field in a Galois algebra has a solution whose field-kernel is isomorphic to the extendable field, then the problem is semidirect.*

PROOF. Let the problem $(K/k, G, \varphi, N)$ be semidirect. Then the algebra $(K; i\colon F_1 \to G)$ is a solution to this problem; here, F_1 is a subgroup that yields a semidirect factorization of G, namely, $G = F_1N$ and $F_1 \cap N = 1$.

Conversely, if the problem $(K/k, G, \varphi, N)$ has a solution $(K; i\colon F_1 \to G)$, where F_1 is a subgroup of G, then F_1 is isomorphic to the Galois group of the field K, which is equal to G/N, and, therefore, G is a semidirect product F_1N. □

§10. Associated problems

A problem related to a given embedding problem and whose solvability is necessary for the solvability of a given problem is called an *associated problem*. We describe two simple procedures which lead to associated problems.

Let $(K/k, G, \varphi, N)$ be a given problem and let N_0 be a normal subgroup of G contained in N. We put $G_1 = G/N_0$, $N_1 = N/N_0$. Then $G_1/N_1 \approx G/N \approx F = \mathrm{Gal}(K/k)$ and the homomorphism φ induces on G_1 (because $\varphi(N_0) = 1$) a homomorphism φ_1 that maps G_1 onto F and has kernel N_1. The embedding problem $(K/k, G_1, \varphi_1, N_1)$ is an associated problem for the initial problem. Indeed, if the algebra A yields a solution of the initial problem, then, obviously, the algebra $A_1 = A^{N_0}$ is a solution of the new problem. The constructed associated problem is called an *associated problem of the first kind.*

We assume that the associated problem $(K/k, G_1, \varphi_1, N_1)$ for the problem $(K/k, G, \varphi, N)$ is solvable, and let an algebra L be a solution of this problem. Then any solution A of the problem $(L/k, G, \varphi_0, N_0)$, where φ_0 is the natural mapping of G onto $G_1 \approx G/N_0$, also is a solution of the initial problem $(K/k, G, \varphi, N)$. Indeed, A contains the field K as a subalgebra and G is mapped onto $F = \mathrm{Gal}(K/k)$ by the homomorphism $\varphi_1\varphi_0 = \varphi$.

Conversely, if A is a solution of the problem $(K/k, G, \varphi, N)$ and L is the corresponding solution of the associated embedding problem $(K/k, G_1, \varphi_1, N_1)$, $L = A^{N_0}$, then A is a solution of the problem $(L/k, G, \varphi_0, N_0)$. Indeed, by definition, L is a subalgebra of the algebra A belonging to the normal subgroup N_0 of G, and G is mapped onto the group $\mathrm{Gal}(L/k)$ identified with $G_1 = G/N_0$ by means of φ_0.

Let N_0 and N_{00} be normal subgroups of G and $N \supset N_{00} \supset N_0$, and let $G_1 = G/N_0$, $G_2 = G/N_{00}$, $N_1 = N/N_0$, $N_2 = N/N_{00}$, $N_{10} = N_{00}/N_0$. Then $G_2 \approx G_1/N_{10}$. The homomorphism φ induces on G_1 a homomorphism φ_1 onto F with kernel N_1. The homomorphisms φ and φ_1 induce on G_2 the same homomorphism φ_2 onto F with kernel $N_2 \approx N_1/N_{10}$. The problems $(K/k, G_1, \varphi_1, N_1)$ and $(K/k, G_2, \varphi_2, N_2)$ are associated problems of the first kind for the problem $(K/k, G, \varphi, N)$, and the problem $(K/k, G_2, \varphi_2, N_2)$ is an associated problem of the first kind for the problem $(K/k, G_1, \varphi_1, N_1)$. Thus, passing twice to an associated problem of the first kind is equivalent to passing once.

Now we consider an associated problem of the second kind. Given a problem $(K/k, G, \varphi, N)$, let G_1 be a subgroup of G that contains the kernel N of the homomorphism φ. We put $F_1 = \varphi(G_1)$ and denote by k_1 the subfield of F_1-invariant elements in K. The group of the extension K/k_1 is F_1. Thus, the embedding problem $(K/k_1, G_1, \varphi_1, N)$ makes sense. It is clear that if the Galois algebra A provides a solution to the initial problem, then it also provides a solution of the problem $(K/k_1, G_1, \varphi_1, N)$ if we regard it as an algebra over k_1. This problem is called an *associated problem of the second kind* for the initial problem.

If $G \supset G_1 \supset G_2 \supset N$, $F_1 = \varphi(G_1)$, $F_2 = \varphi(G_2)$ and if k_1 and k_2 are the subfields of invariants in the field K for the groups F_1 and F_2, then the problem $(K/k_2, G_2, \varphi, N)$ is an associated problem for $(K/k, G, \varphi, N)$ and for $(K/k_1, G_1, \varphi, N)$. Thus, an associated problem of the second kind for an associated problem of the second kind is an associated problem of the second kind for the initial problem.

An associated problem $(K/k_1, G_1/N_0, \varphi_1, N/N_0)$ of the first kind for the associated problem $(K/k_1, G_1, \varphi, N)$ of the second kind for the initial problem is called a *mixed associated problem*.

It is easy to see that the associated problem $(K/k_1, G_1/N_0, \varphi_1, N/N_0)$ of the second kind for the associated problem $(K/k, G/N_0, \varphi, N/N_0)$ of the first kind for the initial problem is at the same time an associated problem of the first kind for the associated problem $(K/k_1, G_1, \varphi, N)$ of the second kind of the initial problem (but, generally speaking, not vice versa because a normal subgroup N_0 of a subgroup G_1 is not necessarily a normal subgroup of G), i.e., it is a mixed associated problem. Hence it follows that the set of mixed associated problems is closed under passage to the associated problems of both the first and the second kind.

§11. An extension of the base field

Let $(K/k, G, \varphi)$ be an embedding problem and let L be an extension of k. We tensor the field K with L (over k). It becomes a Galois algebra $\widetilde{K} = K \otimes L$ over L with group $F = \operatorname{Im}\varphi$. Clearly, the problem $(\widetilde{K}/L, G, \varphi)$ is an associated problem to the problem $(K/k, G, \varphi)$ in the sense that if the initial problem is solvable and A is a Galois algebra providing a solution, then the problem $(\widetilde{K}/L, G, \varphi)$ is also solvable and the algebra $A \otimes L$ is a solution.

In particular, if k is the algebraic number field, then for L we can take the completion of k at a point $\mathfrak{p}$. The obtained problem $(K_\mathfrak{p}/k_\mathfrak{p}, G, \varphi_\mathfrak{p})$ is called a *local associated problem* (here $K_\mathfrak{p} = K \otimes_k k_\mathfrak{p}$).

Here two cases may occur. First, the algebra $\widetilde{K} = K \otimes L$ may be a field if K and L are linearly separated over k. In this case we say that a free extension of the base field holds. But it may happen that $\widetilde{K} = K \otimes L$ is a Galois algebra over L with several components. Let E be an idempotent of the algebra $\widetilde{K}$ that is the unit of a component, we denote by H the subgroup of $F = \operatorname{Gal}(K/k)$ whose elements do not change E and by $\widetilde{H}$ the preimage of it in G. As was clarified above, the field-kernel $\widetilde{K}_1 = \widetilde{K}E$ of the algebra $\widetilde{K}$ has H as the Galois group over L. We denote by k_1 the subfield of K/k belonging to the subgroup H. If $x \in k_1$, then xE is an invariant for the Galois group of the field $\widetilde{K}_1 = \widetilde{K}E$ and, therefore, it belongs to the base field L (more precisely, to the embedding LE of it in the field

$\widetilde{K}_1$). Thus, the field L contains a subfield isomorphic to k_1 and multiplication by E carries out an embedding of k_1 in L. The extension of k to L may be attained in two steps. We first extend k to k_1 (by k-tensor multiplication), then we extend k_1 to L by k_1-tensor multiplication. At the first step the field K is converted to a Galois algebra over k_1 with component K/k_1, and the solution of the embedding problem can be obtained by inducing $\mathrm{ind}_{\widetilde{H}}^{G}$ a solution of the problem $(K/k_1, \widetilde{H}, \varphi)$, i.e., the associated problem of the second kind for the initial problem. At the second step, a free extension with the linearly separated fields K/k_1 and L/k_1 occurs.

§12. Direct multiplication of embedding problems

First we note a simple group-theoretical fact.

LEMMA. *Let N_1 and N_2 be two normal subgroups of the group G such that $N_1 \cap N_2 = 1$. Then G is isomorphic to the direct product of the groups G/N_1 and G/N_2 with amalgamated quotient group G/N_1N_2.*

Since the groups N_1 and N_2 are relatively prime, we have the following commutative diagram with exact rows:

$$\begin{array}{ccccccccc}
 & & & & 1 & & 1 & & \\
 & & & & \downarrow & & \downarrow & & \\
 & & & & N_2 & \xrightarrow{\mathrm{id}} & N_2 & & \\
 & & & & \downarrow & & \downarrow & & \\
1 & \longrightarrow & N_1 & \longrightarrow & G & \xrightarrow{\varphi_1} & G/N_1 & \longrightarrow & 1 \\
 & & \downarrow{\scriptstyle \mathrm{id}} & & \downarrow{\scriptstyle \varphi_2} & & \downarrow{\scriptstyle \varphi_2^*} & & \\
1 & \longrightarrow & N_1 & \longrightarrow & G/N_2 & \xrightarrow{\varphi_1^*} & G/N_1N_2 & \longrightarrow & 1 \\
 & & & & \downarrow & & \downarrow & & \\
 & & & & 1 & & 1 & &
\end{array}$$

where the homomorphism of a group onto a quotient group is natural. The mapping $g \in G \to (\varphi_1 g, \varphi_2 g)$ yields the required isomorphism.

It is obvious that the lemma can be generalized to the case of several normal subgroups $N_1, \dots, N_k$, where each N_i is relatively prime to the product of the remaining subgroups. Here G is isomorphic to the direct product of the quotient groups G/H_i, where $H_i = N_1 \dots N_{i-1}N_{i+1} \dots N_k$, with amalgamated quotient group $G/N_1N_2 \dots N_k$.

Let two embedding problems $(K/k, G_1, \varphi_1)$ and $(K/k, G_2, \varphi_2)$ with the same embeddable field be given. Then $\varphi_1 G_1 = \varphi_2 G_2 = F = \mathrm{Gal}(K/k)$. We denote by N_1 and N_2 the kernels of the homomorphisms φ_1 and φ_2. We denote by G the direct product of the groups G_1 and G_2 with amalgamated quotient group F, i.e., the group that consists of pairs (g_1, g_2), $g_1 \in G_1$, $g_2 \in G_2$, such that $\varphi_1(g_1) = \varphi_2(g_2)$. We define a homomorphism of G onto F by the formula $\varphi(g_1, g_2) = \varphi_1(g_1) = \varphi_2(g_2)$. The direct product $N_1 \times N_2$ of the groups N_1 and N_2 is the kernel of this homomorphism. The embedding problem $(K/k, G, \varphi, N_1 \times N_2)$ is called the *direct product* of given problems.

THEOREM 1.12. *If the problems $(K/k, G_1, \varphi_1, N_1)$ and $(K/k, G_2, \varphi_2, N_2)$ are solvable, then their direct product is also solvable, and, vice versa, if the direct product of embedding problems is solvable, then both factors are solvable.*

PROOF. Let the problems $(K/k, G_1, \varphi_1, N_1)$ and $(K/k, G_2, \varphi_2, N_2)$ be solvable and let algebras A_1 and A_2 be solutions. We denote by A the tensor product of A_1 and A_2 over the field K. The algebra A is an S-algebra over k containing the field K. Let $x_1 \in A_1$, $x_2 \in A_2$, $x_1 \otimes x_2 \in A$ and $(g_1, g_2) \in G$. We put $(x_1 \otimes x_2)^{(g_1,g_2)} = x_1^{g_1} \otimes x_2^{g_2}$. Then for $c \in K$ we have $(cx_1 \otimes x_2)^{(g_1,g_2)} = c^{g_1} x_1^{g_1} \otimes x_2^{g_2}$ and $(x_1 \otimes cx_2)^{(g_1,g_2)} = x_1^{g_1} \otimes c^{g_2} x_2^{g_2}$ but $c^{g_1} = c^{g_2}$. Thus, the elements (g_1, g_2) do not destroy the structure of K-tensor product and are automorphisms of the algebra A. By Theorem 1.6, the algebra A_1 has an N_1-normal basis over K and the algebra A_2 has an N_2-normal basis over K. Then tensor product of these bases is an $N_1 \times N_2$-normal basis of A over K. Therefore, the elements from A invariant under the automorphisms from $N_1 \times N_2$ belong to the field K, hence G-invariant elements belong to the base field k. The dimension of the algebra A over K is $|N_1| \cdot |N_2|$, so that its dimension over k is $|F| \cdot |N_1| \cdot |N_2| = |G|$. Thus, A is a Galois algebra and it is a solution of the embedding problem.

The initial problems are associated problems of the first kind for the direct product of them; namely, the problem $(K/k, G_1, \varphi_1)$ can be obtained from the problem $(K/k, G, \varphi)$ if in the kernel $N_1 \times N_2$ of the homomorphism φ one takes $1 \times N_2$ as the normal subgroup N_0 of G by which one factorizes the problem; moreover, the quotient group $(N_1 \times N_2)/(1 \times N_2)$ is identified with the group N_1. The problem $(K/k, G_2, \varphi_2)$ can be obtained similarly. □

Thus, if both problems are solvable, then their direct product is also solvable; if the direct product of problems is solvable, then both factors are solvable as associated problems.

If, for example, the problem $(K/k, G_2, \varphi_2)$ is solvable, then the problems $(K/k, G_1, \varphi_1)$ and $(K/k, G, \varphi)$ are equivalent in the sense that they both are simultaneously solvable or unsolvable.

We also note that if A_2 is a solution of the second problem, then any solution A of the problem $(A_2/k, G, \overline{\varphi}_1)$ is at the same time a solution of the problem $(K/k, G, \varphi)$ (here φ_1 is a homomorphism of G onto $G_2 = \mathrm{Gal}(A_2/k)$ such that $\overline{\varphi_1}(g_1, g_2) = g_2$); conversely, any solution A of the problem $(K/k, G, \varphi)$ is a solution of the problem $(A_2/k, G, \overline{\varphi}_1)$, where $A_2 = A^{N_1 \times 1}$.

§13. Lifting, descent, and factorization of embedding problems

First we give some preliminary information about extensions of groups.

Let an extension G_1 of a group N by a group F_1 be given, i.e., let an exact sequence

$$1 \longrightarrow N \xrightarrow{\alpha_1} G_1 \xrightarrow{\varphi_1} F_1 \longrightarrow 1$$

be given, as well as an epimorphism θ of a group F onto F_1. Does there exist an extension $1 \longrightarrow N \xrightarrow{\alpha} G \xrightarrow{\varphi} F \longrightarrow 1$ and an epimorphism $\psi\colon G \to G_1$ such that the following diagram

$$(*)\qquad \begin{array}{ccccccccc} 1 & \longrightarrow & N & \xrightarrow{\alpha} & G & \xrightarrow{\varphi} & F & \longrightarrow & 1 \\ & & \big\downarrow{\scriptstyle \mathrm{id}} & & \big\downarrow{\scriptstyle \psi} & & \big\downarrow{\scriptstyle \theta} & & \\ 1 & \longrightarrow & N & \xrightarrow{\alpha_1} & G_1 & \xrightarrow{\varphi_1} & F_1 & \longrightarrow & 1 \end{array}$$

commutes? (Here we do not assume that the groups are finite.)

PROPOSITION 1.13.1. *The required extension exists and is unique up to equivalence of extensions.*

PROOF. Existence is almost obvious; namely, as G one can take the direct product of G_1 and F with amalgamated quotient group F_1, i.e., the set of pairs (g_1, f), where $g_1 \in G_1$, $f \in F$ such that $\varphi_1(g_1) = \theta(f)$ with component-wise multiplication. It should be assumed that $\alpha(a) = (\alpha_1(a), 1)$, $\varphi((g_1, f)) = f$ and $\psi((g_1, f)) = g_1$. The exactness of the upper row and the commutativity of the diagram are obvious.

Now we prove uniqueness. Let a commutative diagram $(*)$ with exact rows be given. We map the elements $g \in G$ to the pairs $\lambda(g) = (\psi(g), \varphi(g)) \in G_1 \times F$. By the commutativity of the diagram, we have $\varphi_1\psi(g) = \theta\varphi(g)$, so that $\lambda(g)$ is contained in the direct product of G_1 and F with amalgamated quotient group F_1. The fact that the mapping λ is a homomorphism is obvious. The kernel of λ is the set of $g \in G$ such that $\psi(g) = 1$ and $\varphi(g) = 1$. From the condition $\varphi(g) = 1$ we conclude that $g = \alpha(a)$ for $a \in N$. Therefore, $\psi(g) = \alpha_1(a)$, and from $\psi(g) = 1$ it follows that $a = 1$ and $g = 1$. Thus, λ is a monomorphism. We now prove that λ is an epimorphism. Let $(g_1, f) = G_1 \times_{F_1} F$ (i.e., the direct product of G_1 and F with amalgamated quotient group F_1), i.e., $\varphi_1(g_1) = \theta(f)$. We find a $g \in G$ such that $\varphi(g) = f$. Then $\varphi_1(g_1) = \theta(f) = \theta\varphi(g) = \varphi_1\psi(g)$ and $\varphi_1(g_1\psi(g)^{-1}) = 1$, so that $g_1\psi(g^{-1}) = \alpha_1(a) = \psi(\alpha(a))$. Consequently, $g_1 = \psi((\alpha a)g)$, but $f = \varphi(\alpha(a)g)$ as well, so that $(g_1, f) = \lambda((\alpha(a))g)$, which proves that λ is an epimorphism and, thus, an isomorphism. Thus, the extension $1 \to N \xrightarrow{\alpha} G \xrightarrow{\varphi} F \to 1$ is equivalent to the extension

$$1 \to N \to G_1 \underset{F_1}{\times} F \to F \to 1. \qquad \square$$

We observe that the group $G \approx G_1 \times_{F_1} F$ has two relatively prime normal subgroups, namely, $(\alpha N, 1) \approx N$ and $(1, \operatorname{Ker}\theta) \approx \operatorname{Ker}\theta$. The passage to the quotient groups of G by $(1, \operatorname{Ker}\theta)$ and of F by $\operatorname{Ker}\theta$ performs the descent of the extension $1 \to N \to G \to F \to 1$ to the extension $1 \to N \to G_1 \to F_1 \to 1$ from which the extension $1 \to N \to G \to F \to 1$ was obtained.

The passage from the extension $1 \to N \to G_1 \to F_1 \to 1$ to the extension $1 \to N \to G \to F \to 1$ (respectively, from $1 \to N \to G \to F \to 1$ to $1 \to N \to G_1 \to F_1 \to 1$) is called *lifting* (respectively, *descent*).

If N is an Abelian group, then the element from $H^2(F, H)$ that corresponds to the extension $1 \to N \to G \to F \to 1$ can be obtained by the lifting of the element from $H^2(F_1, N)$ that corresponds to the extension $1 \to A \to G_1 \to F_1 \to 1$. Indeed, let $\{\overline{f}\}$ be the set of representatives from the preimages in G_1 of the elements $f \in F_1$. Then $z = (f_1, f_2) = \overline{f_1 f_2}^{-1}\overline{f}_1\overline{f}_2$ is a cocycle defining the element from $H^2(F, H)$ that corresponds to the extension of G_1. Next, the pairs $(\overline{f}, l)$, where l ranges over the preimages $\theta^{-1}(f)$ of the elements f in F, form a full set of representatives from preimages in G of the elements in F. A representative of the coset that contains $(\overline{f_1}, l_1)(\overline{f_2}, l_2)$ is $(\overline{f_1 f_2}, l_1 l_2)$. The cocycle that corresponds to this set of representatives is $(\overline{f_1 f_2}^{-1}\overline{f_1}\,\overline{f_2}, (l_1 l_2)^{-1} l_1 l_2) = (\overline{f_1 f_2}^{-1}\overline{f_1 f_2}, 1) = \alpha(\overline{f_1 f_2}^{-1}\overline{f_1}\,\overline{f_2})$ and the cohomology class thus obtained can be obtained from the initial class by lifting.

We observe that here we interpret only a special case of cohomology lifting in terms of extensions. In the general case, lifting is carried out from the group $H^2(F_1, N^T)$ to the group $H^2(F, N)$, where N^T is a subgroup of elements from N left fixed under the action of the elements from the kernel T of the epimorphism $\theta\colon F \to F_1$. In the case under consideration $N^T = N$ because we assume that $a^f = a^{\theta(f)}$, $a \in N$, for $f \in F$, so that $a^t = a^1 = a$ for $t \in T$. In the general case, the interpretation of cohomology lifting in terms of extensions is somewhat more complicated.

PROPOSITION 1.13.2. *In order that the extension* $1 \to N \xrightarrow{\alpha} G \xrightarrow{\varphi} F \to 1$ *lifted from the extension* $1 \to N \xrightarrow{\alpha_1} G_1 \xrightarrow{\varphi_1} F_1 \to 1$ *by the epimorphism* $\theta\colon F \to F_1$ *split* (*i.e., the group* G *be the semidirect product of* N *and the complementary subgroup isomorphic to* F) *it is necessary and sufficient that there exist a homomorphism* ω *of the group* F *to* G_1 *such that* $\varphi_1\omega = \theta$.

PROOF. Let the extension $1 \to N \xrightarrow{\alpha} G \xrightarrow{\varphi} F \to 1$ split. Then there exists a monomorphism $\gamma\colon F \to G$ such that $\varphi\gamma = \mathrm{id}$. Then one can take $\omega = \psi\gamma$, where ψ is an epimorphism of G onto G_1 which arises under lifting. Conversely, let $\omega\colon F \to G_1$ with $\varphi_1\omega = \theta$ exist. We consider G the direct product $G_1 \times F$ with amalgamated quotient group F_1. We map F to the direct product $G_1 \times F$ by putting $\gamma(f) = (\omega(f), f)$. Since $\varphi_1\omega(f) = \theta(f)$, we have $\gamma(f) \in G$. It is clear that γ is a monomorphism of F to G, $\varphi\gamma(f) = f$ for arbitrary $f \in F$. Consequently, the extension $1 \to N \to G \to F \to 1$ splits. □

Now we turn to embedding problems. Let $(K/k, G_1, \varphi_1, N_1)$ be an embedding problem and let L be a field normal over k and containing K. Then L can be regarded as a solution of the embedding problem $(K/k, G_2, \varphi_2, N_2)$, where $G_2 = \mathrm{Gal}(L/k)$. Consider the direct product $(K/k, G, \varphi, N_1 \times N_2)$ of these problems; here G is the direct product of G_1 and G_2 with amalgamated quotient group $\mathrm{Gal}(K/k)$. The constructed problem is equivalent to the problem $(K/k, G_1, \varphi_1, N_1)$, because the second factor is solvable. But, as we have seen in §12, the problem $(K/k, G, \varphi, N_1 \times N_2)$ is equivalent to the problem $(L/k, G, \overline{\varphi_1}, N_1)$, where $\overline{\varphi_1}$ is a homomorphism which maps the pair $(g_1, g_2) \in G$ to the element $g_2 \in \mathrm{Gal}(L/k)$. The subgroup $N_1 \times 1$, which can be identified with N_1, is the kernel of $\overline{\varphi_1}$. The passage from the problem $(K/k, G_1, \varphi_1, N_1)$ to the equivalent problem $(L/k, G, \overline{\varphi_1}, N_1)$ is called *lifting*.

We observe that the extension of groups $1 \to N_1 \xrightarrow{\alpha} G \xrightarrow{\overline{\varphi_1}} F \to 1$ with $F = \mathrm{Gal}(L/k)$ that is related to the problem $(L/k, G, \overline{\varphi_1}, N_1)$ can be obtained by lifting from the extension $1 \to N_1 \xrightarrow{\alpha_1} G_1 \to F_1 \to 1$ with $F_1 = \mathrm{Gal}(K/k)$.

The passage from the problem $(L/k, G, \overline{\varphi_1}, N_1)$ to the problem $(K/k, G_1, \varphi_1, N_1)$ is called *descent*.

If A is a solution of the problem $(K/k, G_1, \varphi_1, N_1)$, then $A \otimes_K L$ is a solution of the problem $(L/k, G, \overline{\varphi_1}, N_1)$.

The group G of the problem obtained by lifting contains the normal subgroup $1 \times N_2$, where N_2 is the kernel of the natural homomorphism of $G_2 = \mathrm{Gal}(L/k)$

onto $F_1 = \mathrm{Gal}(K/k)$. If B is a solution of the lifted problem, then the elements of B invariant for the group $1 \times N_2$ give a solution of the problem $(K/k, G_1, \varphi_1, N_1)$. In general, the following proposition holds.

PROPOSITION 1.13.3. *If the group G in the embedding problem $(L/k, G, \varphi, N_1)$ has a normal subgroup N_2 relatively prime to N_1, then the passage to an equivalent smaller problem $(K/k, G_1, \varphi_1, N_1)$ is possible, where K is the subfield of $\varphi(N_2)$-invariant elements of L, $G_1 = G/N_2$, φ_1 is a natural homomorphism of G_1 onto $F = \mathrm{Gal}(K/k)$. The initial problem can be obtained by lifting from the problem $(K/k, G_1, \varphi_1, N_1)$.*

PROOF. The field L/k can be regarded as a solution of the embedding problem $(K/k, G_2, \varphi_2)$, where $G_2 = \varphi(G) \approx G/N_1$ and the homomorphism φ_2 maps G/N_1 onto $\mathrm{Gal}(K/k) \approx G/(N_1 \times N_2)$.

From the description of lifting it is clear that the passage from the problem $(K/k, G_1, \varphi_1)$ to the problem $(L/k, G, \varphi)$ can be realized by lifting with the help of the multiplication by the problem $(K/k, G_2, \varphi_2)$ having the field L as a solution. Consequently, the problems $(K/k, G_1, \varphi_1)$ and $(L/k, G, \varphi)$ are equivalent. The proof is complete.

Let an embedding problem $(K/k, G, \varphi)$ and a normal subgroup T of G be given. Then $T_0 = \varphi(T)$ is a normal subgroup of $F = \mathrm{Gal}(K/k)$. We denote by K_0 the field K^{T_0} normal over k and by φ_0 the homomorphism of the group G/T onto $\mathrm{Gal}(K_0/k) \approx \varphi(G)/\varphi(T)$ induced by the homomorphism φ. It is easily seen that the problem $(K_0/k, G/T, \varphi_0)$ is associated with the initial problem $(K/k, G, \varphi)$ in the sense that the solvability of the initial problem implies the solvability of $(K_0/k, G/T, \varphi_0)$. Indeed, if A is a solution of the initial problem, then A^T is a solution of the new one. The passage from $(K/k, G, \varphi)$ to $(K_0/k, G/T, \varphi_0)$ is called *factorization* and the group T is called the *kernel of factorization.* It is clear that if the kernel of factorization is contained in the kernel N of the initial problem, then factorization coincides with the passage to an associated problem of the first kind. If the kernel of factorization is relatively prime with the kernel N of the problem, then factorization is descent. In the general case, factorization reduces to descent from some associated problem of the first kind. Indeed, let $T \cap N = N_0$ and let A be a solution of $(K/k, G, \varphi, N)$. Then $A^T = (A^{N_0})^{T/N_0}$ and A^{N_0} is a solution of the problem $(K/k, G/N_0, \overline{\varphi}, N/N_0)$, where $\overline{\varphi}$ is the homomorphism of G/N_0 onto $\mathrm{Gal}(K/k)$ induced by the homomorphism φ, i.e., a solution of the associated problem of the first kind. Then $(N/N_0) \cap (T/N_0) = 1$, so that the passage from A^{N_0} to A^T is obtained by the descent from an embedding problem. □

§14. A splitting field for the embedding problem

Let $(K/k, G_1, \varphi_1, N_1)$ be an embedding problem. A field L normal over k (possibly, an infinite algebraic extension) that contains K is called a *splitting field* of the problem if there exists a homomorphism ω of the group $F = \mathrm{Gal}(L/k)$ to the group G_1 such that $\varphi_1 \omega$ coincides with the natural mapping $F \to F_1 = \mathrm{Gal}(K/k)$.

THEOREM 1.14.1. *In order that the problem $(K/k, G_1, \varphi_1, N_1)$ be solvable it is necessary and sufficient that a splitting field of this problem exist.*

PROOF. Let the problem be solvable, let A be a solution, K_1 the field-kernel of A. The group $H = \mathrm{Gal}(K_1/k)$ can be realized as a subgroup of G_1 that leaves fixed a component K_1e of A. Then $HN_1 = G_1$ (Corollary 4 to Theorem 1.6). Let L be a normal extension of k containing K_1 (in particular, one can take $L = K_1$ or as L one can also take the algebraic separable closure of k). Then there is a natural epimorphism of the group $F = \mathrm{Gal}(L/k)$ onto H and $\omega\varphi_1$ is the required epimorphism. Thus, L is a splitting field of the problem.

Conversely, let a splitting field L exist, let $\mathrm{Gal}(L/k) = F$, and let $\omega\colon F \to G$ be a homomorphism. We denote by K_1 the field of invariants of the group $\operatorname{Ker}\omega$ in L. Then $\mathrm{Gal}(K_1/k) = H$ and the Galois algebra in the standard description $(K_1, H \to G_1)$ yields a solution of the problem. We observe that the inclusion of the Galois group of K_1 in G_1 uniquely determines a solution as an induced G_1-module for the H-module K_1. □

THEOREM 1.14.2. *In order that the field $L \supset K$ normal over k be a splitting field of the problem $(K/k, G_1, \varphi_1, N_1)$, it is necessary and sufficient that the lifted problem $(L/k, G, \overline{\varphi_1}, N_1)$ be semidirect.*

PROOF. The extension of groups $1 \to N_1 \xrightarrow{\alpha} G \xrightarrow{\overline{\varphi_1}} F \to 1$, where $F = \mathrm{Gal}(L/k)$, is obtained by lifting from the extension $1 \to N_1 \xrightarrow{\alpha_1} G_1 \xrightarrow{\varphi_1} F_1 \to 1$ with the help of the epimorphism θ of the group F onto the group $F_1 = \mathrm{Gal}(K/k)$. By Proposition 1.13.2, the extension $1 \to N_1 \to G_1 \to F_1 \to 1$ splits if and only if there exists a homomorphism ω of the group F to G_1 such that $\varphi_1\omega = \theta$. By definition, this means that L is a splitting field for the initial problem. □

COROLLARY. *The embedding problem $(K/k, G_1, \varphi_1, N_1)$ is solvable if and only if it can be lifted to a semidirect problem.*

Before formulating the next theorem, we make some notions more precise.

Let N be a group (non-Abelian, in general) with operators from a group F. A function c_f defined on F with values in N is called a *one-dimensional cocycle* if the relations $c_{f_1}^{f_2} c_{f_2} = c_{f_1 f_2}$ hold. It is easily checked that for any $b \in N$ the function $d_f = b^{-f} c_f b$ is also a cocycle. By definition, the cocycles c_f and d_f belong to the same cohomology class of $H^1(F, N)$. It is clear that if N is Abelian, then this definition leads to the ordinary definition of $H^1(F, N)$. The set $H^1(F, N)$ is not a group if N is non-Abelian.

Recall that in §1.1 we mentioned the solution of the embedding problem "in the strict sense" without making a precise definition. Now we give such a definition. Two solutions A_1 and A_2 of the problem $(K/k, G, \varphi, N)$ are equivalent if there exists an isomorphism $\theta\colon A_1 \to A_2$ that is identity on K and there exist isomorphisms ν_1 and ν_2 of G to the automorphism groups of A_1 and A_2 describing the actions of G on A_1 and A_2 as solutions of the embedding problem. Moreover, we assume that $\theta(x_1^{\nu_1(g)}) = (\theta(x_1))^{\nu_2(g)}$ for $x_1 \in A_1$, $g \in G$.

We study the question on equivalence of solutions of the embedding problem $(K/k, G, \varphi, N)$ in more detail. If A_1 and A_2 are two equivalent solutions of the embedding problem with actions of G with the help of $\nu_1(G) \subset \operatorname{Aut} A_1$ and $\nu_2(G) \subset \operatorname{Aut} A_2$, then A_1 and A_2 are isomorphic as algebras. If we apply to A_2 any isomorphism acting on A_1 together with compatible change of $\nu_2(G)$ we obtain an equivalence of the problems that are realized on one and the same algebra. We first show that the same Galois algebra A yields several different but equivalent solutions of the embedding problem. Indeed, let A be an algebra (together with the action of the elements of G) that yields a solution of the embedding problem. We put $\nu_1(g) = g$ and $\nu_2(g) = b^{-1}gb$ for $b \in N$ and $\theta(x) = x^b$ for $x \in A$. The automorphism θ leaves fixed the elements of the field K, so that it does not destroy the structure of A as a linear space over K. Further, $\theta(x^{\nu_1(g)}) = x^{gb}$ and $(\theta(x))^{\nu_2(g)} = x^{bb^{-1}gb} = \theta(x^{\nu_1(g)})$. Thus, on the algebra A we have constructed two different equivalent solutions of the embedding problem.

Now let A be a solution of the embedding problem with $\nu_1(g) = g$, and let the same algebra be a solution of the same problem equivalent to the initial one, i.e., there are isomorphisms $\theta\colon A \to A$ and $\nu_2\colon G \to G$ such that $\theta(x^g) = (\theta(x))^{\nu_2(g)}$. Let K_1 be the field-kernel of A, let an inclusion of the group $H = \mathrm{Gal}(K_1/k)$ in the group G be given, let A be given as a G-module induced for the H-module K_1, let K_1e be a component of the algebra A left fixed under the action of the automorphisms from H, so that $(ze)^h = z^h e$ for $z \in K_1$, $h \in H$. Then $\theta(K_1e)$ is a component of A, possibly different from K_1e. Let $\theta(K_1e) = K_1e_\rho$, where ρ is a coset in the decomposition of G with respect to H. Since $G = HN$, for representatives of the cosets ρ one can take representatives in the decomposition of N with respect to $N_1 = N \cap H$. The elements from N_1 do not move the components and induce all automorphisms of the field K_1 that leave the elements of K fixed. Therefore, there exists an element $b \in N$ such that for $z \in K_1e$, we have $\theta(z) = z^b$. We put $\theta'(x) = \theta(x^{b^{-1}})$ for $x \in A$. Then θ' is the identity automorphism on K_1e and, consequently, the identity automorphism on $A = \mathrm{ind}_H^G K_1$. So $\theta(x) = x^b$. The relation $(\theta(x))^{\nu_2(g)} = \theta(x^g)$ implies $x^{b(\nu_2(g))} = x^{gb}$ for all $x \in A$, whence $\nu_2(g) = b^{-1}gb$.

In particular, for $h \in H$, $\nu_2(h) = b^{-1}hb$ yields a mapping of H to G corresponding to the second solution of the problem.

Now we turn to the solutions of the embedding problem $(K/k, G, \varphi, N)$ that are related to a given splitting field with the help of the construction described in Theorem 1.14.1. Such solutions are called *solutions belonging to the splitting field* L.

Let L be a field normal over k with Galois group F and $(K/k, G, \varphi, N)$ the embedding problem for which L is a splitting field. We take a solution A_0 belonging to the splitting field L. We denote by K_1 its field-kernel and by H the Galois group of K_1 over k that is realized as a subgroup of G leaving fixed a component K_1e of A_0. Then $NH = G$. Next, there is a homomorphism $\omega\colon F \to G$ whose image is equal to H. The existence of such a homomorphism enables us to regard N as an F-operator group by putting $b^f = (\omega f^{-1})b(\omega f)$ for $b \in N$ and $f \in F$. If N is non-Abelian, then the structure of N as an F-operator group depends on the choice of A_0. If N is Abelian, then $a^f = a^{\varphi_1\omega(f)} = a^{\psi(f)}$, where ψ is the natural epimorphism of the group F onto $F_1 = \mathrm{Gal}(K/k) = G/N$.

THEOREM 1.14.3. *The solutions of the embedding problem $(K/k, G, \varphi, N)$ belonging to the splitting field L with Galois group F are in one-to-one correspondence with one-dimensional cocycles of the group F with values in the group N regarded as an F-operator group with the help of the solution A_0 chosen.*

PROOF. Let A be a solution of the embedding problem. We denote by ω and ω_0 the homomorphisms of F to G that correspond to the solutions A and A_0. Then for every $f \in F$ we have $\omega(f) = (\omega_0(f))c_f$ for $c_f \in N$ because the elements $\omega(f)$ and $\omega_0(f)$ act in the same way on the elements of K. Then $\omega(f_1f_2) = \omega_0(f_1f_2)c_{f_1f_2} = \omega_0(f_1)\omega_0(f_2)c_{f_1f_2}$, but, on the other hand, $\omega(f_1f_2) = \omega(f_1)\omega(f_2) =$

$\omega_0(f_1)c_{f_1}\omega_0(f_2)c_{f_2} = \omega_0(f_1f_2)c_{f_1}^{f_2}c_{f_2}$. Consequently, $c_{f_1}^{f_2}c_{f_2} = c_{f_1f_2}$, i.e., c_f is a cocycle of F with values in N.

Conversely, if c_f is such a cocycle, then the mapping $\omega(f) = \omega_0(f)c_f$ is a homomorphism of F to G, $(\omega(F))N = (\omega_0(F))N = G$, and a solution of the embedding problem corresponds to the homomorphism ω. □

Among the solutions constructed in such a way there are equivalent ones.

THEOREM 1.14.4. *In order that the solutions A_1 and A_2 of the embedding problem $(K/k, G, \varphi, N)$ corresponding to homomorphisms ω_1 and ω_2 of the group $F = \mathrm{Gal}(L/k)$ to the group G be equivalent it is necessary and sufficient that the corresponding cocycles be homologous.*

PROOF. We assume that the solutions corresponding to the homomorphisms ω_1 and ω_2 are equivalent. Then they are isomorphic; in particular, their field-kernels are isomorphic, so that we may assume that the solutions are realized on the same algebra A. The kernels of the homomorphisms ω_1 and ω_2 coincide because both of them coincide with the subgroup to which the subfield K_1 isomorphic to the field-kernel of A belongs. Therefore, if $\omega_1(f) = h \in G$, then $\omega_2(f)$ does not depend on the choice of f and we may assume that $\omega_2(f) = \nu(h)$, where ν is an isomorphic inclusion of H in G. If the extensions are equivalent, then there exists $b \in N$ such that $\theta(x) = x^b$ for $x \in A$ and $\theta(x^g) = (\theta(x))^{\nu_2(g)}$, where ν_2 is an isomorphism of G that coincides with ν on H. Thus, for $g = h \in H$ we have $\theta(x^h) = x^{hb} = (\theta(x))^{\nu(h)} = x^{b\nu(h)}$. Consequently, $\nu(h) = b^{-1}hb$, i.e., $\omega_2(f) = b^{-1}(\omega_1(f))b$. Using the cocycles we obtain $(\omega_0 f)d_f = b^{-1}(\omega_0 f)c_f b = (\omega_0 f)b^{-f}c_f b$, whence $d_f = b^{-f}c_f b$, i.e., the cocycles d_f and c_f are homologous.

Since b can be arbitrary, the equivalence of the corresponding extensions follows from the fact that the cocycles are homologous. □

§15. Look "from above" at the embedding problem

If an embedding problem is solvable and has a splitting field L, then the algebraic separable closure $\overline{k}$ of the base field also is a splitting field. Thus, the following theorems are valid.

THEOREM 1.15.1. *For the solvability of the embedding problem $(K/k, G, \varphi, N)$ it is necessary and sufficient that there exist a homomorphism ω of the group $\mathrm{Gal}(\overline{k}/k)$ to the group G such that $\varphi\omega$ is the canonical homomorphism.*

THEOREM 1.15.2. *The solutions of the embedding problem are in one-to-one correspondence with the cocycles of $\mathrm{Gal}(\overline{k}/k)$ with values in N.*

THEOREM 1.15.3. *The solutions corresponding to cocycles of $\mathrm{Gal}(\overline{k}/k)$ with values in N are equivalent if and only if the cocycles are homologous.*

These theorems immediately follows from Theorems 1.14.1, 1.14.3, and 1.14.4.

CHAPTER 2

The Compatibility Condition

§1. The crossed products

Let K be a normal extension of the field k with group F and $\varphi\colon G \to F$ a homomorphism of a group G onto F with kernel N.

The algebra over k that consists of the formal sums $\sum_{g\in G} u_g x_g$, $x_g \in K$, with multiplication rules $u_{g_1}u_{g_2} = u_{g_1g_2}$ and $xu_g = u_g x^{\varphi(g)}$ for $x \in K$, $g \in G$ is called the *crossed product* $G \times K$ of the group G and the field K. The algebra $G \times K$ is associative. Indeed, $(u_f x u_g y) \cdot u_h z = u_{fg} x^g y u_h z = u_{fgh} x^{gh} y^h z$ and $u_f x \cdot (u_g y u_h z) = u_{fgh} x^{gh} y^h z$. The element u_1 is the unit of the algebra $G \times K$. The dimension of the algebra $G \times K$ over k is equal to $(G{:}1)[K{:}k] = (G{:}N)(N{:}1)(F{:}1) = (N:1)(F:1)^2$ because $(G:N) = (F:1)$.

The algebra $G \times K$ contains the group algebra $K[N]$ of the group N over K as a subalgebra. Indeed, the elements u_t with $t \in N$ commute with the elements of K.

If A is a Galois algebra with group F, then the crossed product $G \times A$ is defined similarly. It contains the group algebra of the group N over A, namely, $\sum^{\oplus} u_t A$, as a subalgebra.

If the embedding problem $(A/k, G, \varphi)$ is given, then the conditions for constructing the crossed product $G \times A$ are satisfied. Therefore, this crossed product is naturally associated with the problem $(A/k, G, \varphi)$.

THEOREM 2.1. *Let a Galois algebra A have the field-kernel K_1 with group H. Then $\Lambda = G \times A$ is isomorphic to the algebra of matrices of order $(F:H)$ over the crossed product $\widetilde{H} \times K_1$, where $\widetilde{H} = \varphi^{-1}(H)$ is the preimage of the group H in the group G.*

PROOF. Recall that $A = \operatorname{ind}_H^G K_1 = \sum_\rho^{\oplus} K_1 e^{\overline{\rho}}$. Here ρ ranges over the cosets of the decomposition of F with respect to H, the $\overline{\rho}$ are their representatives, and the $e^{\overline{\rho}}$ are minimal idempotents that are the units of the components of the algebra A. The spaces $e^{\overline{\rho}}\Lambda$ are right ideals of the algebra Λ, and Λ is a direct sum of them. These ideals are isomorphic as Λ-modules because $e^{\overline{\rho}} = u_{\overline{\rho}}^{-1} e u_{\overline{\rho}}$, so that $e^{\overline{\rho}}\Lambda = u_{\overline{\rho}}^{-1} e u_{\overline{\rho}}\Lambda = u_{\overline{\rho}}^{-1} e\Lambda$ and $e\Lambda = u_{\overline{\rho}} e^{\overline{\rho}}\Lambda$.

By Theorem A.1.2, the algebra Λ is isomorphic to a matrix algebra over the subalgebra $e\Lambda e$, where e is a component of unit in a direct summand of the algebra A.

In our case, $\Lambda = \sum u_g A$, $\Lambda e = \sum u_g A e$, but $Ae = K_1 e$. Further, $e\Lambda e = \sum e u_g K_1 e$. But $e u_g = u_g e^{\varphi(g)}$ and $e^{\varphi(g)} e = 0$ only if $e^{\varphi(g)} \neq e$. We have $e^{\varphi(g)} = e$ if and only if $\varphi(g) \in H$, i.e., $g \in \widetilde{H}$. Thus, $e\Lambda e = \sum_{g\in\widetilde{H}} u_g K_1 e$, which is obviously isomorphic to $\widetilde{H} \times K_1$. □

COROLLARY 1. *If a Galois algebra A with group G splits completely, then $G \times A$ is isomorphic to a matrix algebra over the group algebra of $N = \operatorname{Ker} \varphi$ over k.*

Indeed, in this case $K_1 = k$, $H = 1$, $\widetilde{H} = \varphi^{-1}(H) = N$, and the crossed product $\widetilde{H} \times K_1$ is a group algebra of the group N over k.

COROLLARY 2. *If the characteristic of the field k does not divide the order of the group $N = \operatorname{Ker} \varphi$, then the algebra $\Lambda = G \times A$ is semisimple.*

Indeed, we extend the base field k to a field L in which the algebra A splits completely. Then the algebra $\Lambda \otimes L/L$ is isomorphic to a matrix algebra over the group algebra of group N over L. Therefore, $\Lambda \otimes L/L$ is semisimple, and so is the algebra Λ/k.

§2. The compatibility module

Let K be a normal extension of the field k with group F, and let $\varphi\colon G \to F$ be a homomorphism of a group G onto F with kernel N.

A space P over the field K on which the group G acts in a semi-linear way, i.e., $(px)^g = p^g x^{\varphi(g)}$ for $p \in P$ and $x \in K$, and in which there is an N-normal basis is called a *compatibility module* for the group G and the field K or, equivalently, for the embedding problem $(K/k, G, \varphi)$. We note that the operators from N act linearly because for $t \in N$, $\varphi(t) = 1$. A compatibility module may be regarded as a module for the crossed product $G \times K$ if we assume that $u_g y \in G \times K$ acts on px according to the rule $(px) \circ u_g y = p^g x^{\varphi(g)} y$. It is easily verified that $((px) \circ u_g y) \circ u_h z = px \circ (u_g y u_h z)$ with $p \in P$, $g, h \in G$, $x, y, z \in K$.

The existence of an N-normal basis with respect to the field K in a compatibility module is equivalent to the fact that for the subalgebra $K[N]$ of $G \times K$, a compatibility module is a module of the regular representation. Indeed, if $\{b^t\}$, $t \in N$, form a normal basis of the module over K, then $P = \sum_{t \in N}^{\oplus} b^t K = \sum_{t \in N}^{\oplus} b \circ u_t K = b \circ \sum_{t \in N}^{\oplus} u_t K$, i.e., P is isomorphic to the algebra as a right module with respect to operators in $K[N]$. The converse is obvious.

A compatibility module may exist or may fail to exist. If a compatibility module exists, we say that the group G is compatible with the field K, or the compatibility condition holds for the group G and the field K (for the embedding problem $(K/k, G, \varphi)$).

A compatibility module for a group G and a Galois algebra A with group F is defined similarly. Namely, a compatibility module is a free A-module with an N-normal basis and the action of automorphisms in G according to the rule $(px)^g = p^g x^{\varphi(g)}$ for $p \in P$, $x \in A$, $g \in G$, as before.

If a solution of the embedding problem $(K/k, G, \varphi)$ exists, then every solution of this problem is obviously a compatibility module, so that the compatibility condition is necessary for the solvability of the embedding problem, but, generally speaking, it is not sufficient.

THEOREM 2.2.1. *If a Galois algebra A with group F splits completely, then for each group G with epimorphism $\varphi\colon G \to F$, a unique compatibility module up to $G \times A$-isomorphism exists. In particular, the right ideal $e\Lambda$ is a compatibility module, where $\Lambda = G \times A$ and e is the component of unit in a direct summand of the decomposition of A into a direct sum of fields.*

PROOF. Let P be a compatibility module. By definition, P has an N-normal basis over A, i.e., $P = \sum_{t\in N}^{\oplus} b^t A$ for some b. Further, $A = \sum_{f\in F}^{\oplus} e^f k$. The automorphisms $g \in G$ do not act on the elements of the field k and transform e^f to $e^{f\varphi(g)}$. We show that the elements $(be)^g = b^g e^{\varphi(g)}$, $g \in G$, constitute a G-normal basis of the module P over k. The dimension of P is equal to the order of G, so that the linear independence of the system of elements $(be)^g$, $g \in G$, is what remains to be verified. If $\sum_g (be)^g x_g = 0$ for $x_g \in k$, then we obtain $\sum_{\varphi(g)=f} b^g e^f x_g = 0$ applying the operator e^f. Then we apply the operator $\overline{f}^{-1}$, where $\overline{f} \in \varphi^{-1}(f)$. We obtain $\sum_{t\in N} b^t e x_{t\overline{f}} = 0$. The elements b^t are linearly independent with respect to the algebra A. Consequently, all of $x_{t\overline{f}}$ are equal to 0, i.e., all of x_g are equal to 0.

We clarify how the operators from $G \times A$ act on this basis. The elements in G permute the basis elements in accordance with the regular representation. The elements e^f (which constitute a basis of A) annihilate all elements $(be)^g$ except those for which we have $\varphi(g) = f$, and these elements are not changed by the action of e^f. Thus, the choice of a basis of the compatibility module confirms the uniqueness of the compatibility module up to an operator isomorphism.

Now we consider the right ideal $e\Lambda$ of the algebra $\Lambda = G \times A$. The elements $u_g e^f$, $g \in G$, $f \in F$, or, what is the same up to order, the elements $e^f u_g$, form a basis of the algebra Λ over k. Therefore, eu_g, $g \in G$, constitute a basis of the ideal $e\Lambda$. Multiplied by u_{g_1} these basis elements are permuted in the same way as the elements of the group G multiplied by g_1; multiplied by e^f these basis elements are annihilated if $\varphi(g) \neq f$ and do not change if $\varphi(g) = f$, i.e., they are transformed by the same rules as the elements of the basis chosen above in the compatibility module. Consequently, every compatibility module is Λ-operator-isomorphic to the ideal $e\Lambda$. The theorem is completely proved. □

COROLLARY 1. *If a group G and a Galois algebra A are compatible, then a compatibility module is unique up to isomorphism.*

Indeed, this statement was proved in Theorem 2.2.1 for a splitting Galois algebra. By the usual method of extension of the base field we arrive at this situation and it remains to recall that if two representations of an algebra (here $G \times A$) over a field k become equivalent under an extension of the base field, then they are also equivalent over k.

COROLLARY 2. *If G and A are compatible and the order of the group F is equal to n, then the n-fold direct sum of the compatibility module is equivalent to the algebra $G \times A$ as a right $G \times A$-module.*

Indeed, this is true for a splitting algebra A because $G \times A$ decomposes into a direct sum of right ideals and each of them is isomorphic to the compatibility module.

Consequently, a multiple direct sum of the compatibility module gives a representation of the algebra $G \times A$, which is equivalent to the regular representation under a sufficient extension of the base field. Therefore, this equivalence also occurs over the base field.

COROLLARY 3. *If G and A are compatible, then the compatibility module P has a G-normal basis over k.*

Let $\{q^f\}$ be a normal basis of A over k, $f \in \mathrm{Gal}(A/k)$, so that $A = \sum_f^{\oplus} q^f k$. Then $G \times A = \sum_{g\in G}^{\oplus} u_g A = \sum_{g,f}^{\oplus} u_g q^f k = \sum_f^{\oplus} q^f \sum_g^{\oplus} u_g k$. Thus, $G \times A$ as a right G-module is isomorphic to the direct sum of free G-modules isomorphic to the module $k[G]$ and the number of summands is equal to $m = (A : k)$. The direct sum of m copies of the module P is also isomorphic to $G \times A$ (Corollary 2). By the Krull–Schmidt theorem, the G-modules P and $k[G]$ are isomorphic, and the corollary is proved.

THEOREM 2.2.2. *A group G and an algebra A are compatible if and only if the algebra $G \times A$ is isomorphic to the algebra of matrices of order $n = (F : 1)$ over a subalgebra.*

Indeed, if a compatibility module exists, then $G \times A$ as a right module is isomorphic to the direct sum of n copies of a module isomorphic to the compatibility module. The preimages in $G\times A$ of the direct summands under this isomorphism are isomorphic right ideals. By Theorem A.1.2, an associative algebra with unit that decomposes into a direct sum of n copies of isomorphic right ideals is isomorphic to the algebra of matrices of order n over a subalgebra. The "only if" part is proved.

Further, $G \times A = \sum_{f\in F}^{\oplus} u_{\overline{f}} \sum_{t\in N}^{\oplus} u_t A$, so that $G \times A$ as a right $A[N]$-module decomposes into a direct sum of $n = (F : 1)$ submodules that are isomorphic to a module of the regular representation of the algebra $A[N]$.

Now we assume that $G \times A$ is the algebra of matrices of order n over a subalgebra. Then $G \times A$ decomposes into a direct sum of n isomorphic right ideals. They are also isomorphic as $A[N]$-modules. Thus, the same $A[N]$-module $G \times A$ is represented as a direct sum of n isomorphic modules in two ways. By the Krull–Schmidt theorem, the direct summands of these decompositions are isomorphic. Hence, as $A[N]$-modules they are isomorphic to a module of regular representation. Consequently, as $G \times A$-modules, they are compatibility modules.

§3. Compatibility for a problem with Galois algebra

Let A be a Galois algebra with group F and let a subgroup H of F be the Galois group of the field-kernel K_1 of A. Given an epimorphism $\varphi\colon G \to F$, let $\widetilde{H} = \varphi^{-1}(H)$.

THEOREM 2.3. *The group G is compatible with the Galois algebra A if and only if the group $\widetilde{H}$ is compatible with the field-kernel K_1 of the algebra A.*

PROOF. We have already known that the crossed product $\Lambda = G\times A$ is isomorphic to the algebra of matrices of order $(F : H) = (G : \widetilde{H})$ over the crossed product $\widetilde{H} \times K_1$. If the group $\widetilde{H}$ is compatible with the field K_1, then the algebra $\widetilde{H} \times K_1$ is the algebra of matrices of order $(\widetilde{H} : N) = (H : 1)$ over a subalgebra. Then Λ is isomorphic to the algebra of matrices of order $(F:H)(H:1) = (F:1)$ over the same subalgebra. Consequently, the group G is compatible with the algebra A.

If the group G is compatible with the algebra A, then $\Lambda = G\times A$ is isomorphic to the algebra of matrices of order $(F : 1) = (F : H)(H : 1)$ over a subalgebra. The algebra $G \times A$ is isomorphic to the algebra of matrices of order $(F : H)$ over the crossed product $\widetilde{H} \times K_1$.

By Theorem A.1.4, we conclude that the algebra $H \times K_1$ is isomorphic to the algebra of matrices of order $(H : 1) = (\widetilde{H} : N)$, whence the group $\widetilde{H}$ is compatible with the field K_1. The theorem is proved. □

We easily see direct connections between the compatibility modules for the problems $(A/k, G, \varphi)$ and $(K_1/k, \widetilde{H}, \varphi)$. Namely, if $A = \mathrm{ind}_{\widetilde{H}}^{G} K_1 = \sum_f^{\oplus} K_1 e^{\overline{f}}$, and P and $\widetilde{P}$ are compatibility modules corresponding to the problems given, then we can take $\widetilde{P} = Pe$ if P is given, and if $\widetilde{P}$ is given, we can take $P = \mathrm{ind}_{\widetilde{H}}^{G} \widetilde{P} = \sum_f^{\oplus} \widetilde{P} v_f$ with componentwise multiplication of the elements of $\widetilde{P}$ by the elements of $A = \sum_f^{\oplus} K_1 e^f$.

§4. Inheritance of compatibility for associated problems

We show that if the compatibility condition holds for an embedding problem, then it holds for associated problems.

Given an embedding problem $(K/k, G, \varphi, N)$, we assume that the compatibility condition holds for the group G and the field K. Let N_0 be a normal subgroup of the group G that is contained in N. We denote $N_1 = N/N_0$, $G_1 = G/N_0$. Let P be the compatibility module for G and K. Then P^{N_0} is a compatibility module for the group G_1 and the field K under the natural epimorphism $G_1 \to \mathrm{Gal}(K/k)$ with kernel N_1. Indeed, in passing to N_0-invariant elements of P the field K remains unchanged, an N-normal basis of P over K are converted to an N_1-normal basis, the automorphisms from G of the module P is converted to automorphisms from G_1 of the module P^{N_0}. Thus, compatibility is inherited if we pass to an associated problem of the first kind.

Now let G_1 be a subgroup of G containing N, $F_1 = \varphi(G_1) \subset F = \mathrm{Gal}(K/k)$; let k_1 be the subfield of K that belongs to the subgroup F_1 so that $\mathrm{Gal}(K/k_1) = F_1$. Let P be the compatibility module for the embedding problem $(K/k, G, \varphi, N)$. Then P is obviously a compatibility module for the problem $(K/k_1, G_1, \varphi, N)$. Thus, compatibility is inherited if we pass to an associated problem of the second kind.

In specific cases compatibility for an associated problem implies compatibility for the original problem.

THEOREM 2.4. *Let $(K/k, G, \varphi, N)$ be an embedding problem with kernel N that is a p-group. Then the compatibility condition for it is equivalent to the compatibility condition for the associated problem $(K/k_p, G_p, \varphi, N)$, where G_p is a Sylow p-subgroup of the group G, and k_p is the subfield of K that corresponds to the subgroup $\varphi(G_p)$.*

Let q be an arbitrary prime divisor of the order of G, $p \neq q$, and let G_q be a Sylow q-subgroup of the group G. It is clear that $\varphi(G_q) = F_q$ is isomorphic to G_q. Then the crossed product $G_q \times K \cong F_q \times K$ is isomorphic to the algebra of matrices of order $(G_q : 1)$ over the field $k_q = K^{F_q}$. Thus, $G \times K$ is isomorphic to the algebra of matrices of order $(F_q : 1)$ over a subalgebra of its own (Theorem A.1.1). If the group G_p is compatible with the field K, then, by Theorem 2.2.2, the algebra $G \times K$ is itself isomorphic to the algebra of matrices of order $(F_p : 1)$ over a subalgebra. Then, by Theorem A.1.5, the algebra $G \times K$ is the algebra of matrices of order $(G : N) = (F : 1)$ over a subalgebra of its own, and it remains to use Theorem 2.2.2 once again.

For other reductions of the compatibility condition, see §§5, 6, 9.

§5. Multiplication

Let $(K/k, G_1, \varphi_1, N_1)$ and $(K/k, G_2, \varphi_2, N_2)$ be two embedding problems and let $(K/k, G, \varphi, N_1 \times N_2)$ be their direct product so that G is the direct product of the groups G_1 and G_2 with amalgamated quotient group $F = \mathrm{Gal}(K/k)$, and $\varphi(g_1, g_2) = \varphi_1(g_1) = \varphi_2(g_2)$.

THEOREM 2.5. *Compatibility in each factor is necessary and sufficient for the compatibility of the product of two embedding problems.*

PROOF. Necessity follows from the fact that each of the problems multiplied is an associated problem of the first kind for the product. For the proof of sufficiency we construct a compatibility module for the product by using the compatibility modules for the factors. Let P_1 and P_2 be the compatibility modules for the problems multiplied. We assume that $P = P_1 \underset{K}{\otimes} P_2$ with $(p_1 \otimes p_2)^g = p_1^{g_1} \otimes p_2^{g_2}$ for $p_1 \in P_1$, $p_2 \in P_2$, and $g = (g_1, g_2) \in G$. This does not destroy the structure of the K-tensor product since $z^{g_1} = z^{g_2}$ for $z \in K$.

By the fact that P_1 has an N_1-normal basis over K and P_2 has an N_2-normal basis over K, the tensor product P of the modules P_1 and P_2 has an $(N_1 \times N_2)$-normal basis over K that consists of the tensor products of the basis elements for P_1 and P_2. The theorem is proved. □

Moreover, it is easy to prove that the tensor product over k of the algebras $G_1 \times K$ and $G_2 \times K$ is isomorphic to the full ring of matrices of order $(F:1)$ over the algebra $G \times K$.

COROLLARY. *If N is a nilpotent group, then it is sufficient to verify the compatibility condition for every associated problem $(K/k_p, G_p, \varphi_p, N_p)$, where p is an arbitrary prime divisor of the order of N, G_p is a Sylow p-subgroup of the group G, $N_p = N \cap G_p$, k_p is the subfield of K corresponding to the subgroup $\varphi(G_p)$.*

The assertion easily follows from Theorems 2.4 and 2.5.

§6. Lifting and descent

As was established in §1.13 lifting and descent do not change the embedding condition. The same result is also true for the compatibility conditions.

THEOREM 2.6. *Let the embedding problem $(L/k, \overline{G}, \overline{\varphi}, N)$ be obtained from the problem $(K/k, G, \varphi, N)$ by lifting. Then the compatibility conditions for these problems coincide.*

PROOF. We write F_1 for the Galois group of the extension L/k and N_1 for the kernel of the map $\varphi\colon F_1 \to F = \mathrm{Gal}(K/k)$, i.e., the Galois group $\mathrm{Gal}(L/K)$. Recall that $\overline{G}$ is the product of the groups G and F with the amalgamated quotient F. Consider also the embedding problem $(K/k, \overline{G}, \varphi^*, N \times N_1)$ which is the product of the problems $(K/k, G, \varphi, N)$ and $(K/k, F_1, \varphi_1, N_1)$. The last problem has a solution L, and, by Theorem 2.5, the compatibility conditions for the problems $(K/k, \overline{G}, \varphi^*, N \times N_1)$ and $(K/k, G, \varphi, N)$ coincide.

Let P be the compatibility module for the problem $(K/k, G, \varphi, N)$. Then the module $\overline{P} = P \bigotimes_K L$ admits the action of operators from $\overline{G}$ and is the compatibility module both for the problem $(K/k, \overline{G}, \varphi^*, N \times N_1)$ and the problem $(L/k, G, \overline{\varphi}, N)$.

Conversely, let $\overline{P}$ be the compatibility module for the problem $(L/k, \overline{G}, \overline{\varphi}, N)$. Then $\overline{P}$ is a compatibility module for the associated direct problem $(L/K, N \times N_1, \widetilde{\varphi}, N)$ where $\widetilde{\varphi}$ is the projection of $N \times N_1$ onto N_1. By Corollary 3 in §2, $\overline{P}$ has an $N \times N_1$-normal basis over K and hence is a compatibility module for the problem $(K/k, \overline{G}, \varphi^*, N \times N_1)$. The theorem is proved. □

The compatibility condition is also left unchanged under descent.

§7. Compatibility systems

Let the compatibility condition hold for the problem $(K/k, G, \varphi, N)$ and let P be the compatibility module. Recall that in P a normal basis $\{ b^t \mid t \in N \}$ over the field K exists by definition. We regard P as a module over the algebra $G \times K$. Every element $p \in P$ can uniquely be represented in the form $\sum_{t \in N} b^t x_t = \sum_{t \in N} b \circ u_t x_t = b \circ \sum_{t \in N} u_t x_t$ with $x_t \in K$, $\sum u_t x_t \in K[N] \subset G \times K$. In particular, for each $g \in G$ we have $bu_g = b \circ \lambda_g$, where $\lambda_g \in \sum_{t \in N} u_t K = K[N]$. Then $bu_{g_1 g_2} = bu_{g_1} u_{g_2} = (b \circ \lambda_{g_1}) u_{g_2} = bu_{g_2} u_{g_2}^{-1} \circ \lambda_{g_1} u_{g_2} = b \circ \lambda_{g_2} \lambda_{g_1}^{g_2}$, where $\lambda_{g_1}^{g_2} = u_{g_2}^{-1} \lambda_{g_1} u_{g_2}$. Thus, $\lambda_{g_1 g_2} = \lambda_{g_2} \lambda_{g_1}^{g_2}$. It is clear from the definition that $\lambda_t = u_t$. Let $g_1 = g_2^{-1}$. Then $\lambda_{g_2} \lambda_{g_2^{-1}}^{g_2} = \lambda_1 = u_1$, and u_1 equals the unit element of the algebra $K[N]$. Consequently, λ_g is an invertible element, and $\lambda_{g^{-1}}^{g}$ is the inverse to it. Assume $l_g = \lambda_g^{-1}$. Then

$$l_{g_1 g_2} = l_{g_1}^{g_2} l_{g_2}, \qquad l_t = u_t^{-1}.$$

Note that l_g is a one-dimensional cocycle in $Z^1(G, (K[N])^*)$ which extends the "trivial" cocycle u_t^{-1} of $Z^1(N, (K[N])^*)$. Here $(K[N])^*$ is the group of invertible elements of the algebra $K[N]$.

We call the cocycle l_g, $g \in G$, a *compatibility system* for the embedding problem $(K/k, G, \varphi)$. Thus, the existence of a compatibility system is necessary for the compatibility of the embedding problem $(K/k, G, \varphi, N)$. We show that this condition is also sufficient. Let l_g be a compatibility system. Take the algebra $K[N] = \sum_{t \in N}^{\oplus} u_t K$ and define in it the right operators v_g by the rule $zv_g = l_g^{-1} u_g^{-1} z u_g$. Then $u_1 v_t = l_t^{-1} u_t^{-1} u_1 u_t = u_t$, so that $\{ u_1 v_t \mid t \in N \}$ is an N-normal basis of $K[N]$ over the field K. Further,

$$\begin{aligned}(zv_{g_1})v_{g_2} &= (l_{g_1}^{-1} u_{g_1}^{-1} z u_{g_1}) v_{g_2} = l_{g_2}^{-1} (l_{g_1}^{-1})^{g_2} \cdot u_{g_2}^{-1} u_{g_1}^{-1} z u_{g_1} u_{g_2} \\ &= l_{g_1 g_2}^{-1} (u_{g_1 g_2})^{-1} z u_{g_1 g_2} = zv_{g_1 g_2}.\end{aligned}$$

Thus, the operators v_g yield an action of the group G on $K[N]$. These operators are semilinear. Indeed, let $z \in K[N]$, $x \in K$. Then $(zx)v_g = l_g^{-1} u_g^{-1} z x u_g = l_g^{-1} u_g^{-1} z u_g x^{\varphi(g)} = (zv_g) x^{\varphi(g)}$.

Thus, the operators v_g convert the algebra $K[N]$ to a compatibility module, and, thus, the existence of a compatibility system is sufficient for compatibility.

§8. The structure of the crossed product $G \times K$

Assume that the characteristic of the field K does not divide the exponent n of the kernel N in the embedding problem $(K/k, G, \varphi, N)$ and the field K contains all of the n-th roots of 1. Then, by a theorem of Brauer, all irreducible representations of the group N over K are absolutely irreducible. Moreover, the crossed product

$G \times K$ is a semisimple algebra in this case, as was shown above (Corollary 2 to Theorem 2.1).

We write E_i for the minimal central idempotents of the algebra $\Lambda = G \times K$. Then $\Lambda_i = \Lambda E_i$ are simple components of the algebra Λ. The idempotents E_i belong to the group algebra $K[N]$ since they commute with elements in K. However, they may fail to be minimal in the algebra $K[N]$. Let e be a minimal central idempotent of $K[N]$ that is contained in E_i, so that $E_i e = e$. We write G_e for the subgroup of the group G that consists of all elements in G leaving e unchanged, i.e., $G_e = \{g \in G \mid e^g = u_g^{-1} e u_g = e\}$. Clearly, $N \subset G_e$. Decompose G into the cosets with respect to G_e: $G = \bigcup \rho = \bigcup G_e \overline{\rho}$, where the $\overline{\rho}$ are representatives of the cosets $\overline{\rho}$, the idempotents $e_\rho = u_{\overline{\rho}}^{-1} e u_{\overline{\rho}}$ are different and are contained in E_i because $E_i e_\rho = (E_i e)^{\overline{\rho}} = e^{\overline{\rho}} = e_\rho$. Clearly, $\sum_\rho e_\rho$ is invariant with respect to all $g \in G$, whence this sum is a central idempotent of the algebra $G \times K$. Hence, $\sum_\rho e_\rho = E_i$. Thus, $\Lambda_i = \sum_\rho^\oplus e_\rho \Lambda_i = \sum_\rho^\oplus e_\rho E_i \Lambda = \sum_\rho^\oplus e_\rho \Lambda$.

The ideals $e_\rho \Lambda_i$ are isomorphic because $e_\rho \Lambda_i = u_{\overline{\rho}}^{-1} e \Lambda_i$. Consequently, the algebra Λ_i is isomorphic to the algebra of matrices of order $(G\colon G_e)$ over the algebra $e\Lambda_i e = e\Lambda E_i e = e\Lambda e$.

If $g \notin G_e$, then $e^g \neq e$ and $e u_g e = u_g e^g e = 0$. Therefore, the projection $\Lambda \to e\Lambda e$ annihilates all u_g for $g \notin G_e$, and $\Lambda_e = e\Lambda e = e\overline{\Lambda}_e e = e\overline{\Lambda}_e$, where $\overline{\Lambda}_e = G_e \times K$ is the crossed product of the group G_e and the field K. The algebra Λ_e is a simple component of the algebra $\overline{\Lambda}_e$ and the center of it is the field K_e of elements of K that are invariant with respect to $F_e = \varphi(G_e)$ (projected to the algebra Λ_e by $K \to Ke$).

Let μ be the irreducible representation of the group N that corresponds to the idempotent e, and let d be the degree of this representation. Extend μ to the algebra $K[N]$, assuming that $\mu(xt) = x\mu(t)$, where $x \in K$, $t \in N$. Therefore, μ realizes an isomorphism of the simple algebra $K[N]e$ onto the algebra K_d of matrices of degree d over the field K and a homomorphism $K[N] \to K_d$.

Choose $g \in G$ and construct two other representations of the algebra $K[N]$ connected with the representation μ: $\overline{\mu}_g(z) = \mu(u_g^{-1} z u_g)$ and $\overline{\overline{\mu}}_g(z) = (\mu(z))^{\varphi(g)}$ (in the second case the automorphism $\varphi(g)$ is applied to each element of the matrix $\mu(z)$; here z is an arbitrary element of the algebra $K[N]$).

LEMMA 2.8. *The representations $\overline{\mu}_g$ and $\overline{\overline{\mu}}_g$ are equivalent if and only if $g \in G_e$.*

PROOF. Let $g \in G_e$. The map $\mu y \to (\mu(u_g^{-1} y u_g))^{g^{-1}}$, where $y \in K[N]e$, yields an automorphism of the algebra K_d (isomorphic to $K[N]e$). This automorphism acts identically on the scalar matrices xE, $x \in K$; indeed, $(\mu(u_g^{-1} x u_g))^{g^{-1}} = \mu(x^{\varphi(g)})^{g^{-1}} = (x^{\varphi(g)} E)^{g^{-1}} = xE = \mu(xe)$, therefore, this automorphism is identical on the center of the algebra K_d. Hence, it is an inner automorphism and there exists a matrix $C_g \in K_d$ such that for each $y \in K[N]e$, we have $(\mu(u_g^{-1} y u_g))^{g^{-1}} = C_g^{-1} \mu(y) C_g$. Since $\mu(z) = \mu(ze)$ for $z \in K[N]e$, we obtain $\overline{\mu}_g(z) = \mu(u_g^{-1} z u_g) = (C_g^{-1} \mu(z) C_g)^g = T_g^{-1} \overline{\overline{\mu}}_g(z) T_g$, where $T_g = C_g^{\varphi(g)}$.

Conversely, let $\overline{\mu}_g$ and $\overline{\overline{\mu}}_g$ be equivalent: $\overline{\mu}_g(z) = T_g^{-1}(\mu(z))^{\varphi(g)} T_g$. Let us denote the character of the representation μ by χ: $\chi(z) = \operatorname{Tr} \mu(z)$. Then the characters of the representations $\overline{\mu}_g$, $\overline{\overline{\mu}}_g$ on the elements $t \in N$ are, therefore, equal to $\chi(t^g)$ and $\chi(t)^{\varphi(g)}$, respectively, and are equal, i.e., $\chi(t)^{\varphi(g)} = \chi(t^g)$. As is known, the idempotent e is given by the formula $e = (d/n)\sum_{t\in N}\chi(t^{-1})u_t$ (see [**6**]). Then

$$e^g = \frac{d}{n}\sum_{t\in N}\chi(t^{-1})^{\varphi(g)}u_t^g = \frac{d}{n}\sum\chi(t^{-g})u_{t^g} = e,$$

so that $e^g = e$, whence $g \in G_e$. The lemma is proved. □

We begin the study of the algebra Λ_e with the case where the idempotent e corresponds to the representation μ of degree 1 of the group N (and of the algebra $K[N]$). Therefore, μ maps $K[N]$ to the field K, and, thus, for $z \in K[N]$ we have $ze = \mu(z)e$. The commutator subgroup of N is mapped to the unit element by the representation μ, whence we may regard μ as a representation of the Abelian group $\overline{N} = N/[N,N]$ in the field K.

The group G_e introduced above is an extension of the group N by $F_e = \varphi(G_e)$. For each $f \in F_e$ we take a preimage $\overline{f} \in G_e$. Then $\overline{f}_1\overline{f}_2 = \overline{f_1f_2}h(f_1,f_2)$ for $h(f_1,f_2) \in N$. The elements $h(f_1,f_2)$ define a two-dimensional cocycle (in general, noncommutative). The elements $\overline{h}(f_1,f_2) = \mu(h(f_1,f_2))$ define a two-dimensional cocycle of the group F_e in the group K^* (since $\mu(h)e = he$). The algebra Λ_e is generated by the field K (more precisely, by a subfield K_e of it) and by the elements $u_{\overline{f}}e$ because for $t \in N$ we have $te = \mu(t)e$. Moreover, $xu_{\overline{f}}e = u_{\overline{f}}ex^f$, $u_{\overline{f}_1}eu_{\overline{f}_2}e = u_{\overline{f_1f_2}}e\overline{h}(f_1,f_2)e$, where $x \in K$, $f \in F_e$. Therefore, the algebra Λ_e is the crossed product of the field K and the group $F_e = \mathrm{Gal}(K/k_e)$ with system of factors $\overline{h}(f_1,f_2)$. As was proved above, the algebra $\Lambda_i = \Lambda E_i$ is isomorphic to the algebra of matrices of order $(G:G_e)$ over the subalgebra Λ_e. Hence, Λ_i is the full algebra of matrices (of order $(F:1)$ over k) if and only if the crossed product of K and F_e with a system of factors $\overline{h}(f_1,f_2)$ is the algebra of matrices of order $(F_e:1)$ over a subalgebra of it, and this is possible only if it splits completely, i.e., the image of $\overline{h}(f_1,f_2)$ in the cohomology group $H^2(F_e,K^*)$ is trivial.

Now let an idempotent e correspond to a representation μ of degree $d > 1$ for the group N (and for the algebra $K[N]$), so that $K[N]e$ is isomorphic to the algebra of matrices K_d of degree d over the field K. By the lemma, for every $g \in G_e$ there exist matrices $T_g \in K_d$ such that

$$\mu(u_g^{-1}zu_g) = T_g^{-1}(\mu(z))^{\varphi(g)}T_g, \quad z \in K[N].$$

The matrices T_g are uniquely determined up to scalar multiplier. If $g_1 = gt$, where $t \in N$, then

$$\begin{aligned} T_{gt}^{-1}(\mu(z))^{\varphi(g)}T_{gt} &= \mu(u_{gt}^{-1}zu_{gt}) = \mu(u_t^{-1}u_g^{-1}zu_gu_t) = \mu(t^{-1})\mu(u_g^{-1}zu_g)\mu(t) \\ &= \mu(t)^{-1}T_g^{-1}\mu(z)^{\varphi(g)}T_g\mu(t). \end{aligned}$$

Therefore, we can take the matrices $T_g\mu(t)$ as $T_{g_1} = T_{gt}$. Use this fact and choose matrices T_g such that $T_{\overline{f}t} = T_{\overline{f}}\mu(t)$, where the $\overline{f}$ are the preimages in G_e of the elements $f \in F_e$.

We compute $\mu(u_{g_2}^{-1}u_{g_1}^{-1}zu_{g_1}u_{g_2}) = \mu(u_{g_1g_2}^{-1}zu_{g_1g_2})$ in two ways. From one hand,

$$\mu(u_{g_1g_2}^{-1}zu_{g_1g_2}) = T_{g_1g_2}^{-1}\mu(z)^{\varphi(g_1g_2)}T_{g_1g_2};$$

from the other hand,

$$\mu(u_{g_2}^{-1}u_{g_1}^{-1}zu_{g_1}u_{g_2}) = T_{g_2}^{-1}T_{g_1}^{-g_2}\mu(z)^{\varphi(g_1g_2)}T_{g_1}^{g_2}T_{g_2}.$$

Therefore, we have $T_{g_1g_2} = T_{g_1}^{g_2}T_{g_2}c_{g_1,g_2}$, where the c_{g_1,g_2} are scalar matrices identified with the elements of the field K. Show that the elements c_{g_1,g_2} depend only on $\varphi(g_1)$ and $\varphi(g_2)$ under our choice of the matrices T_g. Indeed, we replace g_1 by g_1t_1 and g_2 by g_2t_2 with $t_1, t_2 \in N$. Then $T_{g_1t_1g_2t_2} = T_{g_1t_1}^{g_2t_2}T_{g_2t_2}c_{g_1t_1,g_2t_2}$. But

$$\begin{aligned}
T_{g_1t_1g_2t_2} &= T_{g_1g_2t_1^{g_2}t_2} = T_{g_1g_2}\mu(t_1^{g_2}t_2) = T_{g_1}^{g_2}T_{g_2}\mu(t_1^{g_2}t_2)c_{g_1,g_2},\\
T_{g_1t_1}^{g_2t_2}T_{g_2t_2} &= (T_{g_1}\mu(t_1))^{g_2t_2}T_{g_2}\mu(t_2) = T_{g_1}^{g_2}T_{g_2}T_{g_2}^{-1}(\mu(t_1))^{\varphi(g_2)}T_{g_2}\mu(t_2)\\
&= T_{g_1}^{g_2}T_{g_2}\mu(u_{g_2}^{-1}t_1u_{g_2})\mu(t_2) = T_{g_1}^{g_2}T_{g_2}\mu(t_1^{g_2}t_2).
\end{aligned}$$

Hence $c_{g_1t_1,g_2t_2} = c_{g_1,g_2}$.

Now let $q_g = \mu^{-1}(T_g) \in K[N]e$ and $w_g = u_gq_g^{-1}$. Then

$$w_{gt} = u_{gt}q_{gt}^{-1} = u_gu_t(\mu^{-1}(T_g\mu(t)))^{-1} = u_gu_tu_t^{-1}q_g^{-1} = w_g,$$

so that the w_g depend only on $\varphi(g)$. Moreover,

$$w_{g_1}w_{g_2} = u_{g_1}q_{g_1}^{-1}u_{g_2}q_{g_2}^{-1} = u_{g_1}u_{g_2}u_{g_2}^{-1}q_{g_1}^{-1}u_{g_2}q_{g_2}^{-1} = w_{g_1g_2}q_{g_1g_2}u_{g_2}^{-1}q_{g_1}^{-1}u_{g_2}q_{g_2}^{-1}.$$

The element $q_{g_1g_2}u_{g_2}^{-1}q_{g_1}^{-1}u_{g_2}q_{g_2}^{-1}$ belongs to $K[N]e$ and its image under the map μ onto K_d is equal to $T_{g_1g_2}T_{g_2}^{-1}(T_{g_1}^{-1})^{g_2}T_{g_2}T_{g_2}^{-1} = T_{g_1g_2}T_{g_2}^{-1}T_{g_1}^{-g_2} = c_{g_1,g_2}$. Hence, $w_{g_1}w_{g_2} = w_{g_1g_2}c_{g_1g_2}$. Therefore, $c_{g_1g_2}$ is a two-dimensional cocycle of the group F_e with values in K^*.

The elements w_g belong to the algebra $e\Lambda_e$ because $g \in G_e$. Let us establish the commutation relations of these elements with the elements of the algebra $K[N]e$. Let $z \in K[N]e$. Then $w_g^{-1}zw_g = q_gu_g^{-1}zu_gq_g^{-1} \in K[N]e$ and

$$\mu(w_g^{-1}zw_g) = T_g\mu(u_g^{-1}zu_g)T_g^{-1} = T_gT_g^{-1}(\mu(z))^{\varphi(g)}T_gT_g^{-1} = \mu(z)^{\varphi(g)}.$$

In particular, we have $\mu(w_g^{-1}xw_g) = x^{\varphi(g)}E$ for $x \in K$, whence $w_g^{-1}xw_g = x^{\varphi(g)}$.

Let D_e be the algebra generated by the field K and the elements w_g; D_e is a simple algebra and is isomorphic to the crossed product of the field K and the group F_e with system of factors $c_{g_1g_2} = c_{\varphi(g_1),\varphi(g_2)}$. The center of this algebra, the field K_e, is the subfield of the field K that belongs to the subgroup F_e. Let us find the centralizer of this algebra in the algebra $\Lambda_e = e\Lambda e$. The elements of $K[N]e$ commute in Λ_e with the elements $x \in K$. Let $z \in K[N]e$ commute with all w_g, $g \in G_e$. Then $\mu(z) = \mu(w_g^{-1}zw_g) = \mu(z)^{\varphi(g)}$, i.e., $\mu(z)$ is a matrix in K_d with entries in the field K_e. Therefore, the required centralizer is isomorphic to the algebra of matrices of order d over the field K_e. Thus, the algebra $e\Lambda e$ is isomorphic to the algebra of matrices of order d over the crossed product $(K, F_e, c_{\varphi(g_1),\varphi(g_2)})$.

We formulate the obtained results as a theorem.

THEOREM 2.8. *In the embedding problem $(K/k, G, \varphi, N)$ let K be a field such that its characteristic does not divide $(N{:}1)$ and all irreducible representations of the group N over K are absolutely irreducible. Denote the minimal central idempotents of the algebra $\Lambda = G \times K$ by $E_1, \dots, E_m$, so that ΛE_i are simple components of the algebra Λ. Let e be a minimal central idempotent of the algebra $K[N]$ that is contained in E_i and let G_e be the group of elements $g \in G$ commuting with e. Then the component ΛE_i of the algebra Λ is isomorphic to the algebra of matrices of order $d \cdot (G : G_e)$ (where d is the degree of the irreducible representation corresponding to the idempotent e) over the crossed product of K and the group $F_e = \varphi(G_e)$ with a* *system of factors* $c_{f_1,f_2} \in Z^2(F_e, K^*)$. *If the idempotent* e *corresponds to a one-dimensional representation* μ, *then one can take* $c_{f_1,f_2} = \mu(h_e(f_1, f_2))$, *where* h_e *is the canonical (standard) cocycle for the extension* G_e *of the group* N *by* F_e.

§9. Reduction of the compatibility condition

As usual, we assume that the characteristic of the base field k does not divide the order of the kernel N. Clearly, if the compatibility condition holds for the original embedding problem $(K/k, G, \varphi, N)$, then it holds for associated problems as well, in particular, for the problem $(K/k, G/N', \varphi', N/N')$, where N' is the commutator subgroup of the group N (and, therefore, the kernel N/N' is Abelian). However, the converse statement is valid in some important special cases. Namely, this occurs if in any finite extension k_1 of the field k, the exponent of the simple central algebra over k_1 in the Brauer group is equal to the index of this algebra (i.e., to the square root of the dimension of the division ring that is contained in it). In particular, this condition is satisfied for local fields and for algebraic number fields.

THEOREM 2.9. *If the field* k *satisfies the condition mentioned above and the compatibility condition holds for the Abelian embedding problem* $(K/k, G/N', \varphi', N/N')$, *where* N' *is the commutator subgroup of* N, *then it holds for the problem* $(K/k, G, \varphi, N)$ *as well.*

PROOF. The lifting procedure enables us to assume that the field K contains all necessary roots of 1, and, thus, all irreducible representations of the group N over K are absolutely irreducible. Let e_0 be a minimal idempotent of the algebra $K[N]$ to which there corresponds a one-dimensional representation μ_0. Since μ_0 acts trivially on the commutator subgroup N' of the group N, we can identify it with a representation $\overline{\mu}_0$ of the group N/N'.

The converse is also valid: every irreducible representation $\overline{\mu}_0$ of the Abelian group N/N' over K is one-dimensional and can be lifted to a homomorphism $\mu_0 \colon N \to K^*$, which is a one-dimensional representation of the group N.

Now let μ be an irreducible representation of the group N (and of the group algebra $K[N]$) over K of degree $d > 1$, and let e be the corresponding idempotent. Consider the one-dimensional representation μ_0 given by the formula $\mu_0 = \det \mu(t)$, where $t \in N$, and let $e_0 = (N:1)^{-1} \sum_{t \in N} \mu_0(t^{-1}) u_t$ be the corresponding idempotent. First we show that $G_e \subset G_{e_0}$, where G_e and G_{e_0} are subgroups of the group G whose elements commute with e and e_0, respectively. Let $g \in G_e$. Then, as shown in the preceding section, there exists a matrix $T_g \in K_d$ such that $\mu(u_g^{-1} z u_g) = T_g^{-1}(\mu(z))^{\varphi(g)} T_g$. Here, by passing to the determinants and assuming that $z = t \in N$, we obtain $\mu_0(t^g) = (\mu_0(t))^{\varphi(g)}$, which means that $e_0^g = e_0$.

The matrices T_g constructed in the preceding section have the properties: $T_{gt} = T_g \mu(t)$ and $T_{g_1 g_2} = T_{g_1}^{g_2} T_{g_2} c_{g_1,g_2}$, where c_{g_1,g_2} is obtained by lifting of a cocycle of $Z^1(F_e, K^*)$ to the group G_e. Let $\delta_g = \det T_g \in K^*$, then $\delta_{g_1 g_2} = \delta_{g_1}^{g_2} \delta_{g_2} c_{g_1,g_2}^d$ and $\delta_{gt} = \det T_{gt} = \det(T_g \mu(t)) = \delta_g \mu_0(t)$. Let $g_1 = \overline{f}_1 t_1$, $g_2 = \overline{f}_2 t_2$, where $f_1, f_2 \in F_e$, $\overline{f}_1$, $\overline{f}_2$ are the preimages of them in G_e, $t_i \in N$. Then $\delta_{g_1 g_2} = \delta_{\overline{f}_1 t_1 \overline{f}_2 t_2} = \delta_{\overline{f_1 f_2} h(f_1,f_2) t_1^{\overline{f}_2} t_2} = \delta_{\overline{f_1 f_2}} \mu_0 h(f_1,f_2) \mu_0(t_1^{\overline{f}_2} t_2)$, where h is the cocycle of the extension G_e of the group N by F_e. Moreover, $\delta_{g_1}^{g_2} = \delta_{\overline{f}_1 t_1}^{\overline{f}_2 t_2} = \delta_{\overline{f}_1}^{f_2} \mu_0(t_1)^{f_2} = \delta_{\overline{f}_1}^{f_2} \mu_0(t_1^{f_2})$ (because $f \in F_e \subset F_{e_0}$) and $\delta_{g_2} = \delta_{\overline{f}_2} \mu_0(t_2)$. Substituting these expressions in the relation

$\delta_{g_1g_2} = \delta_{g_1}^{g_2}\delta_{g_2}c^d_{g_1,g_2}$, we obtain

$$c^d_{g_1,g_2} = \delta_{\overline{f_1f_2}}\delta_{\overline{f}_1}^{-f_2}\delta_{\overline{f}_2}^{-1}\mu_0(h(f_1,f_2)).$$

Since the compatibility condition holds for the Abelian embedding problem $(K/k, G/N', \varphi', N/N')$, all simple components of the algebra $(G/N')\times K$, and, thus, all components of the algebra Λ corresponding to one-dimensional representations, split completely. Therefore, the algebra $e_0\Lambda e_0$ is the full matrix algebra of order $(F_{e_0}:1)$ over its center K_{e_0}. This means that the image of the canonical cocycle $h(f_1,f_2)$ $(f_i \in F_{e_0})$ in the field K under the map μ_0 is homologous to the trivial cocycle. Since $F_e \subset F_{e_0}$, it follows that $\mu_0(h(f_1,f_2))$, $f_1, f_2 \in F_e$, is a coboundary. Then $c^d_{g_1,g_2} = c_{\varphi(g_1),\varphi(g_2)}$, which is different from $\mu_0(h(f_1,f_2))$ by a coboundary, is also a coboundary.

By the preceding theorem, the algebra $\Lambda_e = e\Lambda e$ is the algebra of matrices of order d over the crossed product $D_e = (K, F_e, c_{f_1,f_2})$. Since $c^d_{f_1,f_2}$ is homologous to the unity, the exponent of this crossed product in the Brauer group is a divisor of the number d. Then the index of this algebra is $d' \mid d$. Therefore, D_e is the algebra of matrices of order $(F_e:1)/d'$ over a subalgebra. Hence, Λ_e is isomorphic to the algebra of matrices of order $(F_e{:}1)$ over a subalgebra of it, so every simple component of the algebra Λ is the algebra of matrices of order $(F:1)$ over a subalgebra, and, by Corollary 3 to Theorem 2.2.1, this means that the compatibility condition holds for the embedding problem $(K/k, G, \varphi)$. □

The theorem just proved shows that the compatibility condition is rather weak in the case of a non-Abelian kernel. We show that in the general case (when our assumption about the structure of the Brauer group of the field k does not hold), the compatibility condition cannot nevertheless be reduced to the compatibility condition for the associated Abelian problem.

EXAMPLE. Let N be the quaternion group with generators a, b, so that $a^4 = 1$, $b^2 = a^2$, $ba = a^{-1}b$. As G we take the extension of N by the cyclic group of order 2 with generator f, and the relations $\overline{f}^2 = a^2$, $\overline{f}^{-1}a\overline{f} = a^{-1}$, $\overline{f}^{-1}b\overline{f} = ab$ hold for the preimage $\overline{f}$ in G. The homomorphism φ takes a and b to the unit element of the group $F = G/N$ and $\varphi(\overline{f}) = f \in F$. With this extension we relate the embedding problem $(K/k, G, \varphi, N)$, where K is a quadratic extension of the field k, $K = k(\sqrt{m})$, and consider the compatibility condition for this problem. In the algebra $\Lambda = G \times K$ there are central orthogonal idempotents $E_1 = (1+a^2)/2$, $E_2 = (1-a^2)/2$, $E_1 + E_2 = 1$. The algebra $\Lambda E_1 = \Lambda_1$ is isomorphic to the crossed product $(G/N') \times K$, where $N' = \{a^2\}$ is the commutator subgroup of the kernel. Since G/N' is a semidirect extension of N/N' by F, the algebra Λ_1 is the algebra of matrices of order 2 over its center. Consider the algebra $\Lambda_2 = \Lambda E_2$. As is easily shown, this algebra is simple because its center coincides with the field k and its dimension over k equals 16. We find the structure of Λ_2 by representating it as the tensor product of cyclic algebras. The elements aE_2, bE_2, together with the field K, generate a subalgebra isomorphic to the standard algebra of quaternions over k because $(aE_2)^2 = (bE_2)^2 = -E_2$, $bE_2 \cdot aE_2 = -aE_2 \cdot bE_2$. We find the centralizer of this subalgebra in the algebra Λ_2. Let $(z_1 + u_{\overline{f}}z_2)E_2 \in \Lambda_2$ $(z_i \in K[N])$ commute with aE_2. Then $aE_2(z_1+u_{\overline{f}}z_2)E_2 = (az_1+u_{\overline{f}}a^f z_2)E_2 = (az_1 - u_{\overline{f}}az_2)E_2$, $(z_1 + u_{\overline{f}}z_2)E_2 \cdot aE_2 = (z_1a + u_{\overline{f}}z_2a)E_2$. Therefore, $az_1E_2 = z_1aE_2$, $az_2E_2 = -z_2aE_2$, whence $z_1 = y_{11} + ay_{12}$, $z_2 = by_{21} + bay_{22}$, where $y_{ij} \in K$. Now let

$(z_1 + u_{\bar{f}} z_2)E_2$ commute with bE_2. We have $bE_2(z_1 + u_{\bar{f}} z_2)E_2 = (bz_1 + u_{\bar{f}} abz_2)E_2$, $(z_1 + u_{\bar{f}} z_2)E_2 bE_2 = (z_1 b + u_{\bar{f}} z_2 b)E_2$, i.e., $bz_1 E_2 = z_1 bE_2$, $abz_2 E_2 = z_2 bE_2$. Hence $y_{12} = 0$, $y_{22} = -y_{21}$. Thus, the centralizer of the quaternion subalgebra in the algebra Λ_2 is generated by the field KE_2 and the element $v = u_{\bar{f}}(b - ba)E_2$. The element v anticommutes with $\sqrt{m}E_2$ and $v^2 = u_{\bar{f}}(b - ba)u_{\bar{f}}(b - ba)E_2 = u_{\bar{f}}^2(ab - aba^{-1})(b - ba)E_2 = a^2(ab + b)(b - ba)E_2 = -E_2(-aE_2 - E_2 - E_2 + aE_2) = 2E_2$. Therefore, this centralizer is isomorphic to the algebra of generalized quaternions, in which the squares of anticommuting vectors are equal to m and 2. Consequently, the algebra Λ_2 is isomorphic to the tensor product of the algebras $(-1, -1)$ and $(2, m)$ over the field k. The compatibility condition for our problem means that this tensor product is the algebra of matrices of order 2 over a subalgebra. By the Albert theorem (see Appendix), this is equivalent to the fact that the algebras $(-1, -1)$ and $(2, m)$ contain a common quadratic subfield, i.e., the quadratic form $-x_1^2 - x_2^2 - x_3^2 = 2x_4^2 + mx_5^2 - 2mx_6^2$ must have a nontrivial zero over the field k.

Now let the field k be obtained as a simple transcendental extension $\mathbb{Q}(m)$ of the field of rational numbers $\mathbb{Q}$. We show that the quadratic form indicated above does not have nontrivial zeros. Suppose the contrary. Then, as x_i one can take polynomials in $\mathbb{Q}(m)$ relatively prime in common. Comparing the constant terms, we see that the constant terms of the polynomials x_1, x_2, x_3, x_4 are equal to zero. Then $mx_5^2 - 2mx_6^2 \equiv 0 \pmod{m^2}$ holds, which is impossible if x_5 and x_6 are not divisible by m.

Thus, the compatibility condition does not hold for the embedding problem indicated above.

CHAPTER 3

The Embedding Problem with Abelian Kernel

§1. The Brauer problem with cyclic kernel

Let the kernel N in the embedding problem $(K/k, G, \varphi, N)$ be a cyclic subgroup of order n with generator t. We assume that the characteristic of k does not divide n and K contains a primitive nth root ζ of unity. Such a problem is called the *Brauer problem* if the Galois group F of the field K/k acts on ζ and on t in a "consistent way", i.e., if $g^{-1}tg = t^k$, where k depending on g is defined $\bmod n$, then $\zeta^{\varphi(g)} = \zeta^k$. It is clear that this consistency does not depend on the choice of t and ζ.

THEOREM 3.1. *For the Brauer problem, compatibility is sufficient for solvability.*

PROOF. In a compatibility system $l(g) \in K[N]$, we replace $t \in N$ by ζ. We obtain a system of elements $\lambda(g) \in K$. From $l(g_1)^{g_2} l(g_2) = l(g_1 g_2)$ and $l(t) = t^{-1}$ for the compatibility system we obtain the relations $\lambda(g_1)^{g_2}\lambda(g_2) = \lambda(g_1 g_2)$ and $\lambda(t) = \zeta^{-1}$ by consistency of the action of $g \in G$ on t and on ζ. We let $\mu(g) = \lambda(g)^n$. Then $\mu(g_1)^{g_2}\mu(g_2) = \mu(g_1 g_2)$ and $\mu(t) = 1$. From these relations it follows that $\mu(gt) = \mu(g)$, so that $\mu(g)$ depends only on $f = \varphi(g) \in \mathrm{Gal}(K/k) = F$. By Speiser's theorem $\mu(g) = b^{g-1}$ for some $b \in K$. Consider the algebra $A = K[x]/(x^n - b)$. Let θ be the image of x in this algebra. Then $\theta^n = b$. We show that $\theta \to \theta\lambda(g)$ is an automorphism of A for any $g \in G$ extending the automorphism $\varphi(g)$ of the field K over k. Indeed, $(\theta\lambda(g))^n = \theta^n\lambda(g)^n = b\mu(g) = b^g = b^{\varphi(g)} = (\theta^n)^{\varphi(g)}$. The automorphisms $\widetilde{g}\colon \theta \to \theta\lambda(g)$ form a group isomorphic to G, since $(\theta^{\widetilde{g}})^{\widetilde{h}} = (\theta\lambda(g))^{\widetilde{h}} = \theta^{\widetilde{h}}\lambda(g)^h = \theta\lambda(h)\lambda(g)^h = \theta\lambda(gh) = \theta^{\widetilde{gh}}$ $(g, h \in G)$. The elements of k, and only they, are fixed under the action of these automorphisms. Then $(A:k) = (A:K)(K:k) = n \cdot (F:1) = (G:1)$; therefore, A is a Galois algebra over k (Theorem 1.4).

Note that the element b constructed with the condition $\mu(g) = b^{g-1}$ is determined up to a factor from k^*. Therefore, we obtain a class of solutions for the Brauer problem in the form $A = K[x]/(x^n - bc)$, $c \in k^*$; for $c \in k^{*n}$ we have equivalent solutions.

We show that this construction gives us all solutions of the Brauer problem. Indeed, let A be a solution. We denote by $z \in A$ an element generating a normal basis of A over K. We set $\eta = \sum_{k=0}^{n-1} z^{t^k}\zeta^k$ (the Lagrange resolvent). It is clear that $\eta \neq 0$ by linear independence of the basis elements $z, z^t, \dots, z^{t^{n-1}}$ over K. Then $\eta^t = \zeta^{-1}\eta$. Therefore, $d = \eta^n$ is invariant under the action of t, so that $d \in K$. Further, $d \neq 0$, since A does not contain nilpotent elements, and, therefore, η is an invertible element because $\eta^{-1} = \eta^{n-1}d^{-1}$. For $g \in G$, we have $\eta^g = \sum_{k=0}^{n-1} z^{t^k g}(\zeta^g)^k = \sum_{k=0}^{n-1} z^{g g^{-1}t^k g}(\zeta^g)^k$. The elements $g^{-1}t^k g$, for $k = 0, \dots, n-1$, range over the entire group N, and the elements $(\zeta^g)^k$ range over the corresponding roots of 1 by consistency of the action of the automorphisms $g \in G$ on t and ζ. Thus, $\eta^g = \sum_{k=0}^{n-1}(z^g)^{t^k}\zeta^k$. Therefore $\eta^{gt} = \eta^g\zeta^{-1}$ and $(\eta^g\eta^{-1})^t = \eta^g\eta^{-1}$, so that $\eta^g\eta^{-1} = \mu(g) \in K$. It is clear that $\mu(t) = \zeta^{-1}$, and from the equality $\eta^g = \eta\mu(g)$ it is easy to see that $\mu(g_1)^{g_2}\mu(g_2) = \mu(g_1g_2)$. Raising the equality $\eta^g = \eta\mu(g)$ to the power n, we obtain $\mu(g)^n = d^{g-1}$.

The elements $\mu(g)$ constructed here and $\lambda(g)$ we have constructed earlier differ by a one-dimensional cocycle from $Z^1(F, K^*)$. By Speiser's theorem, we have $\lambda(g) = \mu(g)y^{g-1}$, $y \in K^*$. We replace η by $\theta_1 = \eta y$. It is clear that θ_1 is also a primitive element of A. Moreover, $\theta_1^g = \eta^g y^g = \eta\mu(g)y^g = \eta y\lambda(g) = \theta_1\lambda(g)$. We raise the equality $\theta_1^g = \theta_1\lambda(g)$ to the power n. Taking into account that $\lambda(g)^n = b^{g-1}$, $b \in K$, we obtain $b^{g-1} = (\theta_1^n)^{g-1}$, i.e., $\theta_1^n = bc$, $c \in K^*$. Thus, every solution of the Brauer problem can be obtained in the way described above, and the solutions are in one-to-one correspondence with $k^*/(k^*)^n$. As is very easy to see, this quotient group determines the set of solutions of the semidirect Brauer embedding problem.

Then we observe that if the kernel N of an embedding problem is a cyclic group of the second order, then the problem is Brauer because $(-1)^g = -1$, and if $t^2 = 1$, then for arbitrary $g \in G$ we have $g^{-1}tg = t$. □

§2. The Brauer problem with Abelian kernel

Let a normal extension K/k with group G contain a primitive nth root of unity, and let the kernel N of the embedding problem $(K/k, G, \varphi, N)$ be an Abelian group of exponent n. Clearly, the characteristic of k does not divide n (because of the existence of a primitive root $\sqrt[n]{1}$). By characters χ of the group N we mean the homomorphisms of N to the group of roots of unity contained in K. The characters form a group $\widehat{N}$ isomorphic (not canonically) to N. On $\widehat{N}$ we introduce the action of G by the rule $\chi^g(t) = (\chi(gtg^{-1}))^g$, $g \in G$, $t \in N$. An embedding problem is called Brauer if $\chi^g = \chi$ for all $\chi \in \widehat{N}$, $g \in G$. This means that $(\chi(t))^g = \chi(g^{-1}tg)$. Therefore, if N is a cyclic group, then the problem is Brauer in the sense of the previous section. Let $\zeta \in K$ be a primitive nth root of unity and let $\zeta^g = \zeta^{v(g)}$, $v(g) \in \mathbb{Z}/n\mathbb{Z}$. Then $\chi(t^g t^{-v(g)}) = (\chi(t))^g(\chi(t))^{-v(g)} = 1$. This is valid for all characters χ, so that $t^g = t^{v(g)}$. Therefore, all cyclic subgroups of N are invariant.

We decompose the group N into a direct product of cyclic groups. Then the embedding problem can be represented as a product (Chapter 1, §12) of the problems $(K/k, G_i, \varphi_i, N_i)$, where the N_i are direct factors, G_i is the quotient group of G by the product of all factors of the group N except for N_i, and the φ_i are homomorphisms of G_i onto $\mathrm{Gal}(K/k) = F \approx G/N$ corresponding to φ. Clearly, all these problems are Brauer with cyclic kernels.

As we have seen before, the solvability of the original problem $(K/k, G, \varphi, N)$ is equivalent to the simultaneous solvability of all problems $(K/k, G_i, \varphi_i, N_i)$, and the compatibility condition for the original problem is equivalent to the compatibility conditions for the factors of the problem. This and Theorem 3.1 immediately imply the following result.

THEOREM 3.2. *For the solvability of the Brauer problem with Abelian kernel it is necessary and sufficient that the compatibility condition be fulfilled.*

§3. The compatibility condition and the solvability of the associated Brauer problems

Our next goal is to give an additional statement of the compatibility condition for the embedding problem $(K/k, G, \varphi, N)$ assuming that the kernel N is an Abelian group of exponent n and the field K contains a primitive nth root of unity. We again turn to analysis of the crossed product $\Lambda = G \times K$.

Under these assumptions, the elements $e_\chi = \frac{1}{n}\sum_{t\in N}\chi(t)^{-1}u_t$ are minimal idempotents of the commutative semisimple algebra $K[N]$. Then

$$u_g^{-1}e_\chi u_g = \frac{1}{n}\sum_{t\in N}\chi(t^{-1})^g u_{t^g} = \frac{1}{n}\sum \chi^g(t^{-1})u_t = e_{\chi^g}.$$

We denote by F_x the subgroup of $F = \varphi(G) = \mathrm{Gal}(K/k)$ consisting of the automorphisms leaving e_χ fixed. The idempotents conjugate to e_χ are in one-to-one correspondence with the cosets ρ in the decomposition of F with respect to F_χ. It is clear that $E = \sum_\rho e_{\chi\overline{\rho}}$ is a minimal central idempotent of the algebra $\Lambda = G \times K$, and the corresponding component of this algebra is isomorphic to the matrix algebra of degree $(F : F_\chi)$ over the algebra $\Lambda_\chi = e_\chi(G \times K)e_\chi$ (see A.1.2). If $\varphi(g) \notin F_\chi$, then $e_\chi u_g e_\chi = u_g e_{\chi^g} e_\chi = 0$. Further, $e_\chi = e_\chi\chi(t)$. Let $h(f_1, f_2)$ be a cocycle in the canonical class of the extension $1 \to N \to G \to F \to 1$, so that $u_{\overline{f_1}}u_{\overline{f_2}} = u_{\overline{f_1 f_2}}u_{h(\overline{f_1},\overline{f_2})}$ for some preimages $\overline{f} \in G$ of the elements $f \in F$. Then

$$e_\chi u_{\overline{f_1}}u_{\overline{f_2}}e_\chi = e_\chi u_{\overline{f_1 f_2}}e_\chi\chi(h(f_1, f_2)).$$

Thus, the algebra Λ_χ is isomorphic to the crossed product of the group F_χ and the field K for the cocycle $\chi(h(f_1, f_2))$. The field K_χ of elements in K left fixed by F_χ is the center of Λ_χ. The dimension of Λ_χ over the center is equal to the square of the order of the group F_χ, and the dimension over the center of the corresponding component of the algebra Λ is equal to the square of the order of F. Therefore, for compatibility it is necessary and sufficient that all the algebras Λ_χ split over their centers.

We reformulate this condition in cohomological terms. The group W of nth roots of unity is embedded in the multiplicative group K^* of the field K. This embedding induces a homomorphism μ of the groups $H^2(F_\chi, W)$ to $H^2(F_\chi, K^*)$. The compatibility condition can be stated as follows:

$$\mu\chi(h(f_1, f_2)) = 0$$

for arbitrary χ.

In particular, for a Brauer problem with cyclic kernel all characters are powers of a certain χ_0, all the groups F_χ coincide with F, the fields K_χ are equal to k, and the compatibility condition which is also a solvability condition looks as follows:

$$\mu\chi_0(h(f_1, f_2)) = 0.$$

Now let M_χ be the kernel of the character χ, and let $N_\chi = N/M_\chi$. It is clear that N_χ is a cyclic subgroup isomorphic to the group of values of χ. The character χ is a generator of the character group of N_χ. Further we put $H_\chi = \varphi^{-1}(F_\chi)$ and $G_\chi = H_\chi/M_\chi$. It is clear that G_χ is an extension of the cyclic group N_χ by the group F_χ with respect to the homomorphism φ_χ of G_χ onto F_χ induced by the homomorphism φ of G onto F.

The problem $(K/K_\chi, G_\chi, \varphi_\chi, N_\chi)$ is obviously a Brauer problem with cyclic kernel, associated with the original problem. We note that for the Brauer problem $(K/k, G, \varphi, N)$ with cyclic kernel, we have $F_\chi = F$ for all characters χ of N, so that $K_\chi = k$. All components of the crossed product $G \times K$ are simple central algebras over k, and in the Brauer group all of them are powers of a single one corresponding to the generator of the character group. In order that these algebras split it is sufficient that the algebra corresponding to the generator of the character group split.

Returning to the initial problem $(K/k, G, \varphi, N)$, we see that for the compatibility of this problem it is necessary and sufficient that the associated problems $(K/K_\chi, G_\chi, \varphi_\chi, N_\chi)$ corresponding to the characters of the group N be compatible, and, therefore, solvable, since all of them are Brauer problems. Thus, the following theorem is valid.

THEOREM 3.3. *The compatibility condition for the problem $(K/k, G, \varphi, N)$ with Abelian kernel of exponent n under the assumption that K contains a primitive nth root of unity is equivalent to the solvability of all associated Brauer problems $(K/K_\chi, G_\chi, \varphi_\chi, N_\chi)$ corresponding to the characters.*

It is clear from the above that for the compatibility of the problem with Abelian kernel under the assumptions mentioned above it is necessary and sufficient that all Brauer problems associated with the original one be solvable. Moreover, one can prove that all Brauer problems with cyclic kernels associated with the original problem either correspond to characters or are associated with such problems.

§4. The embedding problem with cyclic kernel

In this section we show that the compatibility condition for the embedding problem with cyclic kernel is sufficient for embeddability if the kernel has odd order, or its order is equal to 4. At the same time, as will be shown below (see §14), even for a cyclic kernel of order 8 the compatibility condition is not sufficient in the general case.

1°. First consider the case of a kernel of order 4 (for a kernel of order 2, the problem is always Brauer). Here we indicate an explicit solution, whenever the compatibility system is given.

Thus, let $(K/k, G, \varphi, N)$ be the problem mentioned above, and the kernel N be generated by an element t of order 4. One may assume that $i = \sqrt{-1}$ is contained in K, otherwise one may pass to an equivalent embedding problem lifting the field K to the field $K(i)$. For any $g \in G$ we have $t^g = t^{\pm 1}$, $i^g = i^{\pm 1}$. Let G_0 be the subgroup of $g \in G$ with $(ti)^g = (ti)^{\pm 1}$. If G_0 coincides with G, we have a Brauer problem, and we have constructed a solution of it in §2. Therefore, we assume that $G_0 \neq G$. Then G_0 has index 2 in G. Consider the function ε_g equal to 1 for $g \in G_0$ and -1 for $g \notin G_0$. It is clear that ε_g is a homomorphism $G \longrightarrow \{\pm 1\}$.

In the group algebra $K[N]$, the unity decomposes into the sum of four orthogonal idempotents E_0, E_1, E_2, E_3, where

$$E_0 = \frac{1+t+t^2+t^3}{4}, \qquad E_1 = \frac{1-t+t^2-t^3}{4},$$
$$E_2 = \frac{1+ti-t^2-t^3i}{4}, \qquad E_3 = \frac{1-ti-t^2+t^3i}{4};$$

moreover, $tE_0 = E_0$, $tE_1 = -E_1$, $tE_2 = -iE_2$, $tE_3 = iE_3$. The idempotents E_0 and E_1 are invariant with respect to the action of $g \in G$, while E_2 and E_3 are invariant with respect to $g \in G_0$ and go one to the other under the action of $g \in G \setminus G_0$.

Let a compatibility system $l_g \in K[N]$ for our problem be given. Recall that $l_{g_1 g_2} = l_{g_1}^{g_2} l_{g_2}$, $l_t = t^{-1}$. The decomposition of l_g into idempotents defines the four functions a_g, b_g, c_g and d_g on G with values in K^*; here a_g and b_g are cocycles, and c_g and d_g are connected by the relations

$$\begin{aligned} c_{g_1 g_2} &= c_{g_1}^{g_2} c_{g_2} \quad \text{if } g_2 \in G_0, \\ c_{g_1 g_2} &= d_{g_1}^{g_2} c_{g_2} \quad \text{if } g_2 \notin G_0; \\ d_{g_1 g_2} &= d_{g_1}^{g_2} d_{g_2} \quad \text{if } g_2 \in G_0, \\ d_{g_1 g_2} &= c_{g_1}^{g_2} d_{g_2} \quad \text{if } g_2 \notin G_0. \end{aligned}$$

Moreover, $a_t = 1$, $b_t = -1$, $c_t = i$, $d_t = -i$. We have $a_{gt} = a_g$ for the elements a_g, i.e., they can be obtained by the lifting of a cocycle from $Z^1(F, K^*)$, where $F = \varphi(G) = \mathrm{Gal}(K/k)$; by Speiser's theorem, we have $a_g = v_0^{\varphi(g)-1}$, where $v_0 \in K^*$, so such elements a_g always exist; but the existence of them gives no information about the problem.

The elements b_g^2 can be obtained by the lifting of a cocycle in $Z^1(F, K^*)$. By Speiser's theorem, there exists an element $y \in K^*$ such that $b_g^2 = y^{g-1}$ (it is more correct to write $y^{\varphi(g)-1}$), and y is uniquely determined up to a factor from k^*.

For the elements c_g and d_g, we have $c_g d_g \in Z^1(G, K^*)$, and $c_t d_t = 1$. Therefore, $c_g d_g$ can be obtained by the lifting of a cocycle from $Z^1(F, K^*)$; so, there exists an element $u \in K^*$ such that $c_g d_g = u^{g-1}$. We set $x_g = c_g$ if $g \in G_0$, and $x_g = c_g^{-1} u^g$ if $g \notin G_0$. Direct verification shows that for each g_1 and g_2 in G, we have $x_{g_1 g_2} = x_{g_1}^{g_2} x_{g_2}^{\varepsilon_{g_1}}$ (where $\varepsilon_g = 1$ if $g \in G_0$, and $\varepsilon_g = -1$ if $g \notin G_0$). Moreover, $x_t = i$. We examine the elements x_g in more detail.

For $g \in g_0$, we see that $x_g \in Z^1(G_0, K^*)$, the x_g^4 are unchanged under the replacement of g by gt, so for $g \in G_0$ we have (by Speiser's theorem) that $x_g^4 = z^{g-1}$ for $z \in K^*$. Consider the elements $x_g^4 z^{-g-1}$ for $g \notin G_0$. Let $g_0 \in G_0$. Then

$$x_{g_0 g}^4 z^{-g_0 g - 1} = (x_{g_0}^g x_g)^4 z^{-g_0 g - 1} = (z^{g_0 - 1})^g x_g^4 z^{-g_0 g - 1} = x_g^4 z^{-g-1};$$

so $a = x_g^4 z^{-g-1}$ does not depend on the choice of $g \notin G_0$, therefore, $x_g^4 = z^{g-1}$ for $g \in G_0$, and $x_g^4 = a z^{g+1}$ for $g \notin G_0$. (It is easy to check that the element a constructed above is contained in the base field k, but we need a stronger assertion).

For x_g^2 with $g \in G_0$ and for b_g, we have the same relations $x_t^2 = b_t = -1$ with $g = t$, and $x_g^2, b_g \in Z^1(G_0, K^*)$. Therefore, $x_g^2 b_g^{-1}$ is obtained by lifting from $Z^1(F_0, K^*)$; so, there exists $w \in K^*$ such that $x_g^2 = b_g w^{g-1}$ for $g \in G_0$. Squaring this equality (and taking into account that $b_g^2 = y^{g-1}$), we obtain $z^{g-1} = (yw^2)^{g-1}$. Therefore, $zy^{-1}w^{-2}$ is invariant with respect to the elements $g \in G_0$, so $z = yw^2 h$, where $h \in K_1 = K^{F_0}$. Hence $x_g^4 = a z^{g+1} = a(yw^2)^{g+1} N(h) = a y^{g-1} y^2 (w^{g+1})^2 N(h) = aN(h)(b_g y w^{g+1})^2$ for $g \notin G_0$. Since the left-hand side is a square in K, we have that $aN(h) = q^2$, where $q \in K^*$. Therefore, for x_g^2 we have the relations

$$(*) \qquad x_g^2 = b_g w^{g-1} \quad \text{for } g \in G_0, \qquad x_g^2 = b_g w^{g+1} y q \delta_g \quad \text{for } g \notin G_0$$

(here $\delta_g = \pm 1$).

We show that $q \in k^*$. For this we substitute relations $(*)$ in the equality $x_{g_1g_2}^2 = (x_{g_1}^2)^{g_2} x_{g_2}^{\varepsilon_{g_1}}$.

If $g_1 \in G_0$, $g_2 \notin G_0$, we obtain (since $g_1g_2 \notin G_0$)

$$b_{g_1g_2} w^{g_1g_2+1} yq\delta_{g_1g_2} = (b_{g_1} w^{g_1-1})^{g_2} b_{g_2} w^{g_2+1} yq\delta_{g_2},$$

whence $\delta_{g_1g_2} = \delta_{g_2}$, i.e., δ_g, $g \notin G_0$, has constant sign. By choosing a suitable q, we may assume that $\delta_g = 1$.

If $g_1 \notin G_0$, $g_2 \in G_0$, we have

$$b_{g_1g_2} w^{g_1g_2+1} yq = (b_{g_1} w^{g_1+1} yq)^{g_2} (b_{g_2} w^{g_2-1})^{-1},$$

whence $q^{g_2-1} = b_{g_1g_2} b_{g_1}^{-g_2} b_{g_2} y^{1-g_2} = b_{g_2}^2 y^{1-g_2} = 1$, i.e., $q^g = q$ for $g \in G_0$. Finally, assuming $g_1 \notin G_0$, $g_2 \notin G_0$, we similarly have $q^g = q$ for $g \notin G_0$.

Thus, $q \in k^*$, so $aN(h)$ is a square in k^*; the latter means that the element a belongs to k^* and is the norm of an element $h_1 \in K_1 = K^{F_0}$. We set $s = h_1 z$. Then $x_g^4 = az^{g+1} = h_1^{g+1} z^{g+1} = s^{g+1}$ for $g \notin G_0$, and $x_g^4 = z^{g-1} = (zh_1)^{g-1} = s^{g-1}$ for $g \in G_0$. Thus, $x_g^4 = s^{g-\varepsilon_g}$ for every $g \in G$. It is clear that the element s is defined up to a factor from k^*.

Now we construct the required extension of the field K. Indeed, we set $L = K(\theta)$, where $\theta^4 = s \in K^*$, and we define the action of automorphisms from G on the primitive element θ by the formula $\theta^g = \theta^{\varepsilon_g} x_g$. In addition, $s^g = (\theta^4)^g = \theta^{4\varepsilon_g} x_g^4 = s^{\varepsilon_g} x_g^4$, so that the elements g act on the field K in a proper way (like $\varphi(g) \in F$), and

$$(\theta^{g_1})^{g_2} = (\theta^{\varepsilon_{g_1}} x_{g_1})^{g_2} = (\theta^{\varepsilon_{g_2}} x_{g_2})^{\varepsilon_{g_1}} \cdot x_{g_1}^{g_2} = \theta^{\varepsilon_{g_1g_2}} x_{g_1}^{g_2} x_{g_2}^{\varepsilon_{g_1}} = \theta^{g_1g_2},$$

i.e., these automorphisms form a group isomorphic to G. Moreover, $\theta^t = \theta x_t = \theta i$. Thus, the field (or the Galois algebra) L is a solution of our embedding problem.

2°. Now let the kernel N of the embedding problem $(K/k, G, \varphi, N)$ be a cyclic group of order $n = p^m$, where $m \geq 1$, $p \neq 2$ a prime. The characteristic of k is different from p. We prove that in this situation compatibility is also sufficient for the solvability of the embedding problem. To this end we lift the problem by adjoining externally the nth roots of unity; we find the associated Brauer problem for the lifted problem and show that, using a special solution of this problem, we can obtain the solution of the original problem.

Consider the algebra $B = K[x]/(\Phi_n(x))$, where Φ_n is the cyclotomic polynomial, $\Phi_n(x) = x^{p^{m-1}(p-1)} + \cdots + x^{p^{m-1}} + 1$. Denote by ζ the image of x under passage to the quotient algebra. It is clear that $\zeta^n = 1$. The algebra B is a Galois algebra over K of dimension $q = \varphi(n) = p^{m-1}(p-1)$ with automorphism group $T = \{\tau\}$, where $\zeta^\tau = \zeta^r$, r is a primitive residue modulo p^m, $\tau^q = 1$. Moreover, one may regard the algebra B as a Galois algebra over k by letting $a^\tau = a$ for $a \in K$, $\zeta^\tau = \zeta^r$ and $f^\tau = f$ for $f = \varphi(g) \in F = \mathrm{Gal}(K/k)$, so that the group $\mathrm{Gal}(B/k)$ is the direct product $F \times T$ of the groups F and T.

One may assume, without loss of generality, that $r^q = 1 + p^m y$ for $(y, p) = 1$. Indeed, for $m = 1$ one can achieve this by adding to r a suitable multiple of p, while for $m \geq 2$ this is always valid. Indeed, let $r^{p-1} = 1 + py_0$. Then $r^{p^k(p-1)} \equiv 1 + p^{k+1} y_0 \pmod{p^{k+2}}$, and if $y_0 \equiv 0 \pmod p$, then $r^{p^{m-2}}(p-1) \equiv 1 \pmod{p^m}$, i.e., r is not a primitive residue for $n = p^m$. Therefore, for a primitive residue r we have $y_0 \not\equiv 0$

(mod p) and $r^q = r^{p^{m-1}(p-1)} \equiv 1 + p^m y_0 \pmod{p^{m+1}}$, i.e.,

$$r^q = 1 + p^m y, \qquad (y, p) = 1.$$

The algebra $K[N] = K[t]$, $t^n = 1$, has a natural homomorphism θ onto the algebra B, namely, $\theta(a) = a$ for $a \in K$ and $\theta(t) = \zeta$. An automorphism of the group N, namely, $t^f = g^{-1}tg = t^{v(f)}$, $v(f) \in \mathbb{Z}/n\mathbb{Z}$, corresponds to each $f = \varphi(g) \in F$. In the group $F \times T$ of automorphisms of B we find a subgroup $\widetilde{F}$ whose elements $\widetilde{f}$ act on t and ζ in a "consistent" way, namely, we put $\widetilde{f} = (f, \zeta) \to \zeta^{v(f)}$. It is clear that $\widetilde{F}$ is isomorphic to F, and $\widetilde{F} \times T = F \times T$. We denote by A the subalgebra of $\widetilde{F}$-invariant elements of the algebra B. The algebra A is a Galois algebra over k with group isomorphic to T.

Let the compatibility condition for the initial problem hold, and let $l(g) \in K[N]$ be a compatibility system: $l(g_1)^{g_2} l(g_2) = l(g_1 g_2)$ and $l(t) = t^{-1}$. We note that the action of the automorphism $g \in G$ in the algebra $K[N]$ depends only on $f = \varphi(g) \in F$. We set $\theta(l(g)) = \lambda(g) \in B$. Then, by the consistency of the action of automorphisms $\widetilde{f}$ on $\zeta \in B$ and on $t \in N$, we have $\lambda(g_1)^{\widetilde{f}_2} \lambda(g_2) = \lambda(g_1 g_2)$, where $f_2 = \varphi(g_2)$, and $\lambda(t) = \zeta^{-1}$. We set $\mu(g) = \lambda(g)^{r-\tau}$. Then $\mu(g_1)^{\widetilde{f}_2} \mu(g_2) = \mu(g_1 g_2)$ and $\mu(t) = 1$. Therefore, $\mu(g)$ depends only on $f = \varphi(g)$ or, what is the same, on $\widetilde{f}$, and redenoting $\mu(g)$ by $\mu(\widetilde{f})$, we have $\mu(\widetilde{f}_1)^{\widetilde{f}_2} \mu(\widetilde{f}_2) = \mu(\widetilde{f}_1 \widetilde{f}_2)$ for $\widetilde{f}_1, \widetilde{f}_2 \in \widetilde{F}$. By Speiser's theorem for Galois algebras, we have $\mu(\widetilde{f}) = b^{\widetilde{f}-1}$ for some $b \in B$. Now we put $M = b^{y'(r^{q-1} + r^{q-2}\tau + \cdots + \tau^{q-1})}$ assuming that $r^q = 1 + yp^m$, $(y, p) = 1$, and $yy' \equiv 1 \pmod{p^m}$.

We adjoin the element $\gamma = \sqrt[n]{M}$ to the algebra B, i.e., consider the algebra $C = B[z]/(z^n - M)$, so that γ is the image of z under the passage to the quotient algebra. We show that the algebra C is a Galois algebra over k with group isomorphic to the direct product $G \times T$.

We have

$$M^{\widetilde{f}} = M \cdot M^{\widetilde{f}-1} = M \mu(g)^{y'(r^{q-1} + r^{q-2}\tau + \cdots + \tau^{q-1})} = M \lambda(g)^{y'(r^q - \tau^q)} = M \Lambda(g)^{p^m},$$

where $\Lambda(g) = \lambda(g)^{y'y}$, and g is an inverse image of $f \in F$ in G. Therefore, the mapping $S(g)\colon \gamma \to \gamma^{\Lambda(g)}$, $u \to u^{\widetilde{f}}$ for $u \in B$ and $f = \varphi(g)$ is an automorphism of the algebra C extending the automorphism $\widetilde{f}$ of the algebra B. It is clear that $\Lambda(g_1)^{g_2} \times \Lambda(g_2) = \Lambda(g_1 g_2)$ and $\Lambda(t) = \zeta^{-1}$. Therefore, the automorphisms $S(g)$ form a group isomorphic to G. The subalgebra of invariant elements for the group $S(t)$ is obviously the algebra B, and the subalgebra of invariant elements for the whole group $S(g)$, $g \in G$, coincides with the subalgebra A of invariant elements of B for the group $\widetilde{F}$.

Further, $M^\tau = M^r M^{\tau - r} = M^r b^{-y'(r^q - \tau^q)} = M^r \Delta_\tau^n$, where $\Delta_\tau = b^{-y'y}$. Therefore, the mapping $S_\tau\colon \gamma \to \gamma^r \Delta_\tau$, $u \to u^\tau$, for $u \in B$ is an automorphism of the algebra C extending the automorphism τ of the algebra B. The automorphisms $S(g)$ and S_τ commute on the algebra B. For the element γ, we also have

$$\begin{aligned} S(g)S_\tau\gamma = S(g)\gamma^r b^{-y'y} &= \gamma^r \Lambda(g)^r b^{-\widetilde{f}y'y} = \gamma^r \lambda(g)^{y'yr} b^{-y'y} \mu(g)^{-y'y} \\ &= \gamma^r \lambda(g)^{y'y\tau} b^{-y'y} = S_\tau \gamma \Lambda(g) = S_\tau S(g)\gamma. \end{aligned}$$

Therefore, the group generated by the automorphisms $S(g)$ and S_τ is isomorphic to the direct product of G and T. The subalgebra L of the algebra C, belonging to the subgroup generated by S_τ, has group over the field k isomorphic to G. The field K is the algebra of invariant elements of the algebra C for the group generated by S_τ and $S(t)$. The elements of $S(g)$ act on the elements of K by the rule $f = \varphi(g)$, so that the algebra L is a solution of the initial embedding problem.

§5. The first Kochendörffer theorem

Let the problem $(K/k, G, \varphi, N)$ with Abelian kernel N be given. The group N can be decomposed into a direct product of Abelian p-groups: $N = N_1 \dots N_k$. It is clear that all direct factors are invariant subgroups of the group G, so that the problem $(K/k, G, \varphi, N)$ itself is a product of the problems with kernels $N_1, \dots, N_k$ in the sense of §12, Chapter 1, and for the solvability of the initial problem it is necessary and sufficient that the problems with kernels $N_1, \dots, N_k$ be solvable. Thus, the following theorem is valid.

THEOREM 3.5. *The solvability of the embedding problem with Abelian kernel is equivalent to the solvability of the problems with kernels that are p-groups, namely, the direct factors of the kernel of the initial problem.*

In particular, for the embedding problem with cyclic kernel of odd order, the compatibility condition is sufficient for the solvability of the embedding problem. Indeed, a cyclic group of odd order is a direct product of groups of order p^m for odd and distinct primes p, and for a cyclic group of order p^m with $p \neq 2$ compatibility is sufficient for the solvability of the embedding problem.

§6. Free normal extensions of normal fields

Let K be a normal extension of the field k with Galois group F of order n, and let L be a normal extension (or a Galois algebra) of the field K with group Γ (not necessarily Abelian).

Construct a Galois algebra A over the field k such that L is contained in A as a subalgebra and A is generated by L and all the conjugates to L in A. To this end we consider the algebras L_f, $f \in F$, conjugate to L; in other words, these algebras differ from L in that in the multiplication table of a basis of L over K all coefficients are replaced by the conjugates by $f \in F$. To a basis of L over K we associate one of the bases of L_f over K with the conjugate multiplication table; we get an isomorphism of L onto L_f over k, that extends the automorphism f of K over k. We denote this isomorphism by $\widetilde{f}$. We assume that for each $h \in F$, $\widetilde{f}$ maps L_h onto L_{hf} isomorphically by the rule: $\widetilde{f}$ is equal to $\widetilde{h}^{-1}\widetilde{hf}$ on L_h. Clearly, all L_f are normal over K and their groups are isomorphic to Γ.

Indeed, if $\gamma \in \Gamma$, then the mapping $\widetilde{f}^{-1}\gamma\widetilde{f}$ is an automorphism of the algebra L_f over K, and all automorphisms of L_f over K can be obtained in this way.

The tensor product A over K of all L_f, $f \in F$, is a Galois algebra over K with group isomorphic to the direct product of n groups isomorphic to Γ. It is natural to regard this direct product as a group of functions ψ defined on F with values in Γ, and we suppose that the function ψ acts on the element $z \in L_f$ as follows: $z^\psi = z^{\widetilde{f}^{-1}\psi(f)\widetilde{f}}$. The set of automorphisms $\widetilde{f}$ can be naturally extended, in turn, to the algebra A as a group of automorphisms $\widetilde{F}$ isomorphic to the group F and representing it as a group of K over k. We clarify the rule of permutation of the automorphisms $\widetilde{f}$ and automorphisms ψ by setting $\psi^f(h) = \psi(hf^{-1})$, $h \in F$.

Let $z \in L_h$. Then $z^{\widetilde{f}}$ and $z^{\psi\widetilde{f}}$ belong to L_{hf}. We have $z^{\psi\widetilde{f}} = z^{\widetilde{h}^{-1}\psi(h)\widetilde{h}\widetilde{f}} = z^{\widetilde{h}^{-1}\psi^f(hf)\widetilde{hf}} = z^{\widetilde{f}\psi^f}$. Thus, for all $z \in L_f$ and hence on the whole algebra A we have $\psi\widetilde{f} = \widetilde{f}\psi^{\widetilde{f}}$. Therefore, the products $\{f\psi\}$ form a group H isomorphic to the wreath product of the group Γ by the regular representation of the group Γ by permutations. Such a wreath product is a group of pairs (f, ψ), where $f \in F$ and the ψ are functions on F with values in Γ, with the multiplication $(f_1, \psi_1)(f_2, \psi_2) = (f_1 f_2, \psi_1^{f_2}\psi_2)$, where $\psi^f(f_1) = \psi(f_1 f^{-1})$. The order of H is equal to nm^n, where n and m are the orders of F and Γ, respectively. The dimension of A over k is also equal to nm^n, and the elements of A fixed under the action of automorphisms ψ and $\widetilde{f}$, form the field k. Therefore, A is a Galois algebra over k and is generated by the subalgebra L and by all conjugates to L in A.

In the case where the fields L_f are separated, i.e., none of them has common subfield with the compositum of the others, the tensor product considered above is a field, namely, the compositum of all the fields L_f.

If K is an algebraic number field, Γ is an Abelian group and the field K contains roots of unity of degree equal to the exponent of the group Γ (one can attain this by lifting and subsequent descent), then it is not hard to achieve the separation of the fields L_f. The Abelian extension L of the field K is a Kummer extension, $L = K(\sqrt[n_1]{a_1}, \sqrt[n_2]{a_2}, \dots, \sqrt[n_k]{a_k})$, where $n_1, \dots, n_k$ are orders of the cyclic direct factors of the group Γ. If we assume that the a_i are divisible by distinct prime noncritical ideals $\mathfrak{P}_i$ of the first order, the a_i are not divisible by $\mathfrak{P}_i^f$ conjugate to $\mathfrak{P}_i$, $f \neq 1$, and the a_j, $j \neq i$, are not divisible by $\mathfrak{P}_i^f$, $f \in F$, then it is easily seen that the fields L_f are separated.

The wreath product is a semidirect product of the group of functions ψ and a group isomorphic to F. In the case where Γ is an Abelian group, the wreath product has the following interesting universality property.

THEOREM 3.6. *Any semidirect extension of an Abelian group Γ by F is a homomorphic image of the wreath product of Γ by F.*

PROOF. Let $G = F \cdot \Gamma$, $\gamma^f = f^{-1}\gamma f \in \Gamma$ for $\gamma \in \Gamma$, $f \in F$. We denote by H the wreath product of Γ by F. Consider the following mapping of H into G:

$$(f, \psi) \to f \prod_{h \in F} (\psi(h))^h.$$

It is clear that this mapping is surjective. We prove that this is a homomorphism. In the group H we have

$$(f_1, \psi_1)(f_2, \psi_2) = (f_1 f_2, \psi_1^{f_2}\psi_2).$$

For the images in G we have

$$f_1 \prod_{h \in F} \psi_1(h)^h \cdot f_2 \prod_{h \in F} \psi_2(h)^h = f_1 f_2 \prod_{h \in F} \psi_1(h)^{hf_2} \prod_{h \in F} \psi_2(h)^h.$$

We set $hf_2 = h'$. Then $\psi_1(h)^{hf_2} = (\psi_1(h' f_2^{-1}))^{h'} = (\psi_1^{f_2}(h'))^{h'}$, so that

$$\begin{aligned} f_1 \prod_{h \in F} \psi_1(h)^h \cdot f_2 \prod_{h \in F} \psi_2(h)^h &= f_1 f_2 \prod_{h' \in F} (\psi_1^{f_2}(h'))^{h'} \prod_{h \in F} \psi_2(h)^h \\ &= f_1 f_2 \prod_{h \in F} ((\psi_1^{f_2}\psi_2)(h))^h. \end{aligned}$$

Thus, the map indicated is really a homomorphism of H onto G. □

It follows from the above that for the semidirect embedding problem with Abelian kernel for algebraic number fields, there exist solutions that are fields.

§7. The generalized wreath product

We consider a group-theoretic construction generalizing the wreath product of groups in exactly the same way as the direct product with amalgamated quotient group generalizes the direct product.

Let $1 \longrightarrow N \longrightarrow G_0 \xrightarrow{\varphi_0} F_0 \longrightarrow 1$ be an exact sequence of finite groups and let F be a finite group containing F_0. We do not assume that N is an Abelian group. We decompose F into the cosets ρ with respect to F_0, $F = \bigcup \rho = \bigcup F_0\overline{\rho}$ with a system of representatives $\overline{\rho}$. Let $\widetilde{G}$ be the set of pairs (f, ψ_f), where $f \in F$ and the ψ_f are functions defined on the cosets ρ with values in the group G_0. We assume that the values of ψ_f satisfy the condition $\varphi_0\psi_f(\rho) = \overline{\rho f^{-1}} f \overline{\rho}^{-1}$. Set $\psi_f^{f_1}(\rho) = \psi_f(\rho f_1^{-1})$. On $\widetilde{G}$, define multiplication by the rule

$$(f_1, \psi_{f_1})(f_2, \psi_{f_2})) = (f_1 f_2, \psi_{f_1}^{f_2}\psi_{f_2}).$$

In order to prove that multiplication is well defined, we must verify that the function $\psi_{f_1}^{f_2}\psi_{f_2}$ satisfies the same condition as $\psi_{f_1 f_2}$. We have

$$\begin{aligned} \varphi_0(\psi_{f_1}^{f_2}\psi_{f_2})(\rho) &= \varphi_0(\psi_{f_1}(\rho f_2^{-1})\psi_{f_2}(\rho)) = \overline{\rho f_2^{-1} f_1^{-1}} f_1 \overline{\rho f_2^{-1}}^{-1} \overline{\rho f_2^{-1}} f_2 \overline{\rho}^{-1} \\ &= \overline{\rho f_2^{-1} f_1^{-1}} f_1 f_2 \overline{\rho}^{-1} = \varphi_0(\psi_{f_1 f_2}(\rho)). \end{aligned}$$

The associativity of multiplication is obvious. The pair $(1, \psi_{10})$ is the unit element, where $\psi_{10}(\rho) = 1$ for all ρ; $(f^{-1}, \psi_{f^{-1}})$ is the inverse to (f, ψ_f) with $\psi_{f^{-1}}(\rho) = (\psi_f(\rho f))^{-1}$.

Clearly, the mapping $\widetilde{\varphi}$ that takes $(f, \psi) \in \widetilde{G}$ to $f \in F$ is a homomorphism. The kernel $\widetilde{N}$ of $\widetilde{\varphi}$ is the set of pairs $(1, \psi_1)$ for which $\varphi_0\psi_1(\rho) = \overline{\rho} \cdot 1 \overline{\rho}^{-1} = 1$, that is, $\psi_1(\rho) \in N$. The order of $\widetilde{N}$ is equal to $(N : 1)^{(F:F_0)}$, whence the order of $\widetilde{G}$ equals $|F| \cdot |\widetilde{N}| = |F| \cdot |N|^{(F:F_0)}$.

Now we establish a relationship between the construction obtained and the embedding problem.

THEOREM 3.7. *Let K/k be a Galois extension with group F; let F_0 be a subgroup of F and let K_0 be the fixed field for F_0. Then the embedding problems $(K/K_0, G_0, \varphi_0, N)$ and $(K/k, \widetilde{G}, \widetilde{\varphi}, \widetilde{N})$ are equivalent, where $1 \longrightarrow N \longrightarrow G_0 \longrightarrow F_0 \longrightarrow 1$ is an exact sequence, the groups $\widetilde{G}$, $\widetilde{N}$ and the homomorphism $\widetilde{\varphi}$ are as above.*

PROOF. Let L_0 be a Galois algebra that yields a solution of the problem $(K/K_0, G_0, \varphi_0, N)$. We construct the algebra L generated by L_0 and its conjugates over k. To this end, we consider the tensor product M over k of $(K : K_0) = (F : F_0)$ copies of L_0, putting in correspondence the factors of M to the cosets ρ in the decomposition of F with respect to F_0. Suppose that the first factor corresponds to the class $\rho_0 = F_0$ with representative $\overline{\rho}_0 = 1$. Let I be the ideal in M generated by the elements $1 \otimes \cdots \otimes x \otimes \cdots \otimes 1 - x^{\overline{\rho}} \otimes 1 \otimes \cdots \otimes 1$, where x is an element of K and it occurs in the first summand at the position corresponding to ρ. The quotient algebra $L = M/I$ is a "twisted" tensor product of the algebras L_0 over K, i.e., the following relations hold in L:

$$xa \otimes a_{\rho_2} \otimes \cdots \otimes a_\rho \otimes \ldots = a_1 \otimes x^{\overline{\rho_2}^{-1}} a_{\rho_2} \otimes \cdots \otimes a_\rho \otimes \ldots$$
$$= a_1 \otimes a_{\rho_2} \otimes \cdots \otimes x^{\overline{\rho}^{-1}} a_\rho \otimes \ldots .$$

Here, $a_1, a_{\rho_2}, \ldots, a_\rho \in L$, $x \in K$. It is clear that the dimension of L over K is equal to $|N|^{(F:F_0)} = |\widetilde{N}|$, and the dimension of L over k equals $|\widetilde{G}|$. □

In the algebra M we define automorphisms from the group $\widetilde{G}$ by setting

$$(1 \otimes 1 \otimes \cdots \otimes a \otimes \cdots \otimes 1)^{(f,\psi_f)} = 1 \otimes \cdots \otimes a^{\psi_f(\rho f)} \otimes \cdots \otimes 1.$$

Here $a \in L$, and on the left-hand side a is located at the position of ρ, while on the right-hand side a is at the position of ρf. We show that the automorphisms defined in this way form a group isomorphic to $\widetilde{G}$. It suffices to prove that to the product of automorphisms there corresponds the product of the corresponding elements of $\widetilde{G}$. Thus,

$$((1 \otimes 1 \otimes \cdots \otimes a \otimes \cdots \otimes 1)^{(f,\psi_f)})^{(g,\psi_g)}$$
$$= (1 \otimes \cdots \otimes a^{\psi_f(\rho f)} \otimes \cdots \otimes 1)^{(g,\psi_g)} = (1 \otimes \cdots \otimes a^{\psi_f(\rho f)\psi_g(\rho f g)} \otimes \cdots \otimes 1)$$
$$= 1 \otimes \cdots \otimes a^{\psi_f^g(\rho f g)\psi_g(\rho f g)} \otimes \cdots \otimes 1 = (1 \otimes \cdots \otimes a \otimes \cdots \otimes 1)^{(f,\psi_f)(g,\psi_g)}.$$

Here, in the first and the fifth expression $a \in L_0$ is at the position of ρ, in the second a is at the position of ρf, and in the third and the fourth a is at the position of $\rho f g$.

Now we prove that the automorphisms (f, ψ_f) take the ideal I into itself. To this end, it suffices to check that the generators $1 \otimes \cdots \otimes x \otimes \cdots \otimes 1 - x^{\overline{\rho}} \otimes 1 \otimes \cdots \otimes 1$ go to elements from I under the action of (f, ψ_f). We have

$$(1 \otimes \cdots \otimes x \otimes \cdots \otimes 1)^{(f,\psi_f)} = 1 \otimes \cdots \otimes x^{\psi_f(\rho f)} \otimes \cdots \otimes 1$$
$$= 1 \otimes \cdots \otimes x^{\varphi_0 \psi_f(\rho f)} \otimes \cdots \otimes 1$$
$$\equiv x^{\overline{\rho f}} \otimes 1 \otimes \cdots \otimes 1 \pmod{I},$$

since $\varphi_0\psi_f(\rho f) = \overline{\rho f f^{-1}} \, \overline{f \rho f}^{-1} = \overline{\rho} f \overline{\rho f}^{-1}$ (here $x \in K$; in the first expression x is at the position of ρ, in the second and the third x is at the position of ρf).

In particular, $(x \otimes 1 \otimes \cdots \otimes 1)^{(f,\psi_f)} \equiv x^f \otimes 1 \otimes \cdots \otimes 1$ and

$$(x^{\overline{\rho}} \otimes 1 \otimes \cdots \otimes 1)^{(f,\psi_f)} \equiv x^{\overline{\rho} f} \otimes 1 \otimes \cdots \otimes 1 \equiv (1 \otimes \cdots \otimes x \otimes \cdots \otimes 1)^{(f,\psi_f)}.$$

Thus, the ideal I goes into itself under the action of the automorphisms from $\widetilde{G}$. Therefore, we can regard $\widetilde{G}$ as the automorphism group of $L = M/I$. It is natural to identify $x \in K$ with $(x \otimes 1 \otimes \cdots \otimes 1)$. We see that $x^{(f,\psi_f)} = x^f$ in L.

Let $\{ u^t \mid t \in N \}$ be a normal basis for L_0 over K. Then $(u \otimes u \otimes \cdots \otimes u)$ generates a normal basis for L over K, since $(u \otimes u \otimes \cdots \otimes u)^{(1,\psi_1)} = u^{\psi_1(\rho_0)} \otimes \cdots \otimes u^{\psi_1(\rho_0)} \otimes \cdots = u^{t_1} \otimes \cdots \otimes u^{t_m}$, where $t_1, \ldots, t_m$ range independently over the group N because they are equal to values of ψ_1. Consequently, L is a Galois algebra over K with group $\widetilde{N}$. The field K is the fixed field for $\widetilde{N}$, and k is the fixed field for $\widetilde{G}$. By Theorem 1.4, L is a Galois algebra over k with group $\widetilde{G}$ because $(L : k) = |\widetilde{G}|$.

Conversely, let L be a solution of the problem $(K/k, \widetilde{G}, \widetilde{\varphi}, \widetilde{N})$. Denote by $\widetilde{G}_0$ the full preimage of the subgroup F_0 of F in the group $\widetilde{G}$. Then the problem $(K/K_0, \widetilde{G}_0, \widetilde{\varphi}, \widetilde{N})$ is an associated problem of the second kind for $(K/k, \widetilde{G}, \widetilde{\varphi}, \widetilde{N})$, and we can regard L as a solution of this problem.

We define a mapping μ from $\widetilde{G}_0$ to G_0 by setting $\mu(f, \psi_f) = \psi_f(\rho_0)$ for $(f, \psi_f) \in \widetilde{G}_0$ (hence $f \in F$). Then for $f, g \in F_0$ we have

$$\begin{aligned}\mu((f,\psi_f)(g,\psi_g)) &= \mu(fg, \psi_f^g \psi_g) = \psi_f^g(\rho_0)\psi_g(\rho_0)\\ &= \psi_f(\rho_0 g^{-1})\psi_g(\rho_0) = \psi_0(\rho_0)\psi_g(\rho_0),\end{aligned}$$

so that μ is a homomorphism. Clearly, it is surjective, i.e., $\mu(\widetilde{N}) = N$ and $\varphi_0\mu = \widetilde{\varphi}$. Therefore, $(K/K_0, G_0, \varphi_0, N)$ is an associated problem of the first kind for $(K/K_0, \widetilde{G}_0, \widetilde{\varphi}, \widetilde{N})$ and, thus, it is solvable.

REMARK. Theorem 3.7 proved above can be established by literally repeating the proof in the case where K is a Galois algebra over k only if the subalgebra of F_0-invariant elements of the algebra K is a field. One can also prove the equivalence of the compatibility conditions for the problems $(K/k, \widetilde{G}, \widetilde{\varphi}, \widetilde{N})$ and $(K/K_0, G_0, \varphi_0, N)$.

The group $\widetilde{G}$ admits a simpler description if the problem $(K/K_0, G_0, \varphi_0, N)$ is an associated problem of the second kind for some problem $(K/k, G, \varphi, N)$, where $\varphi(G) = F = \mathrm{Gal}(K/k)$, G_0 is a subgroup of G, φ_0 is the restriction of φ to G_0, $F_0 = \varphi_0(G_0)$. In this situation, the full φ-preimages of the cosets of F relative to F_0 are cosets in the decomposition of G relative to G_0 and, as their representatives, one can take any φ-preimages of the representatives $\overline{\rho}$ of the cosets ρ in the decomposition of F with respect to F_0. In order not to overload the notation, we keep the notation ρ and $\overline{\rho}$ for the cosets of G with respect to G_0 and their representatives that correspond to the cosets ρ and their representatives (in the previous notation).

We associate the functions ω_f on the cosets ρ with values in the group G with the functions ψ_f on the cosets ρ with values in G_0 by the formulas $\omega_f(\rho) = \overline{\rho f^{-1}}^{-1}\psi_f(\rho)\overline{\rho}$. Then

$$\varphi(\omega_f(\rho)) = \overline{\rho f^{-1}}^{-1}\varphi_0(\psi_f(\rho))\overline{\rho} = \overline{\rho f^{-1}}^{-1}\overline{\rho f^{-1}}\,\overline{f\rho}^{-1}\overline{\rho} = f.$$

We put

$$\omega_f^g(\rho) = \omega_f(\rho g^{-1}) = \overline{\rho g^{-1} f^{-1}}^{-1}\psi_f(\rho g^{-1})\overline{\rho g^{-1}}.$$

Then

$$\begin{aligned}\omega_f^g(\rho)\omega_g(\rho) &= \overline{\rho g^{-1} f^{-1}}^{-1}\psi_f(\rho g^{-1})\overline{\rho g^{-1}}\,\overline{\rho g^{-1}}^{-1}\psi_g(\rho)\overline{\rho}\\ &= \overline{\rho g^{-1} f^{-1}}^{-1}\psi_f^g(\rho)\psi_g(\rho)\overline{\rho}.\end{aligned}$$

By the last formula, the group of pairs (f, ω_f) with multiplication law $(f, \omega_f)(g, \omega_g) = (fg, \omega_f^g\omega_g)$ is isomorphic to the group of pairs (f, ψ_f), i.e., to the group $\widetilde{G}$. Although in the pairs (f, ω_f) the values of functions belong to the group G that is larger than G_0, the restrictions on the values of the functions ω_f, namely, $\varphi(\omega_f) = f$, are of simpler form than the restrictions on the values of ψ_f.

§8. The second Kochendörffer theorem

THEOREM 3.8. *Let $(K/k, G, \varphi, N)$ be an embedding problem and let $(K/K_0, G_0, \varphi_0, N)$ be a problem associated with it, where G_0 is a subgroup of G, φ_0 is the restriction of φ to G_0, $F = \varphi(G) = \mathrm{Gal}(K/k)$, $F_0 = \varphi_0(G_0) = \mathrm{Gal}(K/K_0)$. Let N be an Abelian group whose order n is relatively prime to the index $m = (F : F_0) = (G : G_0)$. Then these two problems are equivalent.*

PROOF. It suffices to prove that the solvability of the associated problem implies the solvability of the original problem. For the problem $(K/K_0, G_0, \varphi_0, N)$ we construct an equivalent "large" problem $(K/k, \widetilde{G}, \widetilde{\varphi}, \widetilde{N})$ by the method described in the preceding section. We regard the group $\widetilde{G}$ as a group of pairs (f, ω_f), where the ω_f are functions on the cosets ρ with values in G, and $\varphi(\omega_f(\rho)) = f$. For every function ω_f we define an "averaging" operation as follows. We assume that $\omega_f(\rho) = g\lambda_f(\rho)$, where $g \in G$ and $\varphi(g) = f$. Then the values of $\lambda_f(\varphi)$ belong to N. We put

$$\mu(\omega_f) = g\Bigl(\prod_\rho \lambda_f(\rho)\Bigr)^{1/m} = g\Bigl(\prod_\rho g^{-1}\omega_f(\rho)\Bigr)^{1/m}.$$

Raising to the power $1/m$ in N is well defined and unique since m and n are relatively prime. The result of averaging does not depend on the choice of g out of the φ-preimages of f. Indeed, if $\varphi(g) = \varphi(g') = f$, then $g' = gt$ for $t \in N$, and the average over g' is equal to

$$gt\Bigl(\prod_\rho t^{-1}\lambda_f(\rho)\Bigr)^{1/m} = gt(t^{-m})^{1/m}\Bigl(\prod_\rho \lambda_f(\rho)\Bigr)^{1/m} = g\Bigl(\prod_\rho \lambda_f(\rho)\Bigr)^{1/m},$$

which coincides with averaging over g. It is clear that for all admissible functions ω_f, their averagings fill up the full φ-preimage of f. We define a mapping of $\widetilde{G}$ to G by assigning $\mu(\omega_f)$ to a pair (f, ω_f). This mapping is clearly surjective. To establish that it is a homomorphism, we need to prove that $\mu(\omega_{f_1}^{f_2}\omega_{f_2}) = \mu(\omega_{f_1})\mu(\omega_{f_2})$. We have

$$\begin{aligned}
\mu(\omega_{f_1}^{f_2}\omega_{f_2}) &= g_1g_2\Bigl(\prod_\rho g_2^{-1}g_1^{-1}\omega_{f_1}(\rho f_2^{-1})\omega_{f_2}(\rho)\Bigr)^{1/m} \\
&= g_1g_2\Bigl(\prod_\rho g_2^{-1}g_1^{-1}\omega_{f_1}(\rho f_2^{-1})g_2\Bigr)^{1/m}\Bigl(\prod_\rho g_2^{-1}\omega_{f_2}(\rho)\Bigr)^{1/m} \\
&= g_1g_2g_2^{-1}\Bigl(\prod_\rho g_1^{-1}\omega_{f_1}(\rho)\Bigr)^{1/m} g_2\Bigl(\prod_\rho g_2^{-1}\omega_{f_2}(\rho)\Bigr)^{1/m} \\
&= g_1g_2g_2^{-1}g_1^{-1}\mu(\omega_{f_1})g_2g_2^{-1}\mu(\omega_{f_2}) = \mu(\omega_{f_1})\mu(\omega_{f_2}).
\end{aligned}$$

Thus, this is a homomorphism of $\widetilde{G}$ onto G, and N is the image of $\widetilde{N}$. Therefore, $(K/k, G, \varphi, N)$ is an associated problem of the first kind for the "large" solvable problem $(K/k, \widetilde{G}, \widetilde{\varphi}, \widetilde{N})$; hence, it is also solvable.

The assumptions of the theorem hold if N is an Abelian p-group and G_0 is a Sylow p-subgroup of G. This special case of the theorem is the second Kochendörffer theorem which in combination with the first theorem reduces the embedding problem with Abelian kernel to the embedding problem for p-groups. □

§9. The Artin–Schreier theorem

This section and the next one are devoted to the embedding problem in the case where the order of the kernel is a power of the characteristic of the fields to be embedded. Here the kernel is not assumed to be Abelian.

THEOREM 3.9 (Artin–Schreier). *Let L/K be a cyclic extension of degree p, and let the characteristic of K also be equal to p. Then in the extension there exists a primitive element θ which satisfies the equation $\theta^p - \theta - c = 0$, where $c \in K$. Conversely, if θ is a root of such an equation, then $K(\theta)$ is a Galois algebra over K with cyclic group of order p. Moreover, $K(\theta)$ is a field if and only if the equation $t^p - t - c = 0$ has no solutions in K.*

PROOF. Let η generate a normal basis for L/K. The element $\operatorname{Tr}\eta = \eta + \eta^a + \cdots + \eta^{a^{p-1}}$, where a is a generator of $\operatorname{Gal}(L/K)$, is contained in K and is different from zero by the linear independence. Therefore, we can assume $\operatorname{Tr}\eta = -1$. Consider the element $\theta = \eta + 2\eta^a + \cdots + (p-1)\eta^{a^{p-2}}$. Then $\theta^a = \eta^a + 2\eta^{a^2} + \cdots + (p-1)\eta^{a^{p-1}}$ and $\theta^a - \theta = -\eta - \eta^a - \cdots - \eta^{a^{p-2}} + (p-1)\eta^{a^{p-1}} = 1$, therefore $\theta^a = \theta + 1$. Since $\theta \notin K$, we have $L = K(\theta)$. It is clear that $(\theta^p - \theta)^a = (\theta+1)^p - (\theta+1) = \theta^p - \theta$, whence $\theta^p - \theta = c \in K$.

Now let θ be a root of the equation $t^p - t - c = 0$, $c \in K$. Then $\theta + 1$, $\theta + 2$, $\ldots$, $\theta + (p-1)$ are also roots of this equation, and $K(\theta)$ is a normal extension of the field K (or a Galois algebra over K) with cyclic Galois group. Since p is prime, this extension either splits completely (if $\theta \in K$) or is a field (if $\theta \notin K$). □

§10. Witt's theorem

Let k be a field of characteristic $p > 0$, K/k a Galois extension with group F, and $1 \to N \to G \xrightarrow{\varphi} F \to 1$ an exact sequence of groups, where N is a p-group.

THEOREM 3.10 (Witt). *The embedding problem $(K/k, G, \varphi, N)$ is always solvable.*

PROOF. We proceed by induction on the order of the group N. First let G be a p-group, and let the kernel be an Abelian group A of order p with generator a.

To the group G there corresponds a cohomology class $h \in H^2(F, A)$. Let $\overline{h}$ be a representative of the class h in the group of cocycles $Z^2(F, A)$. For an arbitrary triple of elements f_1, f_2, $f_3 \in F$, we have $\overline{h}(f_1, f_2)\overline{h}(f_1 f_2, f_3) = \overline{h}(f_1, f_2 f_3)\overline{h}(f_2, f_3)$.

Let $t(f_1, f_2)$ be a function defined on the elements of F with values in the residue field modulo p as follows: $a^{t(f_1, f_2)} = \overline{h}(f_1, f_2)$.

We note that we can choose the generator v of a normal basis for the extension K/k in such a way that $\sum_{f \in F} v^f = 1$.

Let $\{\overline{f}\}_{f \in F}$ be a set of representatives of the cosets of G with respect to A; then each element $g \in G$ is of the form $g = \overline{f}a^i$. For $g = \overline{f}a^i$, we set $x(g) =$

$\sum_{e\in F} t(f,e)v^{e^{-1}} + i$. For $g_1 = \overline{f}_1 a^i$ and $g_2 = \overline{f}_2 a^j$, we have

$$\begin{aligned} x(g_1)^{\varphi(g_2)} + x(g_2) &= \sum_{e\in F} t(f_1,e)v^{e^{-1}f_2} + i + \sum_{e\in F} t(f_2,e)v^{e^{-1}} + j \\ &= \sum_{e\in F} (t(f_1, f_2e) + t(f_2,e))v^{e^{-1}} + i + j \\ &= \sum_{e\in F} (t(f_1,f_2) + t(f_1f_2,e))v^{e^{-1}} + i + j \\ &= t(f_1,f_2) + i + j + \sum_{e\in F} t(f_1f_2,e)v^{e^{-1}} \\ &= x(\overline{f_1f_2h}(f_1f_2)a^{i+j}) = x(g_1g_2). \end{aligned}$$

Now we set $y(g) = x^p(g) - x(g)$. Then, $y(g_1g_2) = y(g_1)^{\varphi(g_2)} + y(g_2)$ and $y(ag) = y(g)$. Therefore, the cocycle $y(g) \in Z^1(G, K^+)$ can be obtained by the lifting a cocycle $y_1 \in Z^1(F, K^+)$. Then there exists an element $c \in K$ such that $y_1(f) = c^f - c$.

Now let θ be a root of the equation $t^p - t - c = 0$, and let L be the algebra obtained by adjoining the element θ to the field K. We define the automorphisms $g \in G$ on L by setting $\theta^g = \theta + x(g)$. Then $(\theta^{g_1})^{g_2} = (\theta + x(g_1))^{g_2} = \theta + x(g_2) + x(g_1)^{\varphi(g_2)} = \theta + x(g_1g_2) = \theta^{g_1g_2}$, $c^g = (\theta^p - \theta)^g = c^{\varphi(g)}$ and $\theta^a = \theta + 1$. Thus, L is a solution of the embedding problem.

Now we proceed to the general case. If the kernel N contains a proper subgroup N_1 invariant in G, then, passing to the associated problem obtained by factorization of the kernel by the subgroup N_1, we find, using the induction assumption, a solution L_1. Then the embedding problem $(L_1/k, G, \varphi_1, N_1)$ is again solvable by inductive assumption, and its solution is a solution of the original problem (here φ_1 is a natural mapping of G onto G/N_1). Therefore, we may assume that the embedding problem is Abelian because in a non-Abelian kernel its commutator subgroup is a proper characteristic subgroup. By the second Kochendörffer theorem, we may assume that G is a p-group. Then, in the kernel A there exists a subgroup A_1 of index p, invariant in G. The passage to the corresponding associated problem yields the case we have considered at the beginning of the proof. To complete the proof of the theorem, it remains to use again the induction assumption. □

§11. Embedding conditions

Let $(K/k, G, \varphi, A)$ be an embedding problem with Abelian kernel related to the exact sequence of groups $1 \longrightarrow A \longrightarrow G \xrightarrow{\varphi} F = \mathrm{Gal}(K/k) \longrightarrow 1$. If $p > 0$ is the characteristic of the field k and the order of A is a multiple of p, then we can represent A as a direct product of a p-group A_1 and a group A_2 whose order is relatively prime to p. By the first Kochendörffer theorem, the solvability of the embedding problem is equivalent to the solvability of the two associated problems obtained by factorization of G by the subgroups A_1 and A_2, respectively. By Witt's theorem, the associated problem $(K/k, G/A_2, \varphi_2, A_1)$ related to the exact sequence $1 \longrightarrow A_1 \longrightarrow G/A_2 \longrightarrow F \longrightarrow 1$ is always solvable. Therefore, it suffices to consider the embedding problems of fields whose characteristic either is equal to zero, or does not divide the order of the kernel. In the sequel we shall consider precisely these problems.

Let n be the period of the kernel A. We assume that the field K contains all the nth roots of unity, since otherwise, adjoining them to the field K, we may pass to the equivalent embedding problem that is obtained by the lifting of an intermediate field. Then the character group $\widehat{A} = \operatorname{Hom}(A, K^*)$ that is naturally a G-module, namely, $\chi^g(a) = \chi(a^{g^{-1}})^{\varphi(g)}$, is well defined. Henceforth, we shall consider modules in multiplicative notation. We denote the unit character by χ_0.

PROPOSITION 3.11.1. *The embedding problem $(K/k, G, \varphi, A)$ is solvable if and only if there exists an extension of G-modules $1 \longrightarrow K^* \longrightarrow M \xrightarrow{\psi_1} \widehat{A} \longrightarrow 1$ such that for $m \in M$ and $a \in A$ the following relation holds*:

$$m^a = m(\psi_1(m)(a)). \tag{1}$$

PROOF. Let the embedding problem be solvable, and let L be a solution. We decompose the space L (over K) into a direct sum of one-dimensional subspaces ("Lagrange resolvents") M_χ, $\chi \in \widehat{A}$. Namely, we introduce K-linear operators π_χ by setting for $z \in L$, $\pi_\chi(z) = \frac{1}{|A|}\sum_{a\in A} z^a\chi(a^{-1})$, $\chi \in \widehat{A}$, and $M_\chi = \operatorname{Im}\pi_\chi \subset L$. It is easy to check that $\pi_{\chi_1} \cdot \pi_{\chi_2} = 0$ for $\chi_1 \neq \chi_2$; $\pi_\chi \cdot \pi_\chi = \pi_\chi$; $\sum_{\chi\in\widehat{A}} \pi_\chi$ is the identity operator on L. The spaces M_χ are nonzero because $\pi(w) \neq 0$, where w is a generator of a normal basis for L/K, each M_χ is elementwise invariant with respect to π_χ and is annihilated by the operator $\pi_{\chi'}$ for $\chi' \neq \chi$. Therefore, L is the direct sum of the subspaces M_χ, and all these spaces are one-dimensional because the number of them is equal to $(L : K)$. Note that $M_{\chi_0} = K$.

The elements of M_χ are multiplied by $\chi(a)$ under the action of the automorphism $a \in A$, and they are characterized by this property. Indeed,

$$\begin{aligned}\left(\frac{1}{|A|}\sum_{b\in A} z^b\chi(b^{-1})\right)^a &= \frac{1}{|A|}\sum_b z^{ba}\chi(b^{-1}) \\ &= \frac{1}{|A|}\sum_{c\in A} z^c\chi(ac^{-1}) = \chi(a)\cdot\frac{1}{|A|}\sum_{c\in A} z^c\chi(c^{-1}).\end{aligned}$$

Conversely, if $y^a = y\chi(a)$, then

$$\pi_\chi(y) = \frac{1}{|A|}\sum_{b\in A} y^b\chi(b^{-1}) = \frac{1}{|A|}\sum_b y\chi(b)\chi(b^{-1}) = y.$$

We show that if $y \in M_\chi$, then $y^g \in M_{\chi^g}$ for $g \in G$. Indeed,

$$(y^g)^a = (y^{gag^{-1}})^g = (y\chi(gag^{-1}))^g = y^g\chi^g(a).$$

Finally, if $y_1 \in M_{\chi_1}$ and $y_2 \in M_{\chi_2}$, then $y_1y_2 \in M_{\chi_1\chi_2}$. Indeed,

$$(y_1y_2)^a = y_1\chi_1(a)y_2\chi_2(a) = y_1y_2(\chi_1\chi_2)(a).$$

By the last property, on the union $\bigcup M_\chi$ with zero excluded multiplication is well defined, and it is associative, commutative, and $1 \in M_{\chi_0}$. We show that, for each $y \in M_\chi$, $y \neq 0$, an inverse element exists. Indeed, if $\chi^m = \chi_0$, then $z = y^m \in M_{\chi_0} = K$ and $z \neq 0$ because the algebra L does not contain the radical.

Therefore, $y^{m-1}z^{-1} = y^{-1} \in M_{\chi^{-1}}$. Thus, $\bigcup M_\chi$ (with zero excluded) is an Abelian group. Therefore, we have the following exact sequence:

$$1 \to K^* = M_{\chi_0} \setminus \{0\} \to \bigcup_\chi (M_\chi \setminus \{0\}) \to \widehat{A} \to 1.$$

By the properties proved above, this sequence is a sequence of G-modules. Therefore, $M = \bigcup_\chi (M_\chi \setminus \{0\})$ is the required extension.

Conversely, let the following extension of G-modules be given:

$$1 \to K^* \to M \xrightarrow{\psi_1} \widehat{A} \to 1,$$

and let $m^a = m \cdot (\psi_1(m)(a))$ for any $m \in M$ and $a \in A$. We put $M_\chi^* = \psi_1^{-1}(\chi)$, $\chi \in \widehat{A}$. Any two elements in M_χ^* differ by a factor from K^*. This enables us to introduce addition on M_χ^* with zero adjoined and make $M_\chi = M_\chi^* \cup \{0\}$ a one-dimensional space over K. Extending by linearity multiplication in $M = \bigcup M_\chi^*$ to the direct sum of the spaces M_χ, we obtain the algebra L that is a solution of the embedding problem $(K/k, G, \varphi, A)$. The proposition is proved. □

Now let $\mathbb{Z}[\widehat{A}]$ be a multiplicatively written free Abelian group with generators c_χ that are put in one-to-one correspondence with the characters $\chi \in \widehat{A}$. We make $\mathbb{Z}[\widehat{A}]$ a G-module by setting

$$\left(\prod_\chi c_\chi^{k_\chi}\right)^g = \prod_\chi c_{\chi^g}^{k_\chi} \quad (k_\chi \in \mathbb{Z}).$$

Let ψ_2 be the natural homomorphism $\mathbb{Z}[\widehat{A}] \to \widehat{A}$ defined by the formula $\psi_2(c_\chi) = \chi$.

PROPOSITION 3.11.2. *For the embedding problem $(K/k, G, \varphi, A)$, the compatibility condition is fulfilled if and only if there exists an extension of G-modules $1 \to K^* \to \overline{P} \xrightarrow{\psi_3} \mathbb{Z}[\widehat{A}] \to 1$ such that, for any $y \in \overline{P}$, $a \in A$, the following relation holds*:

$$y^a = y \cdot (\psi_2\psi_3(y)(a)). \tag{2}$$

PROOF. Let the compatibility condition hold, and let P be the compatibility module. We apply to P the operators of the "Lagrange resolvents" π_χ, putting $\pi_\chi(z) = \frac{1}{|A|}\sum_{a\in A} z^a \chi(a^{-1})$. As in the proof of Proposition 3.11.1, we may see that P is a direct sum of one-dimensional (over K) subspaces P_χ, where $P_\chi = \operatorname{Im}\pi_\chi$, and that $(P_\chi)^g = P_{\chi^g}$. However, we cannot directly define multiplication law in $\bigcup P_\chi$. We do it formally. Namely, in each P_χ we choose a representative $x_\chi \neq 0$ and consider the group of monomials $z\prod_\chi x_\chi^{k_\chi}$ with integer exponents k_χ (assuming $x_\chi^0 = 1$) and with coefficients $z \in K^*$. This group $\overline{P}$ is a G-module.

The mapping $\psi_3 \colon \overline{P} \to \mathbb{Z}[\widehat{A}]$ defined by the formula $\psi_3(z\prod_\chi x_\chi^{k_\chi}) = \prod_\chi c_\chi^{k_\chi}$ is a module homomorphism with kernel K^*. Moreover, for $y = z\prod_\chi x_\chi^{k_\chi}$, we have

$$y^a = \left(z\prod x_\chi^{k_\chi}\right)^a = z\prod x_\chi^{k_\chi}\chi^{k_\chi}(a) = y\prod \chi^{k_\chi}(a) = y(\psi_2\psi_3(y)(a)).$$

The extension $1 \to K^* \to \overline{P} \xrightarrow{\psi_3} \mathbb{Z}[\widehat{A}] \to 1$ is a required extension.

Conversely, let such an extension exist. Let P_χ be the preimage of c_χ with zero adjoined. As in the proof of Proposition 3.11.1, one can make P_χ to a space one-dimensional over K, and $(P_\chi)^g = P_{\chi^g}$ for $g \in G$. It is easy to see that the direct sum $\sum^\oplus P_\chi$ is the compatibility module for the embedding problem. The proposition is proved.

We note that the module $\overline{P}$ is uniquely determined up to G-module isomorphism. Indeed, the compatibility module P is determined up to isomorphism, and if one changes the set of representatives x_χ from P_χ in the process of constructing $\overline{P}$, this implies only changing the coefficients from K^* of monomials. □

PROPOSITION 3.11.3. *If the compatibility condition is fulfilled and $\overline{P}$ is the module constructed above, then the embedding problem is solvable if and only if there exists the following commutative diagram of G-modules with exact rows*:

$$(*)\qquad \begin{array}{ccccccccc} 1 & \longrightarrow & K^* & \longrightarrow & \overline{P} & \longrightarrow & \mathbb{Z}[\widehat{A}] & \longrightarrow & 1 \\ & & \downarrow{\scriptstyle \mathrm{id}} & & \downarrow{\scriptstyle \omega} & & \downarrow{\scriptstyle \psi_2} & & \\ 1 & \longrightarrow & K^* & \longrightarrow & M & \longrightarrow & \widehat{A} & \longrightarrow & 1 \end{array}$$

PROOF. If the embedding problem is solvable, then one can take its solution as a compatibility module and associate with each monomial $z\prod_\chi x_\chi^{k_\chi}$ its value in L. This gives the required mapping ω of the module $\overline{P}$ onto M, where $M = \bigcup_\chi M_\chi \setminus \{0\}$.

Conversely, let the diagram $(*)$ exist. Then ω is surjective, and the image M of the module $\overline{P}$ satisfies the assumption of Proposition 3.11.1 because the monomial $z\prod_\chi x_\chi^{k_\chi}$ is multiplied by $\prod_\chi \chi(a)^{k_\chi}$ under the action of the automorphism $a \in A$, and its image is multiplied by the same factor. The proposition is proved. □

Now we transform the embeddability condition, assuming that the compatibility condition is fulfilled, so that there exists an exact sequence

$$1 \to K^* \to \overline{P} \xrightarrow{\psi_3} \mathbb{Z}[\widehat{A}] \to 1$$

corresponding to a certain element $t \in \mathrm{Ext}^1_G(\mathbb{Z}[\widehat{A}], K^*)$. There is an exact sequence of G-modules $1 \to V \xrightarrow{\alpha} \mathbb{Z}[\widehat{A}] \xrightarrow{\psi_2} \widehat{A} \to 1$; here, the kernel V consists of the elements $\prod_\chi c_\chi^{k_\chi}$ for which $\prod_\chi \chi^{k_\chi} = \chi_0$, the kernel V is a free Abelian group generated by the elements $c_\chi c_{\chi^{-1}}$.

This exact sequence induces the exact sequence

$$\mathrm{Ext}^1_G(\widehat{A}, K^*) \xrightarrow{\psi_2^*} \mathrm{Ext}^1_G(\mathbb{Z}[\widehat{A}], K^*) \xrightarrow{\alpha^*} \mathrm{Ext}^1_G(V, K^*),$$

so that to the embedding problem given above there corresponds (if compatibility holds) the element $u = \alpha^* t$.

It is easy to describe the image of the module $\overline{P}$ under the action of α^*. Namely, this image is the extension

$$1 \to K^* \to R \to V \to 1,$$

where R is the preimage of V in $\overline{P}$. The module R is formed of the monomials $z\prod_\chi x_\chi^{k_\chi} \in \overline{P}$, for which $\prod_\chi \chi^{k_\chi} = \chi_0$.

It is clear that the elements from R are invariant with respect to the automorphisms $\alpha \in A$ because

$$\left(z\prod_{\chi} x_{\chi}^{k_{\chi}}\right)^{a} = z\prod_{\chi} x_{\chi}^{k_{\chi}} \cdot \prod_{\chi} \chi^{k_{\chi}}(a) = z\prod_{\chi} x_{\chi}^{k_{\chi}}.$$

Therefore, R can be regarded as an F-module, just as K^* and V. Thus, one can assume that $u \in \operatorname{Ext}_F^1(V, K^*)$.

If the embedding problem is solvable, then the element t is the image for ψ_2^* by the existence of the diagram $(*)$; so, $u = \alpha^* t = 0$. Conversely, if $u = \alpha^* t = 0$, then $t = \psi_2^* v$, where $v \in \operatorname{Ext}_G^1(\widehat{A}, K^*)$, i.e., the diagram $(*)$ holds, and, by Proposition 3.11.3, the embedding problem is solvable. Thus, we have proved the following theorem.

THEOREM 3.11 (A. V. Yakovlev). *Let the compatibility condition be fulfilled for the embedding problem $(K/k, G, \varphi, A)$. Then there exists a certain element $u \in \operatorname{Ext}_F^1(V, K^*)$ such that the problem is solvable if and only if $u = 0$.*

§12. The semidirect embedding problem with Abelian kernel

In §9 of Chapter 1, it was shown that the semidirect embedding problem is always solvable. We give a description of the set of its solutions in the case of an Abelian kernel. Let K/k be a Galois extension with group F, and let

$$1 \to A \to G \to F \to 1$$

be a semidirect extension with Abelian kernel A so that an embedding of the group F in G exists.

We assume that the field K in the problem $(K/k, G, \varphi, A)$ contains a primitive root of unity of degree equal to the period of the kernel.

THEOREM 3.12 (A. V. Yakovlev). *The solutions of the semidirect embedding problem are in one-to-one correspondence with the elements of the group* $\operatorname{Ext}_F^1(\widehat{A}, K^*)$.

PROOF. Let L be a solution of the problem, then we may assume that the G-module M corresponding to L by Proposition 3.11.1 is also an F-module. Thus, a certain element $z \in \operatorname{Ext}_F^1(\widehat{A}, K^*)$ is given.

Conversely, let $z \in \operatorname{Ext}_F^1(\widehat{A}, K^*)$ and let

$$1 \to K^* \to M \to \widehat{A} \to 1$$

be the corresponding sequence of F-modules. We choose in M some preimages m_χ of the elements $\chi \in \widehat{A}$. We show that m_χ can be chosen in such a way that for all $\chi \in \widehat{A}$ and $f \in F$, the relations $m_\chi^f = m_{\chi^f}$ hold. Let b be an orbit in $\widehat{A}$ under the action of F, we choose $\chi \in b$ and denote by F_χ a subgroup of F acting trivially on χ. For $f \in F_\chi$ we have $m_\chi^f = m_\chi c(f, \chi)$, where $c(f, \chi) \in K^*$. If $f_1, f_2 \in F_\chi$, then $c(f_1 f_2, \chi) = (c(f_1, \chi))^{f_2} c(f_2, \chi)$, i.e., the function $c(f, \chi)$ with χ chosen is a one-dimensional cocycle of the group F_χ in K^*. Therefore, there exist elements z_χ such that $c(f, \chi) = z_\chi^{f-1}$ for $f \in F_\chi$ and χ chosen. We put $\overline{m}_\chi = m_\chi z_\chi^{-1}$. Then for $f \in F_\chi$ we have $\overline{m}_\chi^f = \overline{m}_\chi$. For the other elements χ^f, $f \in F$, of the orbit b, we put $\overline{m}_{\chi^f} = \overline{m}_\chi^f$. We do this for all orbits. Now we put $\overline{m}_\chi^a = \overline{m}_\chi \chi(a)$. We show that, in doing so, we introduce the structure of a G-module on the module M. Indeed, let $g_1 = f_1 a_1$, $g_2 = f_2 a_2$ (here f_1 and f_2 are the images of the elements of F under given embedding of F in G). Then $g_1 g_2 = f_1 f_2 a_1^{f_2} a_2$, $\overline{m}_\chi^{g_1 g_2} = \overline{m}_{\chi^{f_1 f_2}} \chi^{f_1 f_2}(a_1^{f_2} a_2)$, and $(\overline{m}_\chi^{g_1})^{g_2} = \overline{m}_\chi^{g_1 g_2}$. □

§13. The second approach to the description of embedding conditions

In this section, we use "the look from above" at the embedding problem and, applying the techniques of Galois cohomology, we give another proof of Theorem 3.11.

1°. As a preliminary, we prove a group-theoretic theorem dual to Proposition 1.13.1 in a sense.

THEOREM 3.13.1. *Let A_1, A_2 be F-operator Abelian groups, let $1 \longrightarrow A_1 \xrightarrow{\alpha_1} G_1 \xrightarrow{\varphi_1} F \longrightarrow 1$ be an extension of A_1 by F, and let $\psi\colon A_1 \longrightarrow A_2$ be an F-operator homomorphism. Then there exists a unique extension (up to equivalence of extensions) of A_2 by F such that the diagram*

$$(**)\qquad \begin{array}{ccccccccc} 1 & \longrightarrow & A_1 & \xrightarrow{\alpha_1} & G_1 & \xrightarrow{\varphi_1} & F & \longrightarrow & 1 \\ & & \downarrow{\scriptstyle\psi} & & \downarrow{\scriptstyle\gamma} & & \downarrow{\scriptstyle\mathrm{id}} & & \\ 1 & \longrightarrow & A_2 & \xrightarrow{\alpha_2} & G_2 & \xrightarrow{\varphi_2} & F & \longrightarrow & 1 \end{array}$$

is commutative for some homomorphism γ.

One can obtain the canonical class of $H^2(F, A_2)$ with the help of the class from $H^2(F, A_1)$ by applying ψ to the cocycles that constitute the class from $H^2(F, A_1)$.

PROOF. First we prove existence. Let $\{\overline{f}\}$, $f \in F$, $\overline{1} = 1$, be a set of coset representatives in the group G_1 with respect to the subgroup A_1, and $z(f_1, f_2) = \overline{f_1 f_2}^{-1} \overline{f}_1 \overline{f}_2$ a cocycle corresponding to this set of representatives. Then $u(f_1, f_2) = \psi(z(f_1, f_2))$ is also a cocycle for the group F with values in A_2. We associate to the elements $\overline{f}$ the symbols $\overline{\overline{f}}$, and, using these symbols and the elements of A_2, we form a group with the following multiplication law

$$\overline{\overline{f}}_1 a_1 \overline{\overline{f}}_2 a_2 = \overline{\overline{f_1 f_2}}\, u(f_1, f_2) a_1^{f_1} a_2.$$

Here $a_1, a_2 \in A_2$. It is easy to verify that this is actually a group, we denote it by G_2. The homomorphisms which make the diagram $(**)$ commutative and the second row exact may be given by the formulas $\alpha_2(a_2) = \overline{\overline{1}} a_2$, $a_2 \in A_2$; $\varphi_2(\overline{\overline{f}} a_2) = f$; $\gamma(\overline{f} a_1) = \overline{\overline{f}} \psi(a_1)$, $a_1 \in A_1$.

Now we prove uniqueness. Let the diagram $(**)$ be commutative, $\{\overline{f}\}$, $f \in F$, $\overline{1} = 1$, a set of representatives in G_1, and $z(f_1, f_2) = \overline{f_1 f_2}^{-1} \overline{f}_1 \overline{f}_2$ a cocycle corresponding to the first row. Let $\overline{\overline{f}} = \gamma(\overline{f})$, and let $u(f_1, f_2)$ be a cocycle corresponding to this set of representatives. Then $u(f_1, f_2) = \overline{\overline{f_1 f_2}}^{-1} \overline{\overline{f}}_1 \overline{\overline{f}}_2 = \gamma(\overline{f_1 f_2}^{-1} \overline{f}_1 \overline{f}_2) = \gamma z(f_1 f_2)$. The extension $1 \longrightarrow A_2 \longrightarrow G_2 \longrightarrow F \longrightarrow 1$ is uniquely determined by that cocycle.

It is easy to give a more explicit description of the group G_2. Namely, we denote by G_3 the semidirect product of G_1 and A_2, i.e., the set of pairs (g_1, a_2), $g_1 \in G_1$, $a_2 \in A_2$, with multiplication $(g_1, a_2)(g_1', a_2') = (g_1 g_1', a_2^{f_1'} a_2')$, where $f_1' = \varphi_1(g_1')$; we denote by C the subgroup of G_3 formed by the pairs $(\alpha_1(a_1), \psi(a_1^{-1}))$. It is easy to check that C is a normal subgroup of G_3. We put $\alpha'_2(a_2) = (1, a_2)$, $\varphi'_2(g_1, a_2) = \varphi_1(g_1)$, $\gamma'(g_1) = (g_1, 1)$. The group C is contained in the kernel of the homomorphism φ'_2. The group $G_2 = G_3/C$ and the homomorphisms $\alpha_2(a) = (1, a) \bmod C$, $\varphi_2(g_1 a_2) = \varphi_1(g_1)$, $\gamma(g_1) = (g_1, 1) \bmod C$ satisfy the required demands. Indeed, the exactness of the second row and the commutativity of the right square are obvious. Further, $\gamma_1(\alpha_1(a_1)) = (\alpha_1(a_1), 1) \bmod C$ and $\alpha_2\psi(a_1) = (1, \psi(a_1)) \bmod C$, but $(\alpha_1(a_1), 1) \bmod C = (1, \psi(a_1)) \bmod C$, so the left square is also commutative.

2°. Now we consider the embedding problem $(K/k, G, \varphi, A)$. Let $\overline{k}$ be the algebraic separable closure of the field k with profinite Galois group $\overline{F}$. We denote by h the cohomology class from $H^2(F, A)$ corresponding to the extension $1 \to A \to G \to F \to 1$, where $F = \mathrm{Gal}(K/k)$. □

THEOREM 3.13.2. *The embedding problem $(K/k, G, \varphi, A)$ is solvable if and only if the lifting homomorphism λ of the group $H^2(F, A)$ to the group $H^2(\overline{F}, A)$ maps h to the trivial class.*

PROOF. The embedding problem is solvable if and only if there exists a splitting field L for this problem. Let $F_1 = \mathrm{Gal}(L/k)$. The problem $(L/k, G_1, \varphi_1, A)$ that is obtained by the lifting of the original problem becomes semidirect. Therefore, the class $h \in H^2(F, A)$ splits under the lifting of $H^2(F, A)$ to $H^2(F_1, A)$. Conversely, if the class becomes trivial under lifting to the group $H^2(\overline{F}, A)$, then, by the definition of the latter group, it immediately follows that there exists a finite extension L of the field k with Galois group F_1 containing K such that the lifting of the class h to the group $H^2(F_1, A)$ is trivial. Thus, L is a splitting field for the original problem.□

3°. Now we transform the conditions of the solvability of the problem obtained above. We keep the notation of §11. We can make the F-module $\mathbb{Z}[\widehat{A}]$ an $\overline{F}$-module by setting $c_\chi^{\overline{f}} = c_\chi^{\mu(f)}$ for $c_\chi \in \mathbb{Z}[\widehat{A}]$; here, $\overline{f} \in \overline{F}$, μ is a natural homomorphism of $\overline{F}$ onto F.

The exact sequence $1 \to V \to \mathbb{Z}[\widehat{A}] \to \widehat{A} \to 1$ implies, by the completeness of the multiplicative group $\overline{k}^*$ of the field $\overline{k}$, the exact sequence

$$1 \to \mathrm{Hom}(\widehat{A}, \overline{k}^*) \to \mathrm{Hom}(\mathbb{Z}[\widehat{A}], \overline{k}^*) \to \mathrm{Hom}(V, \overline{k}^*) \to 1,$$

but $\mathrm{Hom}(\widehat{A}, \overline{k}^*) = A$.

This exact sequence implies the exact cohomology sequence

$$(***) \qquad H^1(\overline{F}, \mathrm{Hom}(\mathbb{Z}[\widehat{A}], \overline{k}^*)) \to H^1(\overline{F}, \mathrm{Hom}(V, \overline{k}^*)) \xrightarrow{\delta} H^2(\overline{F}, A) \to H^2(\overline{F}, \mathrm{Hom}(\mathbb{Z}[\widehat{A}], \overline{k}^*)).$$

We compute the cohomology groups for the module $\mathrm{Hom}(\mathbb{Z}[\widehat{A}], \overline{k}^*)$. The $\overline{F}$-module $\mathbb{Z}[\widehat{A}]$ is obviously the direct sum of the $\overline{F}$-modules $\mathbb{Z}(b)$, where the b are orbits in $\widehat{A}$ under the action of $\overline{F}$, and, thus, $\mathrm{Hom}(\mathbb{Z}[\widehat{A}], \overline{k}^*)$ is the direct sum of the groups $\mathrm{Hom}(\mathbb{Z}(b), \overline{k}^*)$. Let $\chi \in b$, and let $\overline{F}_\chi$ be a subgroup of the group $\overline{F}$ acting trivially on χ. Further, the generators of the group $\mathbb{Z}(b)$ move under the action of $\overline{F}$ as the cosets of $\overline{F}$ with respect to $\overline{F}_\chi$, so that $\mathrm{Hom}(\mathbb{Z}(b), \overline{k}^*)$ is isomorphic to a group of functions on the cosets of $\overline{F}$ with respect to $\overline{F}_\chi$ with values in $\overline{k}^*$. Using

Lemma A.3.9, we obtain

$$H^i(\overline{F}, \mathrm{Hom}(\mathbb{Z}[\widehat{A}], \overline{k}^*)) \approx \sum_b^{\oplus} H^i(\overline{F}_\chi, \overline{k}^*).$$

The cohomology groups $H^i(\overline{F}_\chi, \overline{k}^*)$ are obtained by lifting of $H^i(F_\chi, K^*)$, where F_χ is the subgroup of $F = \mathrm{Gal}(K/k)$ consisting of the elements fixing the character χ. Since $H^1(F_\chi, K^*) = 0$, we have $H^1(\overline{F}_\chi, \overline{k}^*) = 0$.

In Chapter III, §3 we have shown that the compatibility condition in homology terms is equivalent to the fact that the image of the canonical class $h \in H^2(F, A)$ in the group $H^2(F_\chi, K^*)$ is equal to 0 for all characters χ. Since the lifting of the Brauer group $H^2(F_\chi, K^*)$ to $H^2(\overline{F}_\chi, K^*)$ is injective, the compatibility condition is equivalent to the fact that the image of h is equal to 0 under the composition of the mappings

$$H^2(F, A) \xrightarrow{\lambda} H^2(\overline{F}, A) \to H^2(\overline{F}, \mathrm{Hom}(\mathbb{Z}[\widehat{A}], \overline{k}^*)) = \sum_\chi^{\oplus} H^2(\overline{F}_\chi, \overline{k}^*),$$

i.e., the element $\lambda(h)$ is the image of some element $z \in H^1(\overline{F}, \mathrm{Hom}(V, \overline{k}^*))$ (we use the exactness of the sequence $(***)$).

Now we apply the descent procedure to the group $H^1(\overline{F}, \mathrm{Hom}(V, \overline{k}^*))$. Let F_0 be the kernel of the natural homomorphism of $\overline{F}$ onto F. The group F_0 acts trivially on the elements of V, so the elements of $\mathrm{Hom}(V, \overline{k}^*)$ left fixed under the action of F_0 constitute the group $\mathrm{Hom}(V, K^*)$.

Applying the Hochschild–Serre theorem on cohomology lifting, we obtain the exact sequence

$$0 \to H^1(F, \mathrm{Hom}(V, K^*)) \xrightarrow{l} H^1(\overline{F}, \mathrm{Hom}(V, \overline{k}^*)) \xrightarrow{r} H^1(F_0, \mathrm{Hom}(V, \overline{k}^*)).$$

By the triviality of the action of F_0 on V, the group $\mathrm{Hom}(V, \overline{k}^*)$ is a direct product of several copies of $\overline{k}^*$, and $H^1(F_0, \mathrm{Hom}(V, \overline{k}^*))$ is trivial. Therefore, the lifting l is an isomorphism of the groups $H^1(F, \mathrm{Hom}(V, K^*))$ and $H^1(\overline{F}, \mathrm{Hom}(V, \overline{k}^*))$, so that δl is an embedding of $H^1(F, \mathrm{Hom}(V, K^*))$ in $H^2(\overline{F}, A)$.

Thus, if the compatibility condition is fulfilled for the original problem, then there exists an element z in the group $H^1(F, \mathrm{Hom}(V, K^*))$ such that $\delta l(z) = \lambda(h)$. The embedding problem is solvable if and only if $z = 0$.

Now we show that the additional embedding condition in §11 coincides with the newly-found one.

THEOREM 3.13.3. *There exists a natural isomorphism between* $\mathrm{Ext}^1_F(V, K^*)$ *and* $H^1(F, \mathrm{Hom}(V, K^*))$.

PROOF. Let $1 \to I \to U \to V \to 1$ be a representation of the module V as a quotient module of a free F-module U; then the sequence induces the exact sequences

$$\mathrm{Hom}_F(U, K^*) \to \mathrm{Hom}_F(I, K^*) \to \mathrm{Ext}^1_F(V, K^*) \to \mathrm{Ext}^1_F(U, K^*) = 0,$$

$$\mathrm{Hom}_F(U, K^*) \to \mathrm{Hom}_F(I, K^*) \to H^1(F, \mathrm{Hom}(V, K^*)) \to H^1(F, \mathrm{Hom}(U, K^*)) = 0.$$

Comparing these sequences, we complete the proof of the theorem. □

In Chapter 1, §15, we established a correspondence between the set of solutions of the solvable embedding problem $(K/k, G, A)$ and the set of cocycles $H^1(\overline{F}, A)$, where $\overline{F} = \mathrm{Gal}(\overline{k}/k)$ is the Galois group of an algebraic closure of k.

At the same time, for an Abelian kernel A and a semidirect extension this set of solutions can be written as $\mathrm{Ext}^1_F(\widehat{A}, K^*)$, where $F = \mathrm{Gal}(K/k)$. These two descriptions may not contradict each other.

THEOREM 3.13.4. *There exists a natural isomorphism between $H^1(\overline{F}, A)$ and $\mathrm{Ext}^1_F(\widehat{A}, K^*)$.*

PROOF. We preserve the previous notation. The exact sequence

$$1 \longrightarrow \mathrm{Hom}(\widehat{A}, \overline{k}^*) \longrightarrow \mathrm{Hom}(\mathbb{Z}[\widehat{A}], \overline{k}^*) \longrightarrow \mathrm{Hom}(V, \overline{k}^*) \longrightarrow 1$$

induces the exact cohomology sequence

$$\mathrm{Hom}_{\overline{F}}(\mathbb{Z}[\widehat{A}], \overline{k}^*) \longrightarrow \mathrm{Hom}_{\overline{F}}(V, \overline{k}^*) \longrightarrow H^1(\overline{F}, A) \longrightarrow H^1(\overline{F}, \mathrm{Hom}(\mathbb{Z}[\widehat{A}], \overline{k}^*)) = 0,$$

or

$$\mathrm{Hom}_F(\mathbb{Z}[\widehat{A}], K^*) \longrightarrow \mathrm{Hom}_F(V, K^*) \longrightarrow H^1(\overline{F}, A) \longrightarrow 0.$$

On the other hand, we have the exact sequence

$$\mathrm{Hom}_F(\mathbb{Z}[\widehat{A}], K^*) \longrightarrow \mathrm{Hom}_F(V, K^*) \longrightarrow \mathrm{Ext}^1_F(\widehat{A}, K^*) \longrightarrow 0,$$

induced by the triple $1 \longrightarrow V \longrightarrow \mathbb{Z}[\widehat{A}] \longrightarrow \widehat{A} \longrightarrow 1$. Comparing these sequences, the theorem follows at once. □

4°. Now we show that the group $H^1(F, \mathrm{Hom}(V, K^*))$ completely describes the embedding problems satisfying the compatibility condition. We keep the previous notation.

THEOREM 3.13.5. *For any element $z \in H^1(F, \mathrm{Hom}(V, K^*))$, there exists an embedding problem to which the element z corresponds, and the compatibility condition is fulfilled.*

PROOF. Consider the sequence (***). We set $h = \delta(z) \in H^2(\overline{F}, A)$. Then $\theta(h) = 0$, so that if there exists an embedding problem to which z corresponds, then the compatibility condition is fulfilled for this problem.

Let $\{K_\beta\}$, $K_\beta \supset K$, be an inductive system of finite Galois extensions of the field k with groups F_β, and let the inductive limit of K_β coincide with the separable algebraic closure $\overline{k}$ of the field k. By definition, $H^2(\overline{F}, A) = \varinjlim H^2(F_\beta, A)$, where the inductive limit is taken with respect to the lifting homomorphisms. By the property of the inductive limit, there exists a subscript β_0 such that $h = \lambda_{\beta_0}(h_{\beta_0})$, where $h_{\beta_0} \in H^2(F_{\beta_0}, A)$, and λ_{β_0} is the lifting homomorphism of the group $H^2(F_{\beta_0}, A)$ to $H^2(\overline{F}, A)$. Let

$$1 \longrightarrow A \longrightarrow G_{\beta_0} \xrightarrow{\varphi_{\beta_0}} F_{\beta_0} \longrightarrow 1$$

be a group extension corresponding to h_{β_0}. Then the embedding problem $(K_{\beta_0}/k, G_{\beta_0}, \varphi_{\beta_0}, A)$ is the required one. Indeed, it suffices to establish the isomorphism of the groups $H^1(F, \mathrm{Hom}(V, K^*))$ and $H^1(F_{\beta_0}, \mathrm{Hom}(V, K^*_{\beta_0}))$ with the help of the Hochschild–Serre sequence and to use the arguments of the preceding section. □

§14. Embedding conditions for local and global fields

1°. Let $(K/k, G, \varphi, A)$ be an embedding problem, and let the field K, as before, contain all roots of unity of degree equal to the period of A. Let $\overline{k}$ be the separable algebraic closure of k, and let F, $\overline{F}$ be the Galois groups over k of the fields K and $\overline{k}$, respectively.

Let $\widehat{A} = \mathrm{Hom}(A, K^*)$ be the character group of A with the action of automorphisms from F defined above. We denote by $\widehat{A}_1$ the subgroup of elements of $\widehat{A}$ fixed under the action of all $f \in F$, and by B_0 the intersection of the kernels of all $\chi \in \widehat{A}_1$. Then B_0 is a normal subgroup of G. Indeed, if $\chi \in \widehat{A}_1$, $g \in G$, and $b \in B_0$, then $\chi(g^{-1}bg) = \chi(b^f) = (\chi^{f^{-1}}(b))^f = 1$ (here $f = \varphi(g)$). Let $G_1 = G/B_0$, $A_1 = A/B_0$, and let φ_1 be the homomorphism of the group G_1 onto F induced by φ. It is clear that $\widehat{A}_1$ is the character group for A_1.

The embedding problem $(K/k, G_1, \varphi_1, A_1)$ is an associated (and Brauer) problem for the original problem. It has a solution if the compatibility condition is fulfilled for the original problem.

It is not hard to see that the problem $(K/k, G_1, \varphi_1, A)$ is maximal among the Brauer problems associated with the original problem in the sense that all these problems are associated with $(K/k, G_1, \varphi_1, A_1)$.

We now turn to the embedding problem of local fields.

THEOREM 3.14.1. *The embedding problem $(K/k, G, \varphi, A)$ with local field k and Abelian kernel A is solvable if the compatibility condition is fulfilled.*

PROOF. The problem $(K/k, G_1, \varphi_1, A_1)$ is solvable if the compatibility condition is satisfied. To prove the solvability of the original problem, it suffices to prove that the homomorphism $\overline{\psi}$ of the group $H^2(\overline{F}, A)$ to the group $H^2(\overline{F}, A_1)$ induced by the homomorphism of A onto A_1 is an isomorphism.

Indeed, the diagram

$$\begin{array}{ccc} H^2(\overline{F}, A) & \xrightarrow{\overline{\psi}} & H^2(\overline{F}, A_1) \\ \uparrow \lambda & & \uparrow \lambda_1 \\ H^2(F, A) & \xrightarrow{\psi} & H^2(F, A_1), \end{array}$$

where λ and λ_1 are the lifting homomorphisms, ψ and $\overline{\psi}$ are induced by the homomorphism of A onto A_1, is commutative.

Let $h \in H^2(F, A)$ and $h_1 \in H^2(F, A_1)$ be cohomology classes corresponding to the extensions G and G_1. Theorem 3.13.1 implies that $\psi(h) = h_1$, the solvability of the associated problem implies $\lambda_1(h_1) = \lambda_1(\psi(h)) = 0$, and, thus, $\overline{\psi}\lambda(h) = 0$. If $\overline{\psi}$ is an isomorphism, then $\lambda(h) = 0$, whence the original problem is solvable.

We apply Theorem A.3.2 (Tate local duality, Appendix, §3). This theorem states that for a finite $\overline{F}$-module A, the groups $H^2(\overline{F}, A)$ and $\mathrm{Hom}_{\overline{F}}(A, \overline{k}^*)'$ are isomorphic. It is clear from the definition of the group A_1 that the groups $\mathrm{Hom}_{\overline{F}}(A, \overline{k}^*)$ and $\mathrm{Hom}_{\overline{F}}(A_1, \overline{k}^*)$ coincide, both of them are equal to $\widehat{A}_1$. The isomorphisms α

and α_1 of Theorem A.3.2 make the diagram

$$\begin{array}{ccc} H^2(\overline{F}, A) & \xrightarrow{\overline{\psi}} & H^2(\overline{F}, A_1) \\ \downarrow{\scriptstyle\alpha} & & \downarrow{\scriptstyle\alpha_1} \\ \operatorname{Hom}_{\overline{F}}(A, \overline{k}^*)' & \xrightarrow{\text{id}} & \operatorname{Hom}_{\overline{F}}(A_1, \overline{k}^*)' \end{array}$$

commutative; so, $\overline{\psi}$ is an isomorphism, which completes the proof of the theorem.

A similar result for the field of real numbers has a much simpler proof. □

THEOREM 3.14.1′. *The embedding problem $(K/\mathbb{R}, G, \varphi, A)$ with Abelian kernel A, where $\mathbb{R}$ is the field of real numbers, is solvable if the compatibility condition is fulfilled.*

PROOF. As shown in Chapter I, §6, we may assume that K is a field, and then it makes sense to take $K = \mathbb{C}$, the field of complex numbers. Thus, F is a cyclic group of second order, and let f be a generator of F. The group B_0, in this case, is a subgroup of elements $a^f a$ with $a \in A$, since for $\chi \in A_1$ we have $\chi(a^f a) = \chi(a)^f \chi(a) = \overline{\chi(a)}\chi(a) = 1$. For the elements $\alpha \in A/B_0$, we have $\alpha^f = \alpha^{-1}$, so that the problem $(\mathbb{C}/\mathbb{R},\ G/B_0,\ A/B_0)$ is Brauer. Since the compatibility condition is satisfied for this problem, it is solvable. Since $\mathbb{C}$ is algebraically closed, an embedding of F in G/B_0 exists, and let $\sigma^2 = 1$, $\sigma \in G/B_0$, $\sigma \notin A/B_0$. For a preimage g in G of the element σ, we have $g^2 = a_0^f a_0$ ($a_0 \in A$). We set $g_0 = g a_0^{-1} \in G$, $\varphi(g_0) = f$. Then $g_0^2 = 1$, i.e., G is a semidirect extension of A by F and our problem is solvable. □

2°. Now consider the embedding problem for global fields. In Chapter 1, §11 we gave the definition of the associated local embedding problem. Now we clarify how the compatibility conditions for the global problem are related to its localization.

PROPOSITION 3.14. *The compatibility condition for the global embedding problem $(K/k, G, \varphi, A)$ is equivalent to the solvability of the associated local problems at all points of the field k.*

PROOF. Let the compatibility condition be fulfilled for the global problem. Let $\mathfrak{p}$ be a prime divisor of k, $\mathfrak{P}$ a prime divisor of $\mathfrak{p}$ in K; let F be the Galois group of K/k, and $F_{\mathfrak{P}}$ the decomposition group of the divisor $\mathfrak{P}$ (the Galois group of the extension $K_{\mathfrak{P}}/k_{\mathfrak{p}}$). We denote by $G_{\mathfrak{P}}$ the full preimage of $F_{\mathfrak{P}}$ in G with respect to the homomorphism φ. Then for the associated embedding problem $(K_{\mathfrak{P}}/k_{\mathfrak{p}}, G_{\mathfrak{P}}, \varphi, A)$ the compatibility condition is fulfilled, so it is solvable.

Conversely, for the global embedding problem let the associated local problems be solvable for all points $\mathfrak{p}$ of k including all of the Archimedean $\mathfrak{p}_\infty$. Then the compatibility conditions are valid for all these problems. This means that all simple direct summands of the algebra $G \times K$ split at all points of the fields K_χ that are the centers of the components. Therefore, the algebra $G \times K$ splits over its center, i.e., the compatibility condition is fulfilled.

The restriction homomorphism ρ of $H^i(F, A)$ onto the group $H^i(F_{\mathfrak{P}}, A)$ does not depend on the choice of the prime divisor $\mathfrak{P}$ of $\mathfrak{p}$. For $i = 0$ the statement is verified directly, and for positive integers i it can be proved by the standard method of shifting the dimension. In the sequel we shall write $F_{\mathfrak{p}}$ instead of $F_{\mathfrak{P}}$.

Now we note that if K_1 is a normal extension of the field k, containing K, with Galois group F_1, then the following diagram is commutative:

$$\begin{array}{ccc} H^i(F_1, A) & \xrightarrow{\rho_1} & H^i(F_{1\mathfrak{p}}, A) \\ \uparrow{\scriptstyle\lambda} & & \uparrow{\scriptstyle\lambda_\mathfrak{p}} \\ H^i(F, A) & \xrightarrow{\rho} & H^i(F_\mathfrak{p}, A) \end{array}$$

Here A is an arbitrary F-module, and, thus, A is an F_1-module, λ and $\lambda_\mathfrak{p}$ are the lifting homomorphisms, ρ and ρ_1 are the restriction homomorphisms.

Now let $\{K_\beta\}$ be an inductive system of normal fields over the field k with Galois groups F_β, and let $\varinjlim K_\beta$ coincide with $\overline{k}$. If $\overline{k}_\mathfrak{p}$ is an algebraic closure of the local field $k_\mathfrak{p}$ with Galois group $\overline{F}_\mathfrak{p}$, then the homomorphism of the group $H^i(\overline{F}, A)$ into $H^i(\overline{F}_\mathfrak{p}, A)$ is well defined, and also the corresponding homomorphism of $H^i(\overline{F}, A)$ to the direct product $\prod_\mathfrak{p} H^i(\overline{F}_\mathfrak{p}, A)$ over all points of k is well defined.□

DEFINITION 3.14. The Galois algebra defined over a local field is said to be *unramified* if its field-kernel is an unramified extension.

LEMMA 3.14. *An unramified algebra may be chosen as a solution of the embedding problem $(K/k, G, \varphi, A)$, where K is an unramified Galois algebra over the field k.*

PROOF. Let K_0 be the field-kernel of the algebra K, F_0 the Galois group of the extension K_0/k and G_0 the full preimage of the subgroup F_0 in the group G with respect to the homomorphism φ. Then it suffices to prove that the embedding problem $(K_0/k, G_0, \varphi, A)$ has an unramified solution.

Let k_{unr} be the maximal unramified extension of k with Galois group F_{unr}, h the cohomology class from $H^2(F_0, A)$ that corresponds to the group G_0. The group F_{unr} is a free profinite group with one generator, and for any F_{unr}-module A the groups $H^i(F_{\mathrm{unr}}, A)$ are trivial for $i \geq 2$. By arguments similar to those used in Theorems 1.14.1 and 3.13.2, we may show that the lemma is equivalent to the relation $\lambda(h) = 0$, where λ is the lifting homomorphism of the group $H^2(F, A)$ to $H^2(F_{\mathrm{unr}}, A)$. It remains to note that the group $H^2(F_{\mathrm{unr}}, A)$ is trivial.

Now we return to our global embedding problem and consider the commutative diagram

$$\begin{array}{ccc} H^2(\overline{F}, A) & \xrightarrow{\omega_1} & \prod_\mathfrak{p} H^2(\overline{F}_\mathfrak{p}, A) \\ \uparrow{\scriptstyle\lambda} & & \uparrow{\scriptstyle\overline{\lambda}} \\ H^2(F, A) & \xrightarrow{\omega_2} & \prod_\mathfrak{p} H^2(F_\mathfrak{p}, A) \end{array}$$

It follows from Proposition 3.14 that the compatibility condition is equivalent to the relation $\overline{\lambda}\omega_2(h) = 0$, or, in accordance with the diagram, to the relation $\omega_1\lambda(h) = 0$.

Note that the image of the group $H^2(\overline{F}, A)$ is contained in the subgroup $\sum_\mathfrak{p}^\oplus H^2(\overline{F}_\mathfrak{p}, A)$. Indeed, every element $\widetilde{h}$ of $H^2(\overline{F}, A)$ can be obtained lifting some element h_{β_0} of the group $H^2(F_{\beta_0}, A)$. The corresponding extension K_{β_0}/k is finite and has only finitely many ramification points $\mathfrak{p}_1, \dots, \mathfrak{p}_t$. It follows immediately from Lemma 3.14 that the image $\omega_1(\widetilde{h})$ may have nontrivial components only at finitely many points $\mathfrak{p}_1, \dots, \mathfrak{p}_t$.

Now we apply Tate's global duality (see Appendix, §3). The kernel of the mapping ω_1 is dual to the kernel of the mapping $\omega_3\colon H^1(\overline{F}, \widehat{A}) \to \prod_{\mathfrak{p}} H^1(\overline{F}_{\mathfrak{p}}, \widehat{A})$.

Let $\overline{F}_0$ be a subgroup of $\overline{F}$ acting trivially on the module $\widehat{A}$, F_1 the quotient group $\overline{F}/\overline{F}_0$, and let K_1 be the subfield that corresponds to the subgroup $\overline{F}_0$.

Now we consider the commutative diagram

$$\begin{array}{ccccccc} 0 & \longrightarrow & H^1(F_1, \widehat{A}) & \longrightarrow & H^1(\overline{F}, \widehat{A}) & \longrightarrow & H^1(\overline{F}_0, \widehat{A}) \\ & & \downarrow{\scriptstyle\omega_4} & & \downarrow{\scriptstyle\omega_3} & & \downarrow{\scriptstyle\omega_5} \\ 0 & \longrightarrow & \prod_{\mathfrak{p}} H^1(F_{1\mathfrak{p}}, \widehat{A}) & \longrightarrow & \prod_{\mathfrak{p}} H^1(\overline{F}_{\mathfrak{p}}, \widehat{A}) & \longrightarrow & \prod_{\mathfrak{p}} H^1(\overline{F}_{0\mathfrak{p}}, \widehat{A}) \end{array}$$

where the rows are Hochschild-Serre sequences. We show that the mapping ω_5 is injective. Let $x \in H^1(\overline{F}_0, \widehat{A})$ and $\omega_5(x) = 0$. An Abelian extension of K_1 corresponds to the element x, and this extension splits completely locally at all points, whence the global extension is trivial, i.e., $x = 0$. Thus, the kernels of the mappings ω_3 and ω_4 coincide, and we have proved the following theorem. □

THEOREM 3.14.2. *If the compatibility condition is fulfilled for the global embedding problem $(K/k, G, \varphi, A)$, then some element y of the group $H^1(F_1, \widehat{A})'$ dual to the group $H^1(F_1, \widehat{A})$ corresponds to this problem. The additional embedding condition implies that y lies in the image of the natural mapping*

$$\theta\colon \Big(\prod_{\mathfrak{p}} H^1(F_{1\mathfrak{p}}, \widehat{A})\Big)' \to H^1(F_1, \widehat{A})',$$

where the prime means, as usual, the passage to the dual groups.

We note that K_1 is a subfield of K, and F_1 is a quotient group of F.

COROLLARY. *If F_1 is a cyclic group, then the compatibility condition is sufficient for imbeddability.*

In order to prove it, we note that in this case there exists a divisor $\mathfrak{p}$ of k such that its decomposition group in the extension K_1/k coincides with F_1, i.e., the mapping θ is surjective.

3°. We now show that the compatibility condition is not sufficient for the solvability of the embedding problem for global fields. Theorems 3.13.2 and 3.14.2 imply that it suffices to find an extension K_1 of k with Galois group F and an F_1-module $\widehat{A}$ such that the mapping $\theta_1\colon H^1(F_1, \widehat{A}) \to \prod_{\mathfrak{p}} H^1(F_{1\mathfrak{p}}, \widehat{A})$ is not injective, where $F_{1\mathfrak{p}}$ is the decomposition group of a prime divisor of $\mathfrak{p}$ in k.

Let A be a cyclic group of eighth order with generator a, k the field of rational numbers and K the field obtained by adjoining to k the square roots $\sqrt{-1}$, $\sqrt{2}$, $\sqrt{7}$. Then the Galois group F of the extension K/k is an elementary Abelian 2-group with generators f_1, f_2, f_3. We assume that the action of the group F on the elements of K is defined by the relations:

$$\begin{array}{lll} \sqrt{-1}^{f_1} = -\sqrt{-1}, & \sqrt{-1}^{f_2} = \sqrt{-1}, & \sqrt{-1}^{f_3} = \sqrt{-1}; \\ \sqrt{2}^{f_1} = \sqrt{2}, & \sqrt{2}^{f_2} = -\sqrt{2}, & \sqrt{2}^{f_3} = \sqrt{2}; \\ \sqrt{7}^{f_1} = \sqrt{7}, & \sqrt{7}^{f_2} = \sqrt{7}, & \sqrt{7}^{f_3} = -\sqrt{7}. \end{array}$$

We make A an F-module by setting $a^{f_1} = a$, $a^{f_2} = a$, $a^{f_3} = a^{-1}$. Let ζ be a primitive eighth root of unity, and let χ be a character from $\widehat{A}$ such that $\chi(a) = \zeta$. Then the action of the group F on the module $\widehat{A}$ is defined by the relations

$$\chi^{f_1} = \chi^7, \qquad \chi^{f_2} = \chi^5, \qquad \chi^{f_3} = \chi^7.$$

The subgroup generated by the element $f_1 f_3$ acts trivially on the module A. The subfield $K_1 = k(\sqrt{2}, \sqrt{-7})$ corresponds to this subgroup. The Galois group F_1 of the extension K_1/k has generators $\overline{f}_1$ and $\overline{f}_2$, and their action on the module $\widehat{A}$ is defined as follows: $\chi^{\overline{f}_1} = \chi^7$, $\chi^{\overline{f}_2} = \chi^5$.

Now we note that the decomposition group of any divisor in the extension K_1/k is either a cyclic group or the trivial one. Indeed, it suffices to consider only the divisors 2 and 7. In the field $\mathbb{Q}_2$, (-7) is a square, and in the field $\mathbb{Q}_7$, 2 is a square.

Consider a one-dimensional cocycle $\overline{h}$ from $Z^1(F_1, \widehat{A})$ defined by the relations $\overline{h}(\overline{f}_1) = 1$, $\overline{h}(\overline{f}_2) = \chi^4$. It is clear that it splits locally at all points, but the class with cocycle $\overline{h}$ is not the unit element.

Thus, we have proved the existence of an embedding problem of global fields, for which the compatibility condition is fulfilled, and a nontrivial character from the group $H^1(F_1, \widehat{A})'$ not lying in the image of the homomorphism θ corresponds to this problem.

An explicit example of such an embedding problem $(k_1/k, G, \varphi, A)$ was constructed in [**48**], where k is the field of rational numbers, $k_1 = k(\sqrt{7})$, G is the group of order 16 with two generators g_1 and g_2 and with relations $g_1^8 = 1$, $g_2^2 = g_1^4$, $g_2^{-1} g_1 g_2 = g_1^7$. If F is the Galois group of the extension k_1/k with generator f, then the homomorphism φ is given by the relations $\varphi(g_1) = 1$, $\varphi(g_2) = f$.

A specific form of the additional condition for the solvability of the problem with cyclic group of order 8 as kernel is given in [**51**].

§15. Composition law on the set of solutions of the embedding problem

1°. First we briefly recall some information about the extension of Abelian groups (see [**37**]).

Let A be an Abelian group (written multiplicatively), F a finite group, and $\varkappa\colon F \longrightarrow \operatorname{Aut} A$ a homomorphism (i.e., A is an F-module). The two extensions G_1 and G_2 of the group A by the group F with a given homomorphism $\varkappa$ are said to be *equivalent* if there exists an isomorphism $\eta\colon G_1 \longrightarrow G_2$ for which the following diagram is commutative:

$$\begin{array}{ccccccccc}
1 & \longrightarrow & A & \xrightarrow{i_1} & G_1 & \xrightarrow{\varphi_1} & F & \longrightarrow & 1 \\
 & & \downarrow{\scriptstyle \mathrm{id}} & & \downarrow{\scriptstyle \eta} & & \downarrow{\scriptstyle \mathrm{id}} & & \\
1 & \longrightarrow & A & \xrightarrow{i_2} & G_2 & \xrightarrow{\varphi_2} & F & \longrightarrow & 1
\end{array}$$

Note that there exist isomorphic but not equivalent extensions.

On the classes of equivalent extensions we introduce a composition law called *the Baer product of extensions*. Let $1 \longrightarrow A \longrightarrow G_1 \xrightarrow{\varphi_1} F \longrightarrow 1$ and $1 \longrightarrow A \longrightarrow G_2 \xrightarrow{\varphi_2} F \longrightarrow 1$ be two extensions of the F-module A by F. We consider $\widetilde{G}$, the product of G_1 and G_2 with amalgamated quotient group F. The kernel of the epimorphism $\widetilde{G} \longrightarrow F$ is the group $A \times A$ (we assume that A is embedded in G_i).

In this kernel we consider the subgroup $\widetilde{A}$ consisting of the pairs (a, a^{-1}), $a \in A$. It is clear that $\widetilde{A}$ is normal in $\widehat{G}$ and $(A \times A)/\widetilde{A} \approx A$. Thus, the following exact sequence is given:

$$1 \to A \to \widetilde{G}/\widetilde{A} \xrightarrow{\varphi} F \to 1,$$

where A is embedded in $\widetilde{G}/\widetilde{A}$ by means of $a \to (a, 1) \bmod \widetilde{A}$. The extension $G = \widetilde{G}/\widetilde{A}$ is said to be *the Baer product of the extensions* G_1 *and* G_2 and is denoted by $G_1 \circ G_2$.

It is easy to see that if G_1 is equivalent to G_1', and G_2 is equivalent to G_2', then the products $G_1 \circ G_2$ and $G_1' \circ G_2'$ are also equivalent. Thus, we have defined the composition law $\circ$ on the set of classes of equivalent extensions (the Baer product).

THEOREM 3.15.1. *The classes of equivalent extensions of an F-module A by F form an Abelian group (isomorphic to $H^2(F, A)$) with respect to the product introduced above.*

As is easy to see, the Baer product of the three extensions G_1, G_2, G_3 is the quotient group of the group of triples $\{(g_1, g_2, g_3)\}$, where $g_i \in G_i$, $\varphi_1(g_1) = \varphi_2(g_2) = \varphi_3(g_3)$, by the subgroup $\widetilde{A}$ consisting of triples $\{(a_1, a_2, a_3)\}$ with $a_1a_2a_3 = 1$. This implies the associative property of the composition law. The semidirect extension G_0 of A by F is the unit element. Indeed, let G_0 consist of elements v_ja, where $a \in A$, $v_f \in G_0$, $v_{f_1}v_{f_2} = v_{f_1f_2} (f_i \in F)$. Then there exists an isomorphism of the Baer product $G \circ G_0$ to G given by the formula $(g, v_fa) \bmod \widetilde{A} \to ga$ $(g \in G,$ $a \in A)$, and this isomorphism is an equivalence.

Now we show the existence of the inverse class. Let G be an arbitrary extension. Consider the group $\overline{G}$ consisting of the elements $\overline{g}$ $(g \in G)$ for which $\overline{g}_1\overline{g}_2 = \overline{g_2g_1}$ (the map $g \to \overline{g}^{-1}$ is an isomorphism, whence G and $\overline{G}$ are abstractly isomorphic).

We map the group $\overline{G}$ to F by the epimorphism $\overline{\varphi}$, where $\overline{\varphi}(\overline{g}) = \varphi(g^{-1})$ and we embed A in $\overline{G}$ by $a \to \overline{a}$. Since A is an Abelian group, $\overline{G}$ is the extension of A by F with the same homomorphism $F \to \operatorname{Aut} A$ as for the group G because $\overline{g}^{-1}\overline{a}\,\overline{g} = \overline{gag^{-1}} = \overline{a^{\varphi(g^{-1})}} = a^{\varphi(g^{-1})} = a^{\overline{\varphi}(\overline{g})}$. Now consider the Baer product $G \circ \overline{G}$. Let $v_f (f \in F)$ be representatives of F in G. For $w_f = (v_f, \overline{v_f^{-1}}) \bmod \widetilde{A} \in G \circ \overline{G}$, we have

$$\begin{aligned} w_{f_1}w_{f_2} &\equiv (v_{f_1}, \overline{v_{f_1}^{-1}})(v_{f_2}, \overline{v_{f_2}^{-1}}) \equiv (v_{f_1}v_{f_2}, \overline{v_{f_2}^{-1}v_{f_1}^{-1}}) \\ &\equiv (v_{f_1}v_{f_2}, \overline{(v_{f_1}v_{f_2})^{-1}}) \equiv (v_{f_1f_2}, \overline{v_{f_1f_2}^{-1}}) \bmod \widetilde{A} = w_{f_1f_2}, \end{aligned}$$

for if $v_{f_1}v_{f_2} = v_{f_1f_2}a$ $(a \in A)$ then $\overline{(v_{f_1}v_{f_2})^{-1}} = \overline{(v_{f_1f_2}a)^{-1}} = \overline{a^{-1}v_{f_1f_2}^{-1}} = \overline{v_{f_1f_2}^{-1}}a^{-1}$. Therefore, the elements w_f $(f \in F)$ complement A in $G \circ \overline{G}$, and $G \circ \overline{G}$ is the semidirect extension of A by F. Hence, the classes of equivalent extensions of A by F constitute a group.

For any set of representatives v_f $(f \in F)$ in G, we have $v_{f_1}v_{f_2} = v_{f_1f_2}a(f_1, f_2)$, $a(f_1f_2) \in A$. Associativity implies that $a(f_1, f_2)$ is a two-dimensional cocycle from $Z^2(F, A)$. For any other set of representatives $\{u_f\}$ we have $u_f = v_fb_f$, and the cocycle $a(f_1, f_2)$ is replaced by the homologous cocycle $a'(f_1, f_2) = a(f_1, f_2)b_{f_1f_2}^{-1}b_{f_1}^{f_2}b_{f_2}$. Therefore, to the extension G there corresponds an element of $H^2(F, A)$, namely, the cohomology class of the cocycle $a(f_1, f_2)$. It is clear that to equivalent extensions there corresponds the same cohomology class, and the classes of equivalent extensions are in one-to-one correspondence with the classes from $H^2(F,A)$. Now we show that this correspondence is a homomorphism.

Let G_1 and G_2 be two extensions, u_f and v_f sets of representatives of F in G_1 and G_2, respectively, and $u_{f_1}u_{f_2} = u_{f_1f_2}a(f_1,f_2)$, $v_{f_1}v_{f_2} = v_{f_1f_2}b(f_1,f_2)$, where a, b are cocycles in $Z^2(F,A)$. We put $w_f = (u_f, v_f) \bmod \widetilde{A} \in G_1 \circ G_2$. Then

$$w_{f_1}w_{f_2} = (u_{f_1f_2}a(f_1,f_2), v_{f_1f_2}b(f_1,f_2)) \bmod \widetilde{A} = w_{f_1f_2}a(f_1,f_2)b(f_1,f_2).$$

The theorem is proved.

2°. Let the embedding problem $(K/k, G, \varphi, A)$ be solvable. The knowledge of the algebra L yielding a solution of the problem does not completely determine the solution itself. It is necessary to specify how the elements of the group G act on the algebra L, i.e., to indicate the isomorphism $G \longrightarrow \mathrm{Gal}(L/k)$. This is often essential. For example, when we solve the problem in several steps, we must consider the embedding problems $(L/k, H, B)$, where L is a solution of the problem $(K/k, G, A)$, and the action of the elements from G on L influences the solvability conditions for the problem $(L/k, H, B)$. Thus, to give a solution in the strict sense means to give both an algebra L and an isomorphism $\nu\colon G \longrightarrow \mathrm{Gal}(L/k)$, i.e., to give a pair (L, ν).

We say that two solutions (L_1, ν_1) and (L_2, ν_2) of the same embedding problem $(K/k, G, \varphi, A)$ are equivalent *in the narrow sense* (*S-equivalent*) if there exists a K-isomorphism $\theta\colon L_1 \longrightarrow L_2$ such that for any $x \in L_1$ and any $g \in G$, the relations $(\theta(x))^{\nu_2(g)} = \theta(x^{\nu_1(g)})$ hold.

PROPOSITION 3.15.1. *Let (L_1, ν_1) and (L_2, ν_2) be two S-equivalent solutions of the problem $(K/k, G, \varphi, A)$. Then the solvability conditions for the problems $(L_i/k, H, \psi, B)$ coincide, and the solutions of these problems are in one-to-one correspondence.*

PROOF. Let M_1 be a solution of the problem $(L_1/k, H, \psi, B)$. Then M_1 is isomorphic to the quotient ring of polynomials $L_1[t]$ by the ideal generated by the polynomial $F_1 \in L_1[t]$. Consider the polynomial $F_2 \in L_2[t]$ obtained by applying the isomorphism θ to the coefficients of F_1. It is easily checked that $M_2 = L_2[t]/F_2L_2[t]$ solves $(L_2/k, H, \psi, B)$. Moreover, there also exists a K-isomorphism $M_1 \longrightarrow M_2$ that establishes the S-equivalence of the solutions M_i of the problem $(K/k, H)$.

The inclusion of the solutions into larger equivalence classes is sometimes justified. Namely, we say that the solutions (L_1, ν_1) and (L_2, ν_2) of $(K/k, G, \psi, A)$ are equivalent *in the broad sense*, or *B-equivalent* if there exists an isomorphism $\theta\colon L_1 \longrightarrow L_2$ and an automorphism η of G such that

1) θ is identical on K;
2) η is identical on A;
3) $\theta(x^{\nu_1(g)}) = \theta(x)^{\nu_2\eta(g)}$ for $x \in L_1$, $g \in G$.

In addition, we note that these conditions, together with the commutativity of the diagram

$$\begin{array}{ccc} G & \xrightarrow{\ \nu\ } & \mathrm{Gal}(L/k) \\ & {\scriptstyle\varphi}\searrow & \downarrow \\ & & \mathrm{Gal}(K/k) \end{array}$$

mean that $\varphi\eta(g) = \varphi(g)$.

It is obvious that if two solutions are S-equivalent, then they are also B-equivalent. $\square$

3°. Let an F-module A be given as before. Consider the set Γ consisting of the elements $v_f a$, where the elements v_f are in one-to-one correspondence with the elements $f \in F$, $a \in A$. We make Γ to a space over A by setting $v_f a \cdot b = v_f \cdot ab$, $b \cdot v_f a = v_f(b^f a)$, where $a, b \in A$. Any extension of A by F is isomorphic, as an A-space, to the space Γ.

We say that the Galois algebra L/k containing the field K solves the problem $(K/k, \Gamma)$ if it is a solution of a problem $(K/k, G, \varphi, A)$, where G is an arbitrary extension of the F-module A by F.

For every $f \in F$ we fix an automorphism $v_f \in \mathrm{Gal}(L/k)$ that extends $f \in \mathrm{Gal}(K/k) = F$. This defines the action of the space Γ on the algebra L. Conversely, on the algebra L containing the field k, let an automorphism group isomorphic to the space Γ be given. We assume that the automorphisms (corresponding to the elements $v_f a$) restricted to the field K coincide with the automorphisms $f \in \mathrm{Gal}(K/k) = F$. Then L is a solution of the problem $(K/k, \Gamma)$.

Two such solutions L_1 and L_2 are said to be Γ-*equivalent* if there exists an isomorphism $\theta\colon L_1 \to L_2$, identical on K, such that $(\theta(x))^g = \theta(x^g)$ for all $g \in \Gamma$ and $x \in L_1$ (by x^g (respectively, $(\theta(x))^g$) we mean the action of the automorphism, corresponding to $g \in \Gamma$, on x (respectively, $\theta(x)$)). The set of classes of Γ-equivalent solutions of the problem $(K/k, \Gamma)$ is denoted by R_Γ.

Introduce a composition law on R_Γ. Let L_1 and L_2 be two solutions of the problem $(K/k, \Gamma)$. The set of automorphisms $\widetilde{\Gamma}$ consisting of the pairs $(v_f a_1, v_f a_2)$ with $f \in F$, $a_i \in A$, acts on the tensor product $\widetilde{L} = L_1 \otimes_K L_2$ according to the rule $(x_1 \otimes x_2)^{(v_f a_1, v_f a_2)} = x_1^{v_f a_1} \otimes x_2^{v_f a_2}$, where $x_i \in L_i$. Indeed, for $y \in K$ we have

$$(x_1 y \otimes x_2)^{(v_f a_1, v_f a_2)} = x_1^{v_f a_1} y^f \otimes x_2^{v_f a_2} = x_1^{v_f a_1} \otimes y^f x_2^{v_f a_2} = (x_1 \otimes y x_2)^{(v_f a_1, v_f a_2)}.$$

Let L_3 be the subalgebra of the tensor product corresponding to the subgroup $\widetilde{A}$ of the pairs (a, a^{-1}), $a \in A$. The automorphisms $(v_f a_1, v_f a_2)$ and $(v_f a_3, v_f a_4)$ act in the same way on L_3 if and only if $a_1 a_2 = a_3 a_4$. The mapping $\widetilde{\Gamma} \to \Gamma$ given by the formula $(v_f a_1, v_f a_2) \to v_f a_1 a_2$ corresponds to such an equivalence relation. Thus, the natural action of the space Γ on L_3 is well defined, and L_3 is a solution of the problem $(K/k, \Gamma)$. We denote $L_3 = L_1 \circ L_2$. It is obvious that if L_1' and L_2' are algebras Γ-equivalent to L_1 and L_2, then $L_1' \circ L_2'$ and $L_1 \circ L_2$ are also Γ-equivalent, so the operation $\circ$ is defined on the set R_Γ.

THEOREM 3.15.2. *The set R_Γ is an Abelian group under the operation $\circ$.*

PROOF. Let L_1, L_2, L_3 be three solutions of $(K/k, \Gamma)$. The algebra $L_1 \circ (L_2 \circ L_3)$ is isomorphic to the subalgebra of the tensor product $L_1 \otimes (L_2 \otimes L_3)$ corresponding to the subgroup $\widetilde{A}$ that consists of the triples (a_1, a_2, a_3), where $a_i \in A_i$ and $a_1 a_2 a_3 = 1$. The same is valid for the algebra $(L_1 \circ L_2) \circ L_3$, therefore the solutions $L_1 \circ (L_2 \circ L_3)$ and $(L_1 \circ L_2) \circ L_3$ are Γ-equivalent. Thus, the operation $\circ$ is associative.

Let $L_0 = \sum_{a \in A}^{\oplus} K e_a$, where the e_a are orthogonal idempotents, and $e_a = e_1^a$, $e_1^{v_f} = e_1$, be a completely splitting Galois algebra ($f \in F, v_f \in \Gamma$). It is clear that the automorphism group on L_0 is a semidirect extension of A by F; moreover, $v_{f_1} v_{f_2}$ and $v_{f_1 f_2}$ yield the same automorphisms on L_0. Let L be an arbitrary solution of the problem $(K/k, \Gamma)$. The algebra $L \otimes_K L_0$ is equal to $\sum_{a \in A}^{\oplus} L e_a$, and $L \circ L_0$

consists of the sums $\sum_{a\in A} z^{a^{-1}} e_a$, $z \in L$. The mapping $z \longrightarrow \sum_{a\in A} z^{a^{-1}} e_a$ yields the Γ-equivalence of L and $L \circ L_0$, i.e., L_0 is the unit element in R_Γ.

We need the following lemma before proving the existence of the inverse element. □

LEMMA. *Let L be a solution of the embedding problem $(K/k, G, \varphi, A)$. Then the algebra $L \otimes_K L$ decomposes into a direct sum $\sum^{\oplus}_{a\in A} Le_a$, where e_a are orthogonal idempotents, and e_1 may be chosen in such a way that $e_a = e_1^{(a,1)}$ and $e_1^{(g,g)} = e_1$ for all $g \in G$.*

PROOF. Consider the mapping $\theta\colon L \otimes_K L \longrightarrow L$ given on $x \otimes y$ by the formula $\theta(x \otimes y) = xy$, and let I be the kernel of θ. Let e' be the unit of the ideal I, and let $e_1 = 1 - e'$ be the idempotent orthogonal to I. For any $g \in G$, we have $(x^g y^g \otimes 1 - x^g \otimes y^g)e_1^{(g,g)} = ((xy \otimes 1 - x \otimes y)e_1)^{(g,g)} = 0$, i.e., $e_1^{(g,g)}$ is orthogonal to I, and, thus, $e_1^{(g,g)} = e_1$ for any g.

We denote $e_a = e_1^{(a,1)}$. For $b \in A$ we have $(\theta e_a)^b = \theta(e_a^{(b,b)}) = \theta(e_1^{(b,b)(a,1)}) = \theta e_a$, whence $\theta e_a \in K$; so θe_a is equal either to 0 or to 1.

Further, $0 = \theta(((x \otimes 1 - 1 \otimes x)e_1)^{(a,1)}) = \theta((x^a \otimes 1 - 1 \otimes x)e_1^{(a,1)}) = (x^a - x)\theta e_a$ for any $x \in L$. Therefore, $\theta e_a = 0$ for any $a \neq 1$. Thus, $e_a \in I$, and $e_1 e_a = 0$ for $a \neq 1$. We also have $e_a e_b = (e_1 e_{ba-1})^{(a,1)}$ for $b \neq a$.

Now we put $E = \sum_{a\in A} e_a$. For $\varphi(g_1) = \varphi(g_2)$, we have

$$E^{(g_1,g_2)} = \sum_{a\in A} e_1^{(a,1)(g_1,g_2)} = \sum_{a\in A} e_1^{(g_2,g_2)(g_2^{-1}ag_1,1)} = \sum_{a\in A} e_1^{(g_2^{-1}ag_1,1)} = E,$$

whence $E \in k$. Moreover, $E^2 = E$, $E \neq 0$, i.e., $E = 1$. Thus, $L \otimes_K L = (L \otimes L)E = \sum^{\oplus}_{a\in A}(L \otimes L)e_a = \sum^{\oplus}_{a\in A} Le_a$, and the lemma is proved.

Now let L be an arbitrary solution of the problem $(K/k, \Gamma)$. We consider the algebra $\overline{L}$ abstractly isomorphic to L and consisting of the elements $\overline{x}$, where $x \in L$. We define on $\overline{L}$ the action of the automorphisms from Γ by the formula $\overline{x}^{v_f a} = \overline{x^{v_f a^{-1}}}$. Since $L \otimes_K L$ is a direct sum $\sum^{\oplus}_{a\in A} Le_a$, where the e_a are orthogonal idempotents and, in addition, $e_1^{(g,g)} = e_1$ for all $g \in \Gamma$, $e_a^{(b,b)} = e_a$ for all $b \in A$ (see the lemma), then the algebra $L \otimes_K \overline{L}$ consists of the elements $\sum_{a\in A}(z_a \otimes 1)e_a$, where $z_a \in L$, $e_a \in L \otimes \overline{L}$, $e_a^{(b,b^{-1})} = e_a$, $e_1^{(v_f,v_f)} = e_1$. Thus, the algebra $L \circ \overline{L}$ consists of the elements $\sum_{a\in A}(x_a \otimes 1)e_a$, where $x_a \in K$. The mapping $\sum_{a\in A}(x_a \otimes 1)e_a \longrightarrow \sum_{a\in A} x_a e_a$ establishes an equivalence between $L \circ \overline{L}$ and L_0 that is the unit element in R_Γ. The theorem is proved. □

4°. Let L be a solution to the problem $(K/k, \Gamma)$. This means that there exists an extension G of A by F such that G is isomorphic to the automorphism group given by the automorphisms from Γ on L. Let ν_1, ν_2 be two isomorphisms $G \longrightarrow \mathrm{Gal}(L/k)$ (which are clearly identical on A). Therefore, there exists an automorphism $\eta\colon G \longrightarrow G$, for which $\nu_2(g) = \nu_1\eta(g)$ for all $g \in G$; moreover, η is identical on A. Generally speaking, the algebra L with isomorphisms ν_i $(i = 1, 2)$ determines different solutions of the embedding problem $(K/k, G, \varphi, A)$, but these solutions are equivalent in the broad sense.

Let L_1 and L_2 be two Γ-equivalent solutions of the problem $(K/k, \Gamma)$, and let $\theta\colon L_1 \longrightarrow L_2$ be the isomorphism setting up the equivalence of them. Then both algebras are solutions of the same embedding problem $(K/k, G, \varphi, A)$ for some bijections $\nu_i\colon G \longrightarrow \mathrm{Gal}(L_i/k)$ $(i = 1, 2)$. Since $\nu_2(g) = \nu_1\eta(g)$ for some automorphism $\eta\colon G \longrightarrow G$ identical on A, the solutions (L_1, ν_1) and (L_2, ν_2) are B-equivalent. Thus, we have defined a mapping of the group R_Γ to the set of classes of B-equivalent solutions of the embedding problem $(K/k, G_h, \varphi_h, A)$, where h ranges over the cohomology classes from $H^2(F, A)$, the G_h are the corresponding group extensions for which the problem is solvable. We denote this set of classes of B-equivalent solutions by R_B, and let $\varkappa$ be the corresponding map $R_\Gamma \longrightarrow R_B$. It is clear that $\varkappa$ is surjective.

THEOREM 3.15.3. *The composition law* $\circ$ R_Γ, *induces on* R_B *a composition law with respect to which* R_B *is an Abelian group.*

We define on R_B a composition law by the formula $\varkappa[L_1] \circ \varkappa[L_2] = \varkappa[L_1 \circ L_2]$, where $[L_i]$ is the class of L_i in R_Γ. We check that this operation is well defined. To this end, it suffices to prove that $\varkappa[L_1' \circ L_2] = \varkappa[L_1 \circ L_2]$ if $\varkappa[L_1'] = \varkappa[L_1]$. Let (L_1, ν_1) and (L_1', ν_1') be B-equivalent solutions of the embedding problem $(K/k, G_1, \varphi_1, A)$, and let θ_1 and η_1 be the corresponding isomorphism $L_1 \longrightarrow L_1'$ and the corresponding automorphism $G_1 \longrightarrow G_1$. The isomorphism θ_1 gives rise to an isomorphism $L_1 \otimes_K L_2 \longrightarrow L_1' \otimes_K L_2$. Also, there exists an isomorphism $\theta_3\colon L_3 \longrightarrow L_3'$, where L_3 and L_3' are the subalgebras of these tensor products that correspond to the subgroup $\widetilde{A}$ formed by the pairs (a, a^{-1}), $a \in A$. At the same time, the automorphism η_1 gives rise to an automorphism η_3 on the Baer product of the groups G_1 and G_2 (namely, we put $\eta_3((g_1, g_2) \bmod \widetilde{A}) = (\eta_1(g_1), g_2) \bmod \widetilde{A}$, where $g_1 \in G_1$, $g_2 \in G_2$). Let ν_1, ν_1', ν_2 be the isomorphisms corresponding to the solutions L_1, L_1', L_2. Then the pairs (L_3, ν_3) and (L_3', ν_3'), where $\nu_3((g_1, g_2) \bmod \widetilde{A}) = (\nu_1(g_1), \nu_2(g_2)) \bmod \widetilde{A}$, $\nu_3'((g_1, g_2) \bmod \widetilde{A}) = (\nu_1'(g_1), \nu_2(g_2)) \bmod \widetilde{A}$, are B-equivalent solutions (the equivalence is proved with the help of the isomorphism θ_3 and the automorphism η_3).

Thus, the composition law $\circ$ on R_B determines a surjective homomorphism $R_\Gamma \longrightarrow R_B$, and this proves the theorem.

We note that the unit element of R_B is the class of all completely splitting algebras solving the semidirect embedding problem $(K/k, G_0, \varphi_0, A)$. Indeed, let $L_1 = \sum_{a\in A}^{\oplus} Ke_a$, $L_2 = \sum_{a\in A}^{\oplus} KE_a$ be two solutions of such a problem, and let $e_a = e_1^a$, $E_a = E_1^a$. Let H be a subgroup of the elements of G leaving e_1 fixed. It is clear that H is isomorphic to F, and, thus, it is the complement to A in G, i.e., for any $f \in F$ there exists $h_f \in H$ with $\varphi(h_f) = f$. Similarly, there exists the complement H' isomorphic to F, leaving E_1 fixed and consisting of the elements h_f', $f \in F$. Considering an automorphism $\eta\colon G_0 \longrightarrow G_0$ identical on A that takes h_f to h_f', and also a K-isomorphism $L_1 \longrightarrow L_2$ taking e_a to E_a, we prove the B-equivalence of the algebras L_1 and L_2.

The kernel of the homomorphism $\varkappa$ is formed by the completely splitting solutions (up to Γ-equivalence) of the problem $(K/k, \Gamma)$. A completely splitting solution is determined by specifying a one-dimensional cochain $c(f) \in C^1(F, A)$ using the formula $e_1^{v_f} = e_{c(f)}$, where e_1 is a minimal idempotent of the algebra $\sum^{\oplus} Ke_1^a$, and two such solutions are Γ-equivalent, provided that the corresponding cochains differ by a cochain $b(f) = b^{f-1}$, $b \in A$, i.e., they are cohomologous.

5°. Now we proceed to studying the classes of solutions equivalent in the narrow sense. First we note that if L_1 and L_2 are two solutions of the problem

$(K/k, \Gamma)$ whose images in R_B coincide, then there is no general-purpose method of recognizing whether or not L_1 and L_2 are equivalent in the narrow sense, because the bijection $G \longrightarrow \Gamma$ (where G is isomorphic to the automorphism group on L_i) is not uniquely determined.

For any $h \in H^2(F, A)$ we choose its preimage, a cocycle $\overline{h}(f_1, f_2) \in Z^2(F, A)$. Let the unit cocycle $\overline{h} \equiv 1$ be a preimage for the identity class. Then we consider the groups G_h whose elements are of the form $\{u_f a\}$, $f \in F$, $a \in A$, and $u_{f_1} u_{f_2} = u_{f_1 f_2} \overline{h}(f_1, f_2)$. Thus, the G_h are extensions of A by F taken by one from every equivalence class.

Let $\overline{R}_S$ be the set of those solutions of the problem $(K/k, \Gamma)$ for which the composition of automorphisms $v_{f_1} v_{f_2}$ coincides with the automorphism $v_{f_1 f_2} \overline{h}(f_1, f_2)$, where the h's are the classes in $H^2(F, A)$ for which the corresponding embedding problems $(K/k, G_h, A)$ are solvable. We denote by R_S the image of $\overline{R}_S$ in R_Γ. Let $L \in \overline{R}_S$ correspond to the class h. We define the bijection $\nu\colon G_h \longrightarrow \Gamma$ by the formula $\nu(u_f a) = v_f a$. Then (L, ν) is a solution of the embedding problem $(K/k, G_h, \varphi_h, A)$ (where $\varphi_h(u_f a) = f$). Conversely, let (L, ν) be a solution of the embedding problem $(K/k, G_h, \varphi_h, A)$. We choose in G_h the representatives u_f, $f \in F$, in such a way that $u_{f_1} u_{f_2} = u_{f_1 f_2} \overline{h}(f_1, f_2)$. Then $\nu(u_{f_1}) \nu(u_{f_2}) = \nu(u_{f_1 f_2}) \overline{h}(f_1, f_2)$. The algebra L solves the problem $(K/k, \Gamma)$ and is contained in $\overline{R}_S$. Thus, every solution of the embedding problem $(K/k, G_h, \varphi_h, A)$ has image in $\overline{R}_S$.

PROPOSITION 3.15.2. *Two solutions (L_1, ν_1) and (L_2, ν_2) of the embedding problem $(K/k, G_h, \varphi_h, A)$ are equivalent in the narrow sense if and only if their images in $\overline{R}_S$ are Γ-equivalent.*

PROOF. Let $u_f \in G_h$ be representatives of $f \in F$ in G_h, for which $u_{f_1} u_{f_2} = u_{f_1 f_2} \overline{h}(f_1, f_2)$. Let θ be an isomorphism $L_1 \longrightarrow L_2$ establishing the S-equivalence of the solutions (L_1, ν_1) and (L_2, ν_2). Consider the images of these solutions in $\overline{R}_S$. The mapping θ also establishes that these images are isomorphic. For $v_f \in \Gamma$, $x \in L_1$, we have $\theta(x^{v_f}) = \theta(x^{\nu_1(u_f)}) = \theta(x)^{\nu_2(u_f)} = \theta(x)^{v_f}$. Therefore, these images are Γ-equivalent. The converse is obvious.

Thus, the classes of S-equivalent solutions to the embedding problems $(K/k, G_h, \varphi_h, A)$ are in one-to-one correspondence with the elements in R_S. The set R_S is not necessarily a subgroup of the group R_Γ. But R_S is closed under the multiplication by a certain subgroup of the set R_S. Namely, let R_0 be the subset of R_S corresponding to the semidirect extension G_0. Recall that for this extension $u_{f_1} u_{f_2} = u_{f_1 f_2}$, and for the solutions of the problem $L \in \overline{R}_0$ the composition of the automorphisms v_{f_1} and v_{f_2} coincides with $v_{f_1 f_2}$. □

PROPOSITION 3.15.3. *Let $[L] \in R_S$, $[L_1] \in R_0$. Then $[L] \circ [L_1] \in R_S$ ($[L]$ is the class of $L \in \overline{R}_S$ in R_S).*

Indeed, let $L \in \overline{R}_S$ and let G_h be the corresponding group. The Baer product of the groups G_h and G_0 is naturally isomorphic to the group G_h, and the pairs $(u_f, u_f^0) \bmod \widehat{A}$ ($u_f \in G_h$, $u_f^0 \in G_0$) determine the same cocycle $\overline{h}(f_1, f_2)$ as in G_h. Therefore, $L \circ L_1$ with $L_1 \in \overline{R}_0$ is contained in $\overline{R}_S$.

COROLLARY. *The set R_0 is a subgroup of the group R_Γ.*

Thus, R_S is an R_0-module.

The map $\varkappa\colon R_\Gamma \longrightarrow R_B$ induces a map $R_S \longrightarrow R_B$, and this map is obviously surjective.

THEOREM 3.15.4. *Let the embedding problem $(K/k, G, \varphi, A)$ be solvable and let (L, ν) be a solution of it. Then any solution of this problem is S-equivalent to the composition of L and a solution L_1 of the semidirect problem $(K/k, G_0, \varphi_0, A)$.*

It is clear that we may consider the solutions that are inside the set $\overline{R}_S$. It is obvious that $L \circ L_1$ is a solution of the same embedding problem as the solution L.

First let the solutions $L_1, L_2 \in \overline{R}_S$ be B-equivalent. In this case we show that L_2 is S-equivalent to $L_1 \circ L^0$, where L^0 is a completely splitting solution of the semidirect problem. B-equivalence means the existence of a K-isomorphism $\theta\colon L_1 \longrightarrow L_2$ and an automorphism η on the group G identical on A such that $\theta(x^{\nu_1(g)}) = (\theta(x))^{\nu_2\eta(g)}$ for all $x \in L_1$, $g \in G$. It is clear that $\eta(g) = gb_f$, where $f = \varphi(g)$, $b_f \in A$. Since η is a homomorphism, b_f is a cocycle. In the language of the solutions of the problem $(K/k, \Gamma)$, our condition means that $\theta(x^{v_f a}) = (\theta(x))^{v_f a b_f}$ for $x \in L_1$, $f \in F$, $v_f a \in \Gamma$. Consider the completely splitting solution $L^0 = \sum^{\oplus}_{a\in A} Ke_a$ $(e_a = e_1^a)$, for which $e_a^{v_f} = e_a f_{b_f}^{-1}$ (since b_f is a cocycle, the solution L^0 is contained in $\overline{R}_o$), and the composition $L_1 \circ L^0$. It is clear that $L_1 \circ L^0$ consists of the elements $\sum_{a\in A} x^{a-1} \otimes e_a$, where $x \in L_1$. We define an isomorphism $\theta_1 : L_1 \longrightarrow L_1 \circ L^0$ by the formula $\theta_1(x) = \sum_{a\in A} x^{a-1} \otimes e_a$. Now we have

$$
\begin{aligned}
(\theta_1(x))^{v_f b_f} &= \sum_{a\in A} (x^{a^{-1}} \otimes e_a)^{(v_f, v_f)(b_f,1)} = \sum_{a\in A} (x^{a^{-1} v_f} \otimes e_{a^f b_f^{-1}})^{(b_f,1)} \\
&= \sum_{a\in A} x^{v_f a^{-f} b_f} \otimes e_{a^f b_f^{-1}} = \sum_{a\in A} (x^{v_f})^{a^{-1}} \otimes e_a = \theta_1(x^{v_f}).
\end{aligned}
$$

Therefore, the solutions L_2 and $L_1 \circ L^0$ are equivalent in the narrow sense.

Now let $L, L' \in \overline{R}_S$ be solutions of the embedding problem $(K/k, G, \varphi, A)$. Consider the images of them, $\overline{L}$ and $\overline{L}'$, in the group R_B, and let $\overline{L}_3 = \overline{L}' \circ (\overline{L})^{-1}$; L_3 is the solution of the semidirect problem in $\overline{R}_0$ whose image in R_B equals $\overline{L}_3$. Then we have $\overline{L}' = \overline{L} \circ \overline{L}_3$, whence the B-equivalence of L' and $L \circ L_3$ follows. By the first part of the proof, there exists a completely splitting solution $L^0 \in \overline{R}_0$ for which L' and $(L \circ L_3) \circ L^0$ are S-equivalent; so, $L_3 \circ L^0$ is the required solution. The theorem is proved.

Let $(K/k, G_h, \varphi_h, A)$ be an embedding problem, $\overline{k}$ an algebraic closure of k with Galois group $\overline{F}$. In §15, Chapter 1 we proved that there exists a correspondence between the solutions L of the embedding problem and the homomorphisms $\nu\colon \overline{F} \longrightarrow G$ such that the following triangle is commutative:

$$
\begin{array}{ccc}
\overline{F} & \xrightarrow{\ \nu\ } & G \\
{\scriptstyle\mu}\big\downarrow & \swarrow {\scriptstyle\varphi} & \\
F = \mathrm{Gal}(K/k) & &
\end{array}
$$

where μ is the homomorphism corresponding to the embedding of K in $\overline{k}$.

We note that if ν and ν_0 are the homomorphisms of $\overline{F}$ to G and G_0 corresponding to the solutions L and L_0 (L_0 is a solution of the semidirect problem

$(K/k, G_0, \varphi_0, A))$, then to the composition $L \circ L_0$ there corresponds the homomorphism $\nu_1 \colon \overline{F} \longrightarrow G$ defined according to the relation $\nu_1(\overline{f}) = (\nu(\overline{f}), \nu_o(\overline{f})) \bmod \widetilde{A}$, where $\overline{f} \in \overline{F}$.

Now we find the kernel of the mapping $R_S \longrightarrow R_B$. This kernel is clear to be contained in R_0 and, thus, is a subgroup of R_0. It consists of the classes of S-equivalent completely splitting solutions. Such a solution determines (as shown in the proof of the theorem) a cocycle $b_f \in Z^1(F, A)$. Namely, this cocycle appears in the formula $e_1^{v_f} = e_1^{b_f}$. It is of interest to know what cocycles determine S-equivalent solutions. Let completely splitting solutions L and L' be determined by the cocycles b_f and b'_f, and let the isomorphism $\theta \colon L \longrightarrow L'$ establish the S-equivalence of them. Then $\theta(e_1) = E_c = E_1^c$, where e_1 and E_c are idempotents of the solutions L and L', $c \in A$. The condition $\theta(e_1^{v_f}) = (\theta(e_1))^{v_f}$ means that $cb_f = b'_f c^f$, whence $b'_f = b_f c^{1-f}$, i.e., the cocycles b_f and b'_f are homologous. Conversely, if $b'_f = b_f \cdot c^{1-f}$, then the isomorphism θ taking the idempotents e_a of the algebra L to the idempotents E_{ac} of the algebra L' yields the desired equivalence. Thus, we have proved the following proposition.

PROPOSITION 3.15.4. *The kernel of the homomorphism $R_S \longrightarrow R_B$ is isomorphic to $H^1(F, A)$.*

6°. It is useful to illustrate the results just obtained with a meaningful example.

Let $A = \langle a \rangle$ be a cyclic group of order 4, $F = \langle f \rangle$ a cyclic group of order 2, and let the action of the operators be trivial, i.e., $a^f = a$. In this case there exist two nonequivalent extensions, the direct extension G_0 and the cyclic group G_1 of order 8.

For the field k, we take $GF(p)$, the field with p elements, $p \equiv 1 \pmod 8$ (then k contains a primitive eighth root ε of unity), $K = k(\sqrt{d})$, where $(\frac{d}{p}) = -1$. Here all objects are finite.

The space Γ consists of the following eight elements: $\{1, a, a^2, a^3, v_f, v_f a, v_f a^2, v_f a^3\}$. The solutions of the problem $(K/k, \Gamma)$ are determined by a generating element ω for which $\omega^4 \in K$ and $\omega^a = \omega\varepsilon^2$. For the solutions that correspond to the direct extension, we obtain $\omega^4 = 1$ or $\omega^4 = d$, and $\omega^{v_f} = \omega\varepsilon^{2i}$ $(i = 0, 1, 2, 3)$. For the cyclic group G_1, we have $\omega^4 = \sqrt{d}$ or $\omega^4 = d\sqrt{d}$, and $\omega^{v_f} = \omega\varepsilon^i$ $(i = 1, 3, 5, 7)$. It is easy to see that all these solutions are contained in different Γ-equivalent classes; so, R_Γ is a group with 16 elements. Let the solution L_1 be defined in accordance with the formulas $\omega^4 = 1$, $\omega^{v_f} = \omega\varepsilon^2$ (here $v_f^2 = a^2$). This is a completely splitting solution; moreover, L_1 has order 4 in R_Γ and generates all completely splitting solutions. For L_2, we take the solution corresponding to the group G_1 and defined in accordance with the formulas $\omega^4 = \sqrt{d}$, $\omega^{v_f} = \omega\varepsilon$ $(v_f^2 = a)$. The solution L_2 also has order 4 in R_Γ, and the cyclic groups generated by L_1 and L_2 intersect only at the unit solution. Therefore, R_Γ is a direct product of two cyclic groups of order 4.

The group R_B is a quotient group of R_Γ by the subgroup generated by the class L_1, i.e., is a cyclic group of order 4. We note that the solutions L_2 and $L_2 \circ L_2 \circ L_2$ are isomorphic as fields, but are not B-equivalent.

Now consider the set R_S. We fix the trivial cocycle in the case of the direct extension and the cocycle defined by the formula $v_f v_f = a$ in the case of a cyclic extension. Direct inspection shows that R_S contains the following classes: L_0 (the unit element), L_1^2, L_2, $L_1^2 \circ L_2$, $L_1 \circ L_2^2$, $L_1^3 \circ L_2^2$, $L_1 \circ L_2^3$, $L_1^3 \circ L_2^3$. It is obvious that

R_S is not closed under multiplication in R_Γ. At the same time, the multiplication by the solutions from R_0, i.e., by L_0, L_1^2, $L_1 \circ L_2^2$, $L_1^3 \circ L_2^2$ takes R_S to R_S.

7°. Now we show that it is possible to introduce a group structure on the set of all classes of S-equivalent solutions. As before, let R_B be the group of classes of B-equivalent solutions, D the subgroup of the group $H^2(F, A)$, corresponding to the solvable embedding problems, C the kernel of the natural homomorphism $R_B \to D$, i.e., the group of the classes of B-equivalent solutions of the semidirect problem. Thus, the exact sequences of Abelian groups $1 \to C \to R_B \to D \to 1$, $1 \to H^1(F, A) \to R_0 \to C \to 1$ are given, where R_0 (as before) is the group of classes of S-equivalent solutions of the semidirect problem. The sequence $1 \to C \to R_B \to D \to 1$ induces the exact sequence of the groups $\operatorname{Ext}^1(D, H^1(F, A)) \to \operatorname{Ext}^1(R_B, H^1(F, A)) \to \operatorname{Ext}^1(C, H^1(F, A)) \to 1$, since on Abelian operator-free groups, the functor Ext^1 is exact (see [**37**]). Since R_0 is contained in $\operatorname{Ext}^1(C, H^1(F, A))$, it has a preimage in $\operatorname{Ext}^1(R_B, H^1(F, A))$. Denote it by X. Then the following diagram is commutative:

$$\begin{array}{ccccccccc} 1 & \longrightarrow & H^1(F,A) & \longrightarrow & R_0 & \longrightarrow & C & \longrightarrow & 1 \\ & & \| & & \downarrow & & \downarrow & & \\ 1 & \longrightarrow & H^1(F,A) & \longrightarrow & X & \longrightarrow & R_B & \longrightarrow & 1 \end{array}$$

It is clear that the mapping $R_0 \to X$ is injective and its cokernel coincides with D. At the same time, the group X, as a space over R_0, coincides with the set of classes of S-equivalent solutions of the embedding problems $(K/k, G_h, \varphi_h, A)$, $h \in H^2(F, A)$. This allows us to specify the group structure on this set.

This specification, however, is artificial and not unique, since the choice of the preimage X is determined modulo the image of the group $\operatorname{Ext}^1(D, H^1(F, A))$ in $\operatorname{Ext}^1(R_B, H^1(F, A))$. Examples show that the groups X obtained in such a way need not be abstractly isomorphic.

CHAPTER 4

The Embedding Problem for Local Fields

In this chapter the embedding problem for p-groups is analyzed from the point of view of solvability in Galois algebras and solvability in the proper sense.

§1. The embedding problem for local fields

In this section we consider the embedding problem of p-extensions of local fields in a field (or a Galois algebra) whose Galois group over the base field also is a p-group. The study of this problem depends essentially on whether the numbers of generators of the Galois groups of the initially given extension and the required extension coincide. In the case where these numbers are equal, the answer is sufficiently simple and complete, i.e., such an embedding problem is solvable if and only if the associated Abelian problem is solvable (i.e., for this problem the compatibility condition is fulfilled; see Theorem 3.14.1). In the case where these numbers are not equal, this is not the case, and we restrict our consideration to a counterexample.

1°. We need some auxiliary group-theoretic facts concerning p-groups.

PROPOSITION 4.1.1. *Let G be a finite p-group. Then the Frattini subgroup of G is generated by the commutators and the pth powers of elements of G.*

PROOF. Let H be a maximal proper subgroup of G. Then H has index p in G, and G/H is a cyclic group of order p. Therefore, $[G,G] \subset H$ and $G^p \subset H$, i.e., $[G,G]G^p \subset H$. Since the Frattini subgroup is the intersection of all maximal proper subgroups, $[G,G]G^p$ is contained in the Frattini subgroup of G. The quotient group $G/[G,G]G^p$ is an elementary Abelian group. Let $\beta_1, \dots, \beta_d$ be generators of this quotient group, and let $\overline{\beta}_1, \dots, \overline{\beta}_d$ be their preimages in G. The subgroups H_i generated by the subgroup $[G,G]G^p$ and the elements $\overline{\beta}_1$, $\overline{\beta}_2$, ... , $\overline{\beta}_{i-1}$, $\overline{\beta}_{i+1}$, ... , $\overline{\beta}_d$ are maximal proper subgroups of G, and the intersection of them is $[G,G]G^p$. The proposition is proved. □

COROLLARY. *The number of generators of a p-group G coincides with the number of generators of the elementary Abelian p-group $G/[G,G]G^p = G/\Phi(G)$.*

PROPOSITION 4.1.2. *Let $1 \to B \to G \xrightarrow{\varphi} F \to 1$ be an exact sequence of p-groups and $B_0 = B \cap \Phi(G)$. Then the sequence $1 \to B/B_0 \to G/B_0 \to F \to 1$ obtained by passage to the quotients by B_0 yields a direct extension.*

It is clear that $\varphi([G,G]G^p) = [F,F]F^p = \Phi(F)$, whence the sequence $1 \to B/B_0 \to G/\Phi(G) \to F/\Phi(F) \to 1$ is exact, and this extension is direct (since $G/\Phi(G)$ is an elementary Abelian group).

The following commutative diagram with natural homomorphisms holds:

$$\begin{array}{ccccccccc}
1 & \longrightarrow & B/B_0 & \longrightarrow & G/B_0 & \longrightarrow & F & \longrightarrow & 1 \\
 & & \downarrow{\scriptstyle \mathrm{id}} & & \downarrow & & \downarrow & & \\
1 & \longrightarrow & B/B_0 & \longrightarrow & G/\Phi(G) & \longrightarrow & F/\Phi(F) & \longrightarrow & 1,
\end{array}$$

therefore G/B_0 is a direct product of $G/\Phi(G)$ and F with amalgamated quotient group $F/\Phi(F)$. Therefore, G/B_0 is a direct product of B/B_0 and F.

COROLLARY. *Let $1 \to B \to G \to F \to 1$ be an exact sequence of p-groups, and let $d(G)$ and $d(F)$ be the numbers of generators of G and F, respectively. Then $d(F) \le d(G)$, and $d(F) = d(G)$ if and only if $B \subset \Phi(G)$.*

The following fact, that will be repeatedly used in the sequel, is almost obvious. For the sequence $1 \to B \to G \to F \to 1$, let the relation $d(G) = d(F)$ hold and let B_1 be a subgroup of B, normal in G. Then we have the exact sequences $1 \to B/B_1 \to G/B_1 \to F \to 1$ and $1 \to B_1 \to G \to G/B_1 \to 1$. By $d(F) \le d(G/B_1) \le d(G) = d(F)$, we have $d(G/B_1) = d(F) = d(G)$, so that for both sequences indicated the group and the quotient group have one and the same number of generators.

2°. Suppose a local field k does not contain a primitive pth root of unity. Let an embedding problem $(K/k, G, \varphi, B)$ be given, where B is a p-group.

THEOREM 4.1.1. *The problem just stated is always solvable.*

PROOF. Inductive arguments reduce the proof of the theorem to the case of an Abelian kernel and even to the case where the kernel is a group of order p. Let $\overline{k}$ be the algebraic closure of k with Galois group $\overline{F}$. We denote by h the cohomology class from $H^2(F, B)$, corresponding to the extension G, and by $\overline{h}$ the lifting of it to the group $H^2(\overline{F}, B)$. By Proposition 3.13.2, the theorem is equivalent to the relation $\overline{h} = 0$. By Tate's local duality (see Appendix, §3), $H^2(\overline{F}, B) \approx \mathrm{Hom}_{\overline{F}}(B, \overline{k}^*)'$. Since $\overline{F}$ trivially acts on B and k^* does not contain nontrivial pth roots of unity, this group is trivial. The theorem is proved. □

REMARK. The theorem also follows from the fact that the Galois group of a maximal p-extension of a local field that does not contain a primitive pth root of unity is a free pro-p-group (see [**55**]). On the other hand, this theorem actually proves this fact. Indeed, we have shown that for the diagram

$$\begin{array}{ccccc}
 & & \overline{F} & & \\
 & & \downarrow{\scriptstyle \psi} & & \\
G & \xrightarrow{\varphi} & F & \longrightarrow & 1
\end{array}$$

there exists a homomorphism $\mu\colon \overline{F} \to G$ such that $\varphi\mu = \psi$. Therefore, $\overline{F}$ is free.

3°. Consider the following embedding problem for local fields:

$$(K/k, G, \varphi, B), \tag{1}$$

where G is a finite p-group (whence B and $F = \mathrm{Gal}(K/k)$ are also finite p-groups). The main purpose of this section is to prove the following theorem.

THEOREM 4.1.2. *Let p-groups G and F have the same number of generators. Then the embedding problem* (1) *is solvable if and only if the associated problem* $(K/k, G/B', \varphi', B/B')$ *is solvable, where B' is the commutator subgroup of B.*

We note that the fact that the numbers of generators of G and F are equal is equivalent to $B \subset \Phi(G)$, where $\Phi(G)$ is the Frattini subgroup of G. Thus, any solution of problem (1) is a field, i.e., is a proper solution (Corollary 5 to Theorem 1.6).

The theorem admits the following generalization.

THEOREM 4.1.2$'$. *Except for the case indicated below the embedding problem* $(K/k, G, \varphi, B)$ *for local fields, where G is a finite p-group, is solvable in the proper sense if and only if the associated problem*

$$(K/k, G/\widetilde{B}, \widetilde{\varphi}, B/\widetilde{B}) \tag{2}$$

is solvable in the proper sense, where $\widetilde{B}$ is the mutual commutator group for the groups B and $B_0 = B \cap \Phi(G)$, $\widetilde{\varphi}$ is the homomorphism $G/\widetilde{B} \to G/B = \mathrm{Gal}(K/k)$ induced by the map φ. The exceptional case may occur only if the following conditions are satisfied simultaneously:

a) $p = 2$;

b) -1 *is not a square in* K;

c) *the kernel B is not contained in the Frattini subgroup $\Phi(G)$ of G.*

Theorem 4.1.2 is a consequence of Theorem 4.1.2$'$, because in the case $B \subset \Phi(G)$ we have $\widetilde{B} = B'$, where B' is the commutator subgroup of B. Since the proofs of both theorems have much in common, we shall prove a more general Theorem 4.1.2$'$.

We proceed by induction on the order of the group B. For an Abelian kernel B, the group $\widetilde{B}$ is trivial, which proves the base of induction. Suppose that for all embedding problems whose order of the kernel is less than the order of B the theorem is true. Let the associated embedding problem $(K/k, G/\widetilde{B}, \widetilde{\varphi}, B/\widetilde{B})$ be solvable in the proper sense.

If the field k does not contain a primitive pth root of unity, then, as shown above, the problem

$$(L/k, G, \widetilde{B}), \tag{3}$$

where L is a proper solution of problem (2), is solvable; since $\widetilde{B} \subset \Phi(G)$, the solution of (3) is proper. Thus, problem (1) has a proper solution. Therefore, in the sequel we shall assume that k contains a primitive pth root ζ_0 of unity. For $\widetilde{B} = \{1\}$, the theorem is empty; so, we assume that $\widetilde{B} = [B, B_0]$ is nontrivial.

We prove several auxiliary lemmas.

LEMMA 4.1.1. *Let C be normal in G, and let C be a proper subgroup of B. Then the associated problem $(K/k, G/C, B/C)$ is solvable in the proper sense.*

For the problem $(K/k, \overline{G}, \overline{B})$, where $\overline{G} = G/C$, $\overline{B} = B/C$, the inductive assumption is valid, whence it reduces to the problem $(K/k, \overline{G}/\widetilde{\overline{B}}, \overline{B}/\widetilde{\overline{B}})$, where $\widetilde{\overline{B}} = [\overline{B}, \overline{B} \cap \Phi(\overline{G})]$. The homomorphism $B/\widetilde{B} \to \overline{B}/\widetilde{\overline{B}}$ is surjective, so that this problem is associated with problem (2) that is solvable in the proper sense.

LEMMA 4.1.2. *Let B_1 be a proper subgroup of B, normal in G. If the group $\widetilde{B}_1 = [B_1, B_1 \cap \Phi(G)]$ is nontrivial, problem* (1) *is solvable in the proper sense.*

Consider the embedding problem $(K/k, G/\widetilde{B}_1, B/\widetilde{B}_1)$ that is solvable in the proper sense whenever $\widetilde{B}_1 \neq \{1\}$. Let L_1 be a proper solution of it and $L = L_1^{B_1/\widetilde{B}_1}$. Then L yields a solution to the embedding problem $(K/k, G/B_1, B/B_1)$. Now we consider the problem $(L/k, G, B_1)$. By the inductive assumption, this problem reduces to the problem $(L/k, G/\widetilde{B}_1, B_1/\widetilde{B}_1)$, L is a proper solution of the latter problem. Thus, problem (1) is solvable.

Therefore, in the sequel we assume that $B_1 \cap \Phi(G)$ is contained in the center of a group B_1, where B_1 is any proper subgroup of B, normal in G. In particular, either $B_0 = B$ or B_0 is an Abelian group.

Let B_1 be an arbitrary proper subgroup of B, containing B_0 (if $B_0 \neq B$). Then B_1 is normal in G and, thus, B_0 is contained in the center of B_1. Since $\widetilde{B} \neq \{1\}$, there exists an element $\beta \in B$ that does not commute with all elements of B_0. Therefore, the group B_1 generated by B_0 and by the element β (as a noncommutative group) must coincide with the group B. Since $\beta^p \in B_0$, we have either B_0 has index p in B or $B_0 = B$. If $B_0 \neq B$, then $B' = [B, B] = [B, B_0] = \widetilde{B}$.

Consider the case $B_0 \neq B$.

LEMMA 4.1.3. *Let L be an arbitrary proper solution of problem* (2), *and let L_0 be the subfield of L corresponding to the subgroup $B_0/\widetilde{B}$. If L_0 is not a cyclotomic extension of the field K, then problem* (1) *has a proper solution.*

It is obvious that the field L_0 is a solution of the embedding problem $(K/k, G/B_0, B/B_0)$. We note that this problem corresponds to a direct extension of groups (Proposition 4.1.2), whence $L_0 = K \otimes_k k_1$, where k_1 is a normal extension of k of degree p, i.e., $k_1 = k(\sqrt[p]{a})$ where $a \in k$. Consider the embedding problem $(L_0/k, G, B_0)$. Since $B_0 \subset \Phi(G)$, any solution of this problem (if it exists) is proper. Moreover, by the reduction we have just made, B_0 is an Abelian group; therefore, the criterion for solvability of this problem consists of the solvability of all associated Brauer problems of the form $(L_0/k, G/\operatorname{Ker}_\chi, B_0/\operatorname{Ker}_\chi)$, where $\chi \in \operatorname{Hom}_G(B_0, L_0^*)$ (see Theorem 3.14.1). Let $\chi \in \operatorname{Hom}_G(B_0, L_0^*)$. Since L_0 is not a cyclotomic extension of K, it follows that $\chi \in \operatorname{Hom}_G(B_0, K^*)$. For $\alpha \in B_0$, $\beta \in B$ we have $\chi(\alpha^\beta) = \chi(\alpha)^\beta = \chi(\alpha)$, whence $\chi(\beta^{-1}\alpha\beta\alpha^{-1}) = 1$ and, thus, $\widetilde{B} = [B, B_0] \subset \operatorname{Ker}_\chi$. Therefore, all Brauer problems of the form $(L_0/k, G/\operatorname{Ker}\chi, B_0/\operatorname{Ker}\chi)$, where $\chi \in \operatorname{Hom}_G(B_0, L_0^*)$, are associated with the problem $(L_0/k, G/\overline{B}, B_0/\overline{B})$ for which L is a solution. Thus, the embedding problem $(L_0/k, G, B)$ has a (proper) solution that also is a solution of problem (1).

Thus, if $B_0 \neq B$, we reduce the problem to the case, where the field $L_0 = K(\sqrt[p]{\zeta})$, where ζ is a root of unity which is contained in the field k (the fields K and $k(\sqrt[p]{\zeta})$ are linearly separated over k).

Consider the subgroup $A = [G, B]B^p$ of B. It is obvious that $A \subset B_0 \subset B$, and B/A is an elementary Abelian p-group, which is contained in the center of G/A.

LEMMA 4.1.4. *If $B_0 \neq B$, then either A coincides with B_0 or A has index p in B_0 and coincides with the center of the kernel B.*

Suppose that $(B_0 : A) \geq p^2$. Since B_0/A is an elementary Abelian group, there exist two subgroups A_1 and A_2 of B of index p containing A. Let B_i be the subgroups generated by A_i and an element $\beta \in B$, $\beta \notin B_0$. Then $B_i \cap \Phi(G) =$

A_i, whence (by reduction) β commutes with the elements of A_1 and A_2, i.e., it commutes with B_0, a contradiction to the fact that $\widetilde{B} = [B, B_0]$ is a nontrivial group. If A has index p in B_0, then the same arguments applied to B_1 generated by A and by an element $\beta \in B$, $\beta \notin B_0$, show that $[A, B_1] = \{1\}$. Since $B_0 B_1 = B$, the group A is contained in the center of B and coincides with it.

LEMMA 4.1.5. *Let $A \neq B_0 \neq B$. Then the embedding problem* (1) *has a proper solution.*

Let L be a proper solution of problem (2), and let $L_0 = L^{B_0/\widetilde{B}}$, $L_1 = L^{A/\widetilde{B}}$. The reduction we have made allows us to assume that L_0 is the composite of K and the field that is a cyclotomic extension of k of degree p, linearly separated of K. The field L_1 is an extension of L_0 of degree p. Since B/A is an elementary Abelian group of order p^2, L_1 is not a cyclotomic extension of L_0. Consider the embedding problem $(L_1/k, G, A)$. The solvability of this problem (hence, the solvability in the proper sense) reduces to verification of the solvability of the associated Brauer problems $(L_1/k, G/\operatorname{Ker}\chi, A/\operatorname{Ker}\chi)$, where $\chi \in \operatorname{Hom}_G(A, L_1^*)$. We have proved that A is the center of B. Therefore, for any $\alpha \in A$, $\beta \in B$ we have $\chi(\alpha) = \chi(\alpha^\beta) = \chi(\alpha)^\beta$, whence $\chi(\alpha) \in L_1^{B/A} = K$. Thus, $\chi(\alpha) \in k^*$, so that

$$\operatorname{Hom}_G(A, L_1^*) = \operatorname{Hom}_G(A, k^*) = \operatorname{Hom}_{\mathbb{Z}}(A, k^*).$$

Consider a fixed element $\beta \in B$ which is not contained in B_0. For $g \in G$, we have $\beta^g = \beta\alpha(g)$, where $\alpha(g) \in A$. It is clear that $\alpha(g_1 g_2) = \alpha(g_1)^{g_2}\alpha(g_2)$, where $g_i \in G$. Moreover, $\alpha(g)$ is determined by the image of g in $G/A = \operatorname{Gal}(L_1/k)$. Let $u(\chi, g) = \chi(\alpha(g))$, where $\chi \in \operatorname{Hom}_G(A, L_1^*)$. Then $u(\chi, g)$ is a cocycle in $Z^1(G/A, L_1^*)$ and, by Speiser's theorem, $u(\chi, g) = v_\chi^{1-g}$, where $v_\chi \in L_1^*$ (it is more correct to write $v_\chi^{1-\overline{\varphi}(g)}$, where $\overline{\varphi}$ is the map $G \to G/A$). Consider the element $c_\chi = \chi(\beta^p)v_\chi^p$. Then $c_\chi^g = \chi(\beta^p)^g v_\chi^{pg} = \chi(\beta^{gp})v_\chi^{gp} = \chi(\beta^p\alpha(g)^p)v_\chi^{gp} = \chi(\beta^p)u(\chi, g)^p v_\chi^{gp} = c_\chi$, whence $c_\chi \in k^*$. But $\chi(\beta^p) \in k^*$, and $v_\chi^p \in k^*$. Therefore, $v_\chi^g = v_\chi\zeta_0^{l(g)}$, where $\zeta_0 \in k^*$ is a primitive pth root of 1, and $l(g)$ is an integer defined modulo p. It is clear that $l(g)$ is a homomorphism $G \to \mathbb{Z}/p\mathbb{Z}$; consequently, for $g \in \Phi(G)$ we have $l(g) \equiv 0 \pmod p$ and $v_\chi^g = v_\chi$. Therefore, for $g \in \Phi(G)$ the element $\alpha(g) = \beta^g\beta^{-1}$ belongs to the kernel of χ and $\widetilde{B} = [B, B_0] \subset \operatorname{Ker}\chi$ for $\chi \in \operatorname{Hom}_G(A, L_1^*)$. Hence, for such χ the embedding problem $(L_1/k, G/\operatorname{Ker}\chi, A/\operatorname{Ker}\chi)$ is associated with the problem $(L_1/k, G/\widetilde{B}, A/\widetilde{B})$ for which L is a solution. Thus, the problem $(L_1/k, G, A)$ is also solvable and, in addition, in the proper sense, and its solution is also a solution of problem (1). The lemma is proved.

Therefore, it remains to consider only the cases $B_0 = B$ and $B_0 = A$.

LEMMA 4.1.6. *Let $B_0 = B$. If the index of A in B is different from p, then the embedding problem* (1) *has a proper solution.*

Let B_1 be a subgroup of the group B of index p, which is normal in G. Then the quotient group B/B_1 has order p and is central in G/B_1. Therefore, $A = [G, B]B^p$ is contained in B_1, whence A is a proper subgroup of B. Suppose that $(B : A) \geq p^2$. We choose in B a subgroup B_1 such that

(a) $A \subset B_1 \subset B$;

(b) $(B : B_1) = p$;

(c) the field $L_1 = L^{B_1/\widetilde{B}}$ (where L is a proper solution of problem (2)) is not a cyclotomic extension of K.

Such a subgroup B_1 exists, because there is at most one subgroup satisfying the first two conditions and not satisfying the third. Moreover, the group B_1 is normal in G in view of $A \subset B_1 \subset B$.

Consider the embedding problem $(L_1/k, G, B_1)$. As before, we may assume that the kernel B_1 is an Abelian group. Let $\chi \in \mathrm{Hom}_G(B_1, L_1^*)$. Since L_1 is not a cyclotomic extension of K, the values of χ belong to K. Thus, for any $\alpha \in B_1$, $\beta \in B$, we have $\chi(\alpha^\beta) = \chi(\alpha)^\beta = \chi(\alpha)$, whence $\chi(\beta^{-1}\alpha\beta\alpha^{-1}) = 1$. Therefore, $[B, B_1]$ is contained in the kernel of any character χ indicated above. But B/B_1 is a cyclic group of order p, whence $[B, B_1] = [B, B]$. Therefore, for any $\chi \in \mathrm{Hom}_G[B_1, L_1^*]$ the Brauer embedding problem $(L_1/k, G/\operatorname{Ker}\chi, B_1 \operatorname{Ker}\chi)$ is associated with the problem $(L_1/k, G/\widetilde{B}, B_1/\widetilde{B})$ for which L is a solution. Thus, the embedding problem $(L_1/k, G, B_1)$, as well as problem (1), is solvable. Since $B_1 \subset B \subset \Phi(G)$, problem (1) is solvable in the proper sense.

Thus, in either case $B_0 \neq B$ and $B_0 = B$, we may assume that the group $A = [G, B]B^p$ has index p in B and the field $L_1 = L^{A/\widetilde{B}}$, where L is a proper solution of problem (2), is a cyclotomic extension of K. Note that $B' = [B, B] = [B, A] = \widetilde{B}$.

LEMMA 4.1.7. *Suppose the group $C = [G, \widetilde{B}]\widetilde{B}^p$ is nontrivial. Then the embedding problem* (1) *has a proper solution.*

Let L_0 be a proper solution of the problem $(K/k, G/C, B/C)$ and $L_1 = L_0^{A/C}$. Then L_1 is a solution of the embedding problem $(K/k, G/A, B/A)$. Consider the embedding problem $(L_1/k, G, A)$. For $\chi \in \mathrm{Hom}_G(A, L_1^*)$ we have $\chi(\alpha^\beta) = \chi(\alpha)^\beta$, where $\alpha \in A$, $\beta \in B$. Since L_1 is an extension of K of degree p, the values $\chi(\alpha^p)$ belong to K. Therefore, $(\chi(\alpha^\beta))^p = (\chi(\alpha)^\beta)^p = \chi(\alpha^p)^\beta = \chi(\alpha^p) = \chi(\alpha)^p$ and $\chi(\alpha^\beta) = \chi(\alpha)\zeta_0$, with $\zeta_0^p = 1$. Thus, $\chi(\alpha^\beta\alpha^{-1}) \in k^*$ and $\chi(\alpha^\beta\alpha^{-1})^p = 1$. Therefore, the values of χ on $\widetilde{B} = [B, A]$ belong to k and have period p. If $\gamma \in \widetilde{B}$, $g \in G$, then $\chi(\gamma^p) = 1$ and $\chi(\gamma^g) = \chi(\gamma)^g = \chi(\gamma)$. Therefore, the values of χ on C are equal to 1. Thus, any Brauer problem of the form $(L_1/k, G/\operatorname{Ker}\chi, A/\operatorname{Ker}\chi)$ is associated with the problem $(L_1/k, G/C, A/C)$ for which L_0 is a solution, and the problem $(L_1/k, G, A)$ is solvable in the proper sense. Thus, problem (1) is also solvable.

We have thereby reduced problem (1) to the case where the group $\widetilde{B} = B'$ is central in G and has exponent p.

If $p = 2$, it is necessary to require that K contains $\sqrt{-1}$. If $B_0 = B$, this requirement may be satisfied without loss of generality. Indeed, if -1 is not a square in K, we may consider the problem obtained by the lifting of K to $K(\sqrt{-1}) = \overline{K}$. The group $\overline{F} = \mathrm{Gal}(\overline{K}/k)$ is a direct product of $F = \mathrm{Gal}(K/k)$ and the group of order 2; in just the same way, the direct product $\overline{G}$ of G and $\overline{F}$ with amalgamated quotient group F is a direct product of G and $\mathbb{Z}/2\mathbb{Z}$. The condition $B \subset \Phi(G)$ remains valid, whence lifting leads to a problem equivalent to the original one in the proper sense. The case $p = 2$, $\sqrt{-1} \notin K$, and $B \neq B_0$ is exceptional and is indicated in the statement of the theorem. We consider this case in the next section.

Now we complete the proof of the theorem. By reduction performed above, we assume that $L_1 = L^{A/\widetilde{B}}$, where L is a proper solution of the embedding problem (2), is a cyclotomic extension of degree p of the field k and is a solution of the problem $(K/k, G/A, B/A)$. Consider the embedding problem $(L_1/k, G, A)$ that has a proper solution if and only if all associated Brauer problems $(L_1/k, G/\operatorname{Ker}\chi, A/\operatorname{Ker}\chi)$ are solvable, where $\chi \in \operatorname{Hom}_G(A, L_1^*)$. We fix some χ.

Consider $\alpha = \beta^p$ with $\beta \in B$ and show that $\chi(\alpha) \in K^*$. This is obvious for $\beta \in A$; if $\beta \notin A$, we have $\alpha^\beta = \alpha$, whence $\chi(\alpha)^\beta = \chi(\alpha^\beta) = \chi(\alpha)$. Therefore, $\chi(\alpha) \in L_1^{B/A} = K$.

Now let $\alpha = \beta^g\beta^{-1} \in A$, where $\beta \in B$, $g \in G$. Since the commutator subgroup B' is contained in the center of B, we have $\alpha^l = (\beta^g)^l\beta^{-l}\gamma^{\binom{l}{2}}$, where γ is the commutator of β^g and β (l is a positive integer, $\binom{l}{2} = \frac{l(l-1)}{2}$). Since the commutator subgroup has period p, we have $\alpha^p = (\beta^g)^p\beta^{-p}$ for odd p; if $p = 2$, then $\alpha^4 = (\beta^g)^4\beta^{-4}$. We have already shown that $\chi(\beta^p) \in K^*$. Since G is a p-group, $\chi(\beta^p)^g = \chi(\beta^p)^{1+pm}$ for integer m. Therefore, we have $\chi(\alpha)^p = \chi(\beta^p)^g\chi(\beta^{-p}) = \chi(\beta^p)^{pm}$ for odd p and $\chi(\alpha)^4 = \chi(\beta^4)^{4m}$ for $p = 2$. Thus, $\chi(\alpha) = \chi(\beta^p)^m\zeta_0$, where ζ_0 is a pth root of 1 for $p \neq 2$; if $p = 2$, then $\chi(a) = \chi(\beta^4)\varepsilon$, where ε is a root of degree 4 of 1. Since ζ_0 (respectively, ε) belongs to K, we have $\chi(a) \in K^*$.

Thus, the values of χ on the elements from B^p and $[B, G]$ are contained in K. Since $A = [B, G]B^p$, $\chi \in \operatorname{Hom}_G(A, K^*)$, we have $\chi(\alpha^\beta) = \chi(a)^\beta = \chi(\alpha)$, $\alpha \in A$, $\beta \in B$, whence the mutual commutator subgroup of A and B is contained in the kernel of χ. But $\widetilde{B} = [A, B]$ and the embedding problem $(L_1/k, G/\operatorname{Ker}\chi, A/\operatorname{Ker}\chi)$ is associated with the solvable problem $(L_1/k, G/\widetilde{B}, A/\widetilde{B})$ (L_3 is a solution of this problem). Therefore, the problem $(L_1/k, G, A)$ is also solvable, and a field must be a solution of this problem. Hence, the problem $(K/k, G, \varphi, B)$ is solvable in the proper sense. Theorem 4.1.2$'$ is proved.

Therefore, Theorem 4.1.2 is also proved.

COROLLARY. *The above theorem proves the equivalence of the following statements*:

(A) *The embedding problem* (1) *under the above assumptions is solvable.*

(B) *The compatibility condition holds for problem* (1).

(C) *The compatibility condition holds for problem* (2).

(D) *The compatibility conditions hold for all Brauer problems* $(K/k, G/\operatorname{Ker}\chi, B/\operatorname{Ker}\chi)$, *with* $\chi \in \operatorname{Hom}_G(B, K^*)$.

Indeed, the implications $A \implies B \implies C \implies D$ are obvious. Condition (D) is equivalent to the solvability of problem (2) and, by the theorem just proved, this implies (A).

4°. We present an example showing that in Theorem 4.1.2 we cannot eliminate the requirement that the numbers of generators in the group and in the quotient group be equal.

Let k be a local number field containing a primitive pth root ζ of unity, but not containing $\sqrt[p]{\zeta}$ (p is odd). The group k^*/k^{*p} is finite and is isomorphic to a finite-dimensional linear space over the field with p elements. On this space the Hilbert symbol gives rise to a skew-symmetric bilinear form. Therefore, the dimension ν of this space is even (Lemma 5.3.2 proves that $\nu = 2$ whenever the characteristic of the residue field of k is not equal to p; for the case $\nu = n + 2$ with $n = (k : \mathbb{Q}_p)$, see [**27**]). In the space k^*/k^{*p} we choose a basis $e_1, e_2, \dots, e_\nu$ on which the form under consideration takes the values $(e_1, e_2) = 1$, $(e_3, e_4) = 1$, $\dots$, $(e_{\nu-1}, e_\nu) = 1$ (respectively, $(e_2, e_1) = (e_4, e_3) = (e_\nu, e_{\nu-1}) = -1$), and all remaining values of the form at basis vectors are zero. Let a_i be the preimages of e_i in the field k. Then for the (multiplicative) Hilbert symbols (a_i, a_j) we have $(a_1, a_2) = (a_3, a_4) = \cdots =$

$(a_{\nu-1}, a_\nu) = \zeta$ and $(a_2, a_1) = (a_4, a_3) = \cdots = (a_\nu, a_{\nu-1}) = \zeta^{-1}$, and the other values of (a_i, a_j) are equal to unity. Without loss of generality we may assume that $a_1 = \zeta$.

Now let the field K be obtained from k by adjoining the radicals $\sqrt[p]{a_1}$, $\sqrt[p]{a_2}, \ldots$, $\sqrt[p]{a_\nu}$. Then $F = \mathrm{Gal}(K/k)$ is an elementary Abelian p-group with ν generators $f_1, f_2, \ldots, f_\nu$, where $(\sqrt[p]{a_i})^{f_i} = \sqrt[p]{a_i}\zeta$, $(\sqrt[p]{a_i})^{f_j} = \sqrt[p]{a_i}$ for $j \neq i$. Let G_0 be an Abelian p-group with 2ν generators $\alpha_1, \ldots, \alpha_\nu, g_1, \ldots, g_\nu$ and $\alpha_i^p = 1$, $g_1^{p^2} = 1$, $g_i^p = g_1^p$ $(i = 2, \ldots, \nu)$; we denote the element g_i^p by γ. Finally, the group G is generated by the group G_0 and an element β such that $\beta^p = 1$, $\beta^{-1}\gamma\beta = \gamma$, $\beta^{-1}\alpha_i\beta = \alpha_i\gamma$, $\beta^{-1}g_i\beta = g_i\alpha_i$. The Frattini subgroup of G is generated by the elements $\gamma, \alpha_1, \ldots, \alpha_\nu$, whence G has $(\nu+1)$ generators.

Now we define the homomorphism $\varphi\colon G \longrightarrow F$ by the formulas $\varphi(g_i) = f_i$, $\varphi(\alpha_i) = 1$, $\varphi(\beta) = 1$ and consider the embedding problem $(K/k, G, \varphi, B)$, where $B = \mathrm{Ker}\,\varphi$ is generated by the elements α_i and β. The commutator subgroup B' of the kernel B is generated by the element γ and is a cyclic group of order p. The group F can obviously be embedded in the quotient group G/B' (namely, the images of g_i in G/B' generate a group isomorphic to F), therefore the Abelian embedding problem $(K/k, G/B', B/B')$ associated with problem (1) is semidirect and solvable.

PROPOSITION 4.1.3. *The embedding problem* (1) *indicated above is unsolvable.*

We put $A = \langle \gamma, \alpha_1, \ldots, \alpha_\nu \rangle$; A is the Frattini subgroup of G and is contained in B, whence the extension $1 \longrightarrow B/A \longrightarrow G/A \longrightarrow F \longrightarrow 1$ is direct. Therefore, any solution of the problem $(K/k, G/A, B/A)$ is of the form $L_x = K \otimes k(\sqrt[p]{x})$, where $x \in k^*$. Here we assume that $\sqrt[p]{x}^{\overline{\beta}} = \sqrt[p]{x}\zeta$, $\sqrt[p]{x}^{\overline{g}_i} = \sqrt[p]{x}$, where $\overline{\beta}$ and $\overline{g}_i$ are images of β and g_i in G/A.

We try to clarify the properties of $x \in k^*$ for which the embedding problem $(L_x/k, G/B', A/B')$ is solvable, where $B' = \langle \gamma \rangle$ is the commutator subgroup of B. It is clear that $(L_x/k, G/B', A/B')$ is solvable for $x = 1$ (because $(K/k, G/B', B/B')$ is a semidirect problem). Consider the characters $\chi_i \in \mathrm{Hom}_{G/B'}(A/B', K^*)$ $(i = 1, \ldots, \nu)$ given by the formulas $\chi_i(\alpha_i) = \zeta$ and $\chi_i(\alpha_j) = 1$, $j \neq i$. The Brauer problems corresponding to them are of the form $(L_x/k, G/\mathrm{Ker}\,\overline{\chi}_i, A/\mathrm{Ker}\,\overline{\chi}_i)$, where $\overline{\chi}_i$ is an extension of χ_i to A. Let S_i be a subgroup of $G/\mathrm{Ker}\,\overline{\chi}_i$ generated by the elements $g_1, \ldots, g_{i-1}, g_{i+1}, \ldots, g_\nu$. The group S_i is normal in $G/\mathrm{Ker}\,\overline{\chi}_i$ and intersects $A/\mathrm{Ker}\,\overline{\chi}_i$ only by the unit element. The image of S_i in F is the subgroup $F_i = \langle f_1, \ldots, f_{i-1}, f_{i+1}, \ldots, f_\nu \rangle$ of F and $L_x^{F_i} = k(\sqrt[p]{a_i}, \sqrt[p]{x}) = L_i$. By descent with respect to S_i, we reduce every Brauer problem to equivalent Brauer problems of the form $(L_i/k, H_i, A/\mathrm{Ker}\,\overline{\chi}_i)$, where H_i is a group of order p^3 and of period p with generators $\overline{\alpha}_i$, $\overline{\beta}$, $\overline{g}_i$; here, $\overline{\alpha}_i$ is a central element and $\overline{\beta}^{-1}\overline{g}_i\overline{\beta} = \overline{g}_i\overline{\alpha}_i$. The component of the crossed product $(H_i \times L_i)e$, where e is an idempotent element corresponding to the homomorphism $\alpha_i \longrightarrow \zeta$, is a tensor product of algebras over k generated by the pairs $(\overline{\beta}e, \sqrt[p]{x})$ and $(\overline{g}_i\sqrt[p]{x^{-1}}e, \sqrt[p]{a_i})$. The first of them is a matrix algebra (since $\overline{\beta^p} = 1$), and the second is defined by the Hilbert symbol (x^{-1}, a_i). Therefore, the solvability of the embedding problem $(L_x/k, G/\mathrm{Ker}\,\overline{\chi}_i, A/\mathrm{Ker}\,\overline{\chi}_i)$ is equivalent to the triviality of the Hilbert symbol (x^{-1}, a_i). Since $a_1, \ldots, a_\nu$ are generators of the group k^*/k^{*p}, by the nonsingularity of the Hilbert symbol we obtain $x \in k^{*p}$, i.e., we may assume that $x = 1$. This is a necessary and sufficient condition for the solvability of the problem $(L_x/k, G/B', A/B')$.

Every solution of problem (1) (if it exists) is a solution of the problem $(L_x/k, G, A)$, where L_x is a solution of the direct embedding problem indicated above. Since we have shown that $x \sim 1$, we consider the equivalent problem $(K/k, G_0, A)$ over the field-kernel (recall that G_0 is an Abelian group generated by the elements α_i, g_i, $i = 1, \dots, \nu$). For all characters $\chi \in \operatorname{Hom}(A, K^*)$ with $\chi(\gamma) = 1$, this problem is solvable, because it determines a direct extension. Consider now a character $\psi(\gamma) = \zeta$, $\psi(a_i) = 1$ for $i = 1, \dots, \nu$. The corresponding Brauer problem is of the form $(K/k, G/\operatorname{Ker}\psi, A/\operatorname{Ker}\psi)$. Denote by S_0 the subgroup of $G/\operatorname{Ker}\psi$ generated by $g_i g_j^{-1}$ $(i, j = 1, \dots, \nu)$; S_0 is an elementary Abelian p-group whose image F_0 is generated by $f_i f_j^{-1}$. The F_0-invariant subfield of K is a cyclic extension of k of degree p obtained by adjoining the radical $\sqrt[p]{a_1 a_2 \dots a_\nu}$ to k. By descent with respect to S_0, we reduce the embedding problem $(K/k, G/\operatorname{Ker}\chi, A/\operatorname{Ker}\chi)$ to an equivalent embedding problem in which the extension $k(\sqrt[p]{a_1 a_2 \dots a_\nu})$ of degree p is embedded in a cyclic extension of degree p^2 (because $g_i^p = \gamma \neq 1$). A simple computation shows that the solvability of this problem, as well as the solvability of the original problem (1), is equivalent to the triviality of the Hilbert symbol $(\zeta, a_1 a_2 \dots a_\nu)$. But $\zeta = a_1$, whence $(\zeta, a_i) = 1$ for $i \neq 2$, and $(\zeta, a_2) \neq 1$. Thus, problem (1) is unsolvable.

REMARK. By Theorem 2.9, the compatibility condition is fulfilled for the problem indicated, because the corresponding Abelian problem is semidirect.

§2. Proper solutions of the embedding problem for local fields

As shown in Chapter 3, §6, if the embedding problem with Abelian kernel over an algebraic number field has a positive solution for Galois algebras, then there also exists a solution that is a field. In the sequel we shall prove this fact for the case of a nilpotent kernel (Theorem 5.5.5).

At the same time, for the embedding problems given over extensions of local fields, the solvability in the sense of algebras does not guarantee the solvability in the proper sense (i.e., for fields). This is due to the finiteness of the quotient group k^*/k^{*p}. Therefore, the question of the solvability in the proper sense of the embedding problems for local fields is of significance by itself. In this section we discuss this question in the situation where both the Galois group of a given extension and the group of the required extension are p-groups.

1°. As shown in the previous section (Theorem 4.1.2′), the embedding problem $(K/k, G, \varphi, B)$ under consideration reduces (excluding the exceptional case) to the case where the group $B \cap \Phi(G)$ is contained in the center of the group B. In the case $B \subset \Phi(G)$ the problem amounts to the case of an Abelian kernel and the exceptional case is lacking. We show that in the general case, we may neither ignore the exceptional case nor reduce the problem to a problem with Abelian kernel.

EXAMPLE 1. Let k be the 2-adic number field $\mathbb{Q}_2$. Note that k^*/k^{*2} is a space of dimension 3 over the field with two elements, and as a basis for this space, we can take the classes modulo k^{*2} generated by the numbers -1, 5, 2. The corresponding Hilbert symbols are equal to $(-1,-1) = -1$, $(-1,5) = 1$, $(-1,2) = 1$, $(5,5) = 1$, $(5,2) = -1$, $(2,2) = 1$. Let $K_1 = k(\sqrt{5})$ and $K = K_1(\theta)$ with $\theta^2 = 2\sqrt{5} - 5$. Then K is a normal extension of the field k with Galois group F isomorphic to a cyclic group of order 4. We choose a generator g in F in such a way that $\theta^g = \theta(2 + \sqrt{5})$ (respectively, $\sqrt{5}^g = -\sqrt{5}$ and $\theta^{g^2} = -\theta$).

Let B be a group of order 16 generated by the elements α_1, α_2, β, and $\alpha_1^4 = 1$, $\alpha_2^2 = 1$, $\beta^2 = 1$, $\alpha_1\alpha_2 = \alpha_2\alpha_1$, $\beta^{-1}\alpha_1\beta = \alpha_1^{-1}$, $\beta^{-1}\alpha_2\beta = \alpha_2$. The commutator subgroup B' of the group B consists of the elements $1, \alpha_1^2$.

As G, we take the semidirect extension of the group by F with $\alpha_1^g = \alpha_1\alpha_2$, $\alpha_2^g = \alpha_2$, $\beta^g = \beta\alpha_1$. The group $B_0 = B \cap \Phi(G)$ is generated by the elements α_1 and α_2, and $\widetilde{B} = [B, B_0] = B'$.

Consider the embedding problem $(K/k, G, \varphi, B)$, where φ is the standard mapping of G onto F. Suppose it has a proper solution L. Let $L_1 = L^{B'}$, $L_2 = L^{B_0}$. Then, L_2 is a solution of $(K/k, G/B_0, B/B_0)$, L_1 is a solution of $(L_2/k, G/B', B_0/B')$, and L solves $(L_2/k, G, B_0)$.

The embedding problem $(K/k, G/B_0, B/B_0)$ is given by a direct extension, so $L_2 = k(\sqrt{x})$ with $x \in k^*$. For the embedding problem $(L_2/k, G/B', B_0/B')$ the compatibility condition is fulfilled. For $\chi \in \operatorname{Hom}_{G/B'}(B_0/B', L_2^*)$ we have $\chi(\overline{a}) = \pm 1 \in k$ (here $\overline{\alpha}$ is the image of $\alpha \in B_0$ in the quotient group $\overline{B}_0 = B_0/B'$) because $\overline{B}_0$ is a group of period 2. Thus, $\chi(\alpha_1\alpha_2) = \chi(\alpha_1^g) = \chi(\alpha_1)^g = \chi(\alpha_1)$, whence $\chi(\alpha_2) = 1$. Therefore, $\operatorname{Hom}_{G/B'}(\overline{B}_0, L_2^*)$ is a group of order 2. Let $\chi(\overline{\alpha}_1) = -1$, $\chi(\overline{\alpha}_2) = 1$, $e_\chi = \frac{1-\overline{\alpha}_1}{2} \cdot \frac{1+\overline{\alpha}_2}{2}$ be the corresponding idempotent element. The component $(\overline{G} \times L_2)e_\chi$ of the crossed product contains the subalgebra $(F \times K)e_\chi$ isomorphic to the matrix algebra over k (of order 4), and its centralizer is generated by the elements $\sqrt{x}e_\chi$ and $\overline{\beta}\sqrt{5}e_\chi$ anticommuting with each other. Therefore, the symbol $(x, 5)$ must be trivial, whence $x \sim \pm 1, \pm 5$. Since L_2 is a field, $x \sim -1$ and $x \sim -5$ are the only possibilities. (The fields $L_2' = K(\sqrt{-1})$ and $L_2'' = K(\sqrt{-5})$ are abstractly isomorphic, since 5 is a square in K, but they determine distinct solutions of the embedding problem because in the first case $\sqrt{-1}^g = \sqrt{-1}$ and in the second $\sqrt{-1}^g = -\sqrt{-1}$.) This argument shows that the embedding problem $(K/k, G/B', B/B')$ is solvable in the proper sense. Indeed, for $x = -1$, $x = -5$ the compatibility condition is fulfilled for the problem $(L_2/k, G/B', B_0/B')$, whence it is solvable, and each of the solutions of it is proper.

Now consider the problem $(L_2/k, G, B_0)$. For $x \sim -1$, the mapping $\chi \colon B_0 \to L_2^*$ given by the formulas $\chi(\alpha_1) = i = \sqrt{-1}$, $\chi(\alpha_2) = 1$ is a G-operator mapping. Let $e_\chi = \frac{1}{8}(1 - \alpha_1 i - \alpha_1^2 + \alpha_1^3 i)(1 + \alpha_2)$ be the corresponding idempotent, so that $\alpha_1 e_\chi = ie_\chi$, $a_2 e_\chi = e_\chi$. The algebra $(G \times L_2)e_\chi$ is the tensor product of the matrix algebra $(F \times K)e_\chi$ and the algebra generated over k by the anticommuting elements ie_χ and $\beta\theta(i - 2 - \sqrt{5})e_\chi$. Since $(\beta\theta(i - 2 - \sqrt{5}))^2 = \theta^2((2 + \sqrt{5})^2 + 1) = (2\sqrt{5} - 5)(10 + 4\sqrt{5}) = -10$, the invariant of the algebra $(G \times L_2)e_\chi$ equals $(-1, -10) = -1$. Thus, for $x \sim -1$ the problem $(L_2/k, G, B_0)$ is unsolvable.

Now let $x \sim -5$. In this case, the homomorphism $\chi \colon B_0 \to L_2^*$ defined by $\chi(\alpha_1) = i = \sqrt{-1} = \sqrt{5}/\sqrt{-5}$, $\chi(\alpha_2) = 1$ is a G-operator mapping. Then the centralizer of the algebra $(F \times K)e_\chi$ in the algebra $(G \times L_2)e_\chi$ is generated over k by the anticommuting elements $\sqrt{-5}e_\chi$ and $\beta(1 + i)e_\chi$. But $(\beta(1 + i)e_\chi)^2 = 2e_\chi$, and the Hilbert symbol $(-5, 2) = (-1, 2)(5, 2) = -1$. Thus, the embedding problem $(K/k, G, \varphi, B)$ has no proper solutions, while the problem $(K/k, G/\widetilde{B}, B/\widetilde{B})$ is solvable in the proper sense. Thus, the exceptional case of Theorem 4.1.2′ in the previous section cannot be removed.

Now we show that the solvability in the proper sense of $(K/k, G, B)$ and the solvability in the proper sense of $(K/k, G/B', B/B')$ (where B' is the commutator subgroup of B) are also not equivalent for p odd.

EXAMPLE 2. Let k be a local field, and let $p \neq 2$ be the characteristic of its residue field. Suppose k contains a primitive pth root of unity. As in the example of the previous section, we see that the group k^*/k^{*p} is a space over the field with p elements. Let $e_1, e_2, \dots, e_\nu$ be a basis of it, where ν is even. We again choose the numbers of the generators in such a way that only the Hilbert symbols (e_1, e_2), $(e_3, e_4), \dots, (e_{\nu-1}, e_\nu)$ are nontrivial. Let the numbers $a_i \in k$ be preimages of the basis elements e_i. As K we consider the field obtained from k by adjoining the radicals $\sqrt[p]{a_3}, \sqrt[p]{a_4}, \dots, \sqrt[p]{a_\nu}$. Then the Galois group $F = \mathrm{Gal}(K/k)$ is an elementary Abelian p-group with $\nu - 2$ generators $g_3, g_4, \dots, g_\nu$, and we assume that $\sqrt[p]{a_i}^{g_j} = \sqrt[p]{a_i}$ for $j \neq i$ and $\sqrt[p]{a_i}^{g_i} = \sqrt[p]{a_i}\zeta$, where ζ is a primitive pth root of unity.

Now we construct the group G. As kernel B, we consider the group of period p with generators α_{ij}, β_i, where $i = 1, 2$; $j = 3, 4, \dots, \nu$; α_{ij} are central elements; $[\beta_1, \beta_2] = \gamma \neq 1$, and γ is a central element that cannot be expressed via α_{ij} (whence $(B : 1) = p^{3+2(\nu-2)}$).

Let G be a semidirect extension of the kernel B by the group $F = \mathrm{Gal}(K/k)$; here γ and α_{ij} are central elements of G and $g_j^{-1}\beta_i g_j = \beta_i\alpha_{ij}$ ($i = 1, 2$; $j = 3, 4, \dots, \nu$). It is clear that $A = \Phi(G)$ is generated by the elements γ and α_{ij} and is contained in B, and the commutator subgroup B' is generated by γ.

Consider the embedding problem $(K/k, G, \varphi, B)$, where φ is the natural mapping $G \to G/B = \mathrm{Gal}(K/k)$. We are interested in proper solutions of this problem (a semidirect problem is solvable in the sense of algebras). The associated problem $(K/k, G/A, B/A)$ with $A = \Phi(G) \subset B$ is given by a direct extension, whence each of its solutions is $L = K(\sqrt[p]{x_1}, \sqrt[p]{x_2})$, where $x_1, x_2 \in k^*$. This solution is proper if and only if the images of the elements x_1, x_2 in the space k^*/k^{*p} are linearly independent over the subspace $(k^* \cap K^{*p})/k^{*p}$; in other words, the elements x_1, x_2, $a_3, \dots, a_\nu$ must generate multiplicatively the entire group k^*.

Now consider the embedding problem $(L/k, G, A)$ with $L = K(\sqrt[p]{x_1}, \sqrt[p]{x_2})$. As is easily seen, for the character χ_{ij} ($i = 1, 2$; $j = 3, 4, \dots, \nu$) defined by $\chi_{ij}(\alpha_{ij}) = \zeta$, where ζ is a fixed primitive pth root of unity, and trivial on the other generators α_{kl} ($(k, l) \neq (i, j)$) and on γ, the invariant of the corresponding Brauer problem is equal to the Hilbert symbol (a_j, x_i). Thus, the embedding problem $(L/k, G/B', A/B')$ with $B' = [B, B] = \langle\gamma\rangle$ is solvable if and only if all symbols (a_j, x_i) are trivial. We can easily meet this requirement by setting, for example, $x_1 = a_1$, $x_2 = a_2$, and in this case the solution is proper. Thus, the problem $(K/k, G/B', B/B')$ has a proper solution.

Now consider the character $\psi \in \mathrm{Hom}_G(A, L^*)$ defined nontrivially on the element $\gamma \in B'$ and trivially on all of the α_{ij}. The invariant of the corresponding problem is equal to (x_1, x_2). Now let $(a_j, x_i) = 1$ and $(x_1, x_2) = 1$ ($i = 1, 2$; $j = 3, 4, \dots, \nu$). The conditions $(a_j, x_i) = 1$ mean that the decomposition of the element $\overline{x}_i$ in the basis $e_1, \dots, e_\nu$ (where $\overline{x}_i$ is the image of x_i in the space k^*/k^{*p}) has zero coordinates at $e_3, e_4, \dots, e_\nu$. Therefore, the elements x_1 and x_2 are generated by the elements a_1 and a_2 over k^{*p}. Let $x_1 \equiv a_1^{l_1}a_2^{m_1} \pmod{k^{*p}}$ and $x_2 \equiv a_1^{l_2}a_2^{m_2} \pmod{k^{*p}}$. Since the symbol (a_1, a_2) is not trivial, the condition $(x_1, x_2) = 1$ implies $\begin{vmatrix} l_1 & m_1 \\ l_2 & m_2 \end{vmatrix} \equiv 0 \pmod p$, i.e., $x_2 \equiv x_1^n \pmod{k^{*p}}$, where n is an integer. Then L is not a proper solution. Thus, the embedding problem indicated above is unsolvable in the proper sense, but the corresponding Abelian problem has a proper solution.

2°. As we have seen, Theorem 4.1.2$'$ cannot be applied whenever $p = 2$ and the field K does not contain $\sqrt{-1}$. The following theorem may serve as a substitute.

THEOREM 4.2.1. *The embedding problem $(K/k, G, \varphi, B)$ has a proper solution if the associated problem $(K/k, G/B_0', B/B_0')$, where B_0' is the commutator subgroup of $B_0 = B \cap \Phi(G)$, has a proper solution.*

Let L be a proper solution of the associated problem, and $L_1 = L^{B_0/B_0'}$. Then $\mathrm{Gal}(L_1/k) \approx G/B_0$, and we consider the embedding problem $(L_1/k, G, B_0)$. Since $B_0 \subset \Phi(G)$, Theorem 4.1.2 of the previous section can be applied to this problem, whence it is equivalent to the associated problem $(L_1/k, G/B_0', B/B_0')$, and L is a proper solution of the latter. Thus, the problem $(K/k, G, B)$ has a proper solution. Therefore, in considering the embedding problems for local fields and p-groups in the proper sense, one can reduce the problem to the case where $B_0 = B \cap \Phi(G)$ is an Abelian group.

In the situation where Theorem 4.1.2$'$ can be applied, we may significantly simplify the problem.

THEOREM 4.2.2. *Let $(K/k, G, \varphi, B)$ be an embedding problem with local fields K and k, where G is a p-group and $B_0 = B \cap \Phi(G)$ is contained in the center of B. Let C be the intersection of the kernels of all characters $\chi \in \mathrm{Hom}_G(B_0, K^*)$. Then the embedding problem* (1) *is solvable in the proper sense if and only if the associated embedding problem $(K/k, G/C, B/C)$ has a proper solution.*

PROOF. Let L be a proper solution of this problem, and let $L_1 = L^{B_0/C}$. Consider the embedding problem $(L_1/k, G, B_0)$. Its solvability (and thus solvability in the proper sense) is equivalent to the solvability of all associated Brauer problems $(L_1/k, G/\mathrm{Ker}\,\chi, B_0/\mathrm{Ker}\,\chi)$, where $\chi \in \mathrm{Hom}_G(B_0, L_1^*)$. For $\alpha \in B_0$, $\beta \in B$, we have $\chi(\alpha^\beta) = \chi(\alpha)^\beta$, but B_0 is a central subgroup in B, whence $\chi(\alpha)^\beta = \chi(\alpha)$. Thus, $\chi(\alpha) \in K^*$ independently of whether or not the field L_1 contains a cyclotomic extension of K. Therefore, $\mathrm{Hom}_G(B_0, L_1^*) = \mathrm{Hom}_G(B_0, K^*)$. For $\gamma \in C$ we have $\chi(\gamma) = 1$ for any χ, therefore the Brauer problem corresponding to each $\chi \in \mathrm{Hom}_G(B_0, L_1^*)$ is associated with the embedding problem $(L_1/k, G/C, B_0/C)$ for which L is a solution. The theorem is proved. □

REMARK. The group B_0 is an F-module, $F = \mathrm{Gal}(K/k)$, because B_0 is contained in the center of B. Therefore,

$$\mathrm{Hom}_G(B_0, K^*) = \mathrm{Hom}_F(B_0, K^*) \approx \mathrm{Hom}_Z(B_0/C, K^*).$$

Moreover, in the group B_0/C the relations $\beta_0^f = \beta_0^{l(f)}$ hold, where $\beta_0 \in B_0/C$, $f \in F$, and the numbers $l(f)$ are defined by $\zeta^f = \zeta^{l(f)}$, where ζ is a primitive root of the maximal p-power of 1 contained in K (obviously, $l(f) \equiv 1 \pmod p$). The theorem proved reduces the embedding problem to the case where the elements f act on all elements $\beta_0 \in B_0$ in the same way. In particular, each subgroup of B_0/C is normal in G/C.

3°. Now we find conditions for solvability in the proper sense for the embedding problem in terms of the consistency of a system of equations. We assume that we have made the necessary reduction, i.e., for the embedding problem $(K/k, G, \varphi, B)$, the group $B_0 = B \cap \Phi(G)$ is contained in the center of B and every homomorphism $B_0 \to K^*$ is a G-operator homomorphism (hence an F-operator homomorphism, where $F = \mathrm{Gal}(K/k)$). Since G/B_0 is a direct product of the groups B/B_0 and F, there exists a subgroup G_0 of G such that $\varphi(G_0) = F$ and $G_0 \cap B = B_0$ (the group G_0 is not uniquely determined, but we fix one of them). This enables us to regard the group B as an F-operator group by setting $\beta^f = \overline{f}^{-1}\beta\overline{f}$, where $\beta \in B$, $f \in F$, and $\overline{f}$ is a representative of f in G_0 (this action is well defined because $B_0 = B \cap \Phi(G)$ is contained in the center of B). Let $\beta_1, \beta_2, \dots, \beta_l$ be the elements of B whose images in the quotient group B/B_0 form a basis for this group (regarded as a space over the field with p elements).

Any solution of the direct embedding problem $(K/k, G/B_0, B/B_0)$ is of the form $K \otimes k(\sqrt[p]{x_1}, \dots, \sqrt[p]{x_l})$, where $x_i \in k^*$, and the adjoined radicals are transformed by the elements from G (from G/B_0, to be more precise) by the formulas $\sqrt[p]{x_i}^h = \sqrt[p]{x_i}$ for $h \in G_0$, $\sqrt[p]{x_i}^{\beta_j} = \sqrt[p]{x_i}\zeta_0^{\delta_{ij}}$, where ζ_0 is a fixed pth root of 1, δ_{ij} is the Kronecker symbol. This solution is proper if and only if the images of the elements $x_1, \dots, x_l$ in the space k^*/k^{*p} are linearly independent over the subspace $(k^* \cap K^{*p})/k^{*p}$.

In order to solve the problem $(K/k, G, \varphi, B)$ in the proper sense, we need to find a proper solution L of the intermediate direct problem such that the embedding problem $(L/k, G, B_0)$ is solvable, whence each of the solutions of it is proper. Therefore, we need to clarify for which $x_1, \dots, x_l \in k^*$ all associated Brauer problems $(L/k, G/\operatorname{Ker}\chi, B_0/\operatorname{Ker}\chi)$ are solvable, where $\chi \in \operatorname{Hom}_G(B_0, L^*)$. Since for such χ we have $\chi(\alpha) = \chi(\alpha^\beta) = \chi(\alpha)^\beta$, where $\alpha \in B_0$, $\beta \in B$, it follows that $\chi \in \operatorname{Hom}(B_0, K^*) = \widehat{B}_0$ independent of whether or not the field L contains a cyclotomic extension of K.

Let the elements $\gamma(i, f) \in B_0$ be defined by $\beta_i^f = \beta_i\gamma(i, f)$ ($\beta_1, \dots, \beta_l$ are generators of B over B_0, $f \in F$). Then $\gamma(i, f_1f_2) = \gamma(i, f_1)^{f_2}\gamma(i, f_2)$. We set $u(i, f, \chi) = \chi(\gamma(i, f))$, $\chi \in \widehat{B}_0$. Then $u(i, f, \chi)$ is a one-dimensional cocycle from $Z^1(F, K^*)$, and, by Speiser's theorem, we have $u(i, f, \chi) = v(i, \chi)^{1-f}$, where $v(i, \chi) \in K^*$. Let e_χ be a minimal idempotent of the group algebra $K[B_0]$ corresponding to the character χ: $e_\chi = (B_0 : 1)^{-1}\sum_{\alpha \in B_0} \alpha\chi(\alpha^{-1})$; so, $\alpha e_\chi = \chi(\alpha)e_\chi$ for $\alpha \in B_0$ and $e_\chi^f = e_\chi$ for $f \in F$. Now we have $(\beta_i v(i, \chi)e_\chi)^f = \beta_i\gamma(i, f)v(i, \chi)^f e_\chi = \beta_i u(i, f, \chi)v(i, \chi)^f e_\chi = \beta_i v(i, \chi)e_\chi$, whence the elements $\beta_i v(i, \chi)e_\chi$ are invariant with respect to F. Thus, the elements $c(i, \chi) = \chi(\beta_i^p)v(i, \chi)^p \in K^*$ are also invariant under F; therefore, $c(i, \chi) \in k^*$. The elements $v(i, \chi)$ are defined uniquely up to factors from k^*, therefore the elements $c(i, \chi)$ are defined uniquely up to factors from k^{*p}. For simplicity, the images of these elements in the space k^*/k^{*p} are also denoted by $c(i, \chi)$.

Now compute the invariants of the algebras $(G \times L)e_\chi$, where $G \times L$ is the crossed product of the group G and the field $L = K \otimes_k k(\sqrt[p]{x_1}, \dots, \sqrt[p]{x_l})$ (since for $\alpha \in \operatorname{Ker}\chi$ we have $\alpha e_\chi = e_\chi$, these components are isomorphic to the algebras $(G/\operatorname{Ker}\chi) \times L$). The algebra $(G_0 \times K)e_\chi$, which is independent of the choice of x_i, is isomorphic to the crossed product of K and the group $F = \mathrm{Gal}(K/k)$ with factors $\chi(\alpha(f_1, f_2))$, where $\alpha(f_1, f_2)$ is the cocycle of the extension G_0. Let inv_χ be the invariant of this algebra in the multiplicative notation. The algebra generated over k by the elements $\sqrt[p]{x_i}$ and $\beta_i v(i, \chi)e_\chi$ ($i = 1, \dots, l$) is the centralizer of the algebra $(G_0 \times K)e_\chi$ in $(G \times L)e_\chi$. The commutators of the elements β_i and β_j are elements from B_0 of period g and are central in G. If any one of these commutators is different from 1 and is not contained in the kernel of χ, the elements $\beta_i v(i, \chi)e_\chi$ and $\beta_j v(j, \chi)\sqrt[p]{x_i}^m e_\chi$ can be made commuting, and the exponents $m = m(i, j, \chi)$ will be determined modulo p. The centralizer of the algebra $(G_0 \times K)e_\chi$ in $(G \times L)e_\chi$

proves to be represented as a tensor product of cyclic algebras over k generated by the elements $\sqrt[p]{x_i}$ and $\beta_i v(i,\chi)\prod_{k<i}\sqrt[p]{x_k}^{m(k,i,\chi)}e_\chi$, each of these algebras is the algebra of generalized quaternions whose invariant is equal to the invariant of the cyclic algebra $(x_i, c(i,\chi))\prod_{k<i}x_k^{m(k,i,\chi)}$. Thus, the invariant of the algebra $(G\times L)e_\chi$ is equal to

$$\operatorname{inv}_\chi \prod_{i=1}^{l}(x_i, c(i,\chi))\prod_{k<i}(x_i, x_k)^{m(k,i,\chi)}.$$

The solvability condition for the problem $(L/k, G, B_0)$, where L is a field, is equivalent to the compatibility condition for this problem, i.e., to the fact that all the invariants indicated are trivial. Thus, we have obtained the following theorem.

THEOREM 4.2.3. *The problem $(K/k, G, \varphi, B)$ has a proper solution if and only if there exist elements $x_1, x_2, \dots, x_l \in k^*$ (l is the number of generators of B/B_0) such that*

(a) *the images of $x_1, \dots, x_l$ in the space k^*/k^{*p} are linearly independent over the space $(k^*\cap K^{*p})/k^{*p}$;*

(b) *for any $\chi\in\widehat{B}_0$, the following relations hold:*

$$\operatorname{inv}_\chi \prod_{i=1}^{l}(x_i, c(i,\chi))\prod_{k<i}(x_i, x_k)^{m(k,i,\chi)} = 1.$$

It is easy to see that if these products are equal to 1 for two characters $\chi_1,\chi_2\in\widehat{B}_0$, then this also occurs for the product $\chi_1\chi_2$. Therefore, it is sufficient to verify these relations on the set of generators of the group $\widehat{B}_0$. Moreover, if χ acts trivially on the commutator subgroup of B, then all the exponents $m(k,i,\chi)$ are equal to zero, whence the equations indicated above are linear.

4°. Now consider the case where the kernel B is Abelian. Moreover, we assume that the initial embedding problem $(K/k, G, \varphi, B)$ has a solution (not necessarily proper). Then there exist $a_1, \dots, a_l$ (not necessarily linear independent over the space $k^*\cap K^{*p}$) for which $\operatorname{inv}_\chi\prod_{i=1}^{l}(a_i, c(i,\chi)) = 1$ with $\chi\in\widehat{B}_0$. We set $x_i = a_i y_i$. Then the solvability of the problem $(K/k, G, \varphi, B)$ in the proper sense is equivalent to the existence of $y_1, \dots, y_l \in k^*$ such that

(a) the images of the elements $a_i y_i$ in the space k^*/k^{*p} are linearly independent over the subspace $(k^*\cap K^{*p})/k^{*p}$;

(b) for any $\chi\in\widehat{B}_0$ the following relation is valid:

$$\prod_{i=1}^{l}(y_i, c(i,\chi)) = 1.$$

We denote the number of generators of a p-group H by $d(H)$. The symbol $d(k^*)$ designates the dimension of the space k^*/k^{*p} (as is known, $d(k^*) = d(\overline{F})$, where $\overline{F}$ is the Galois group of the maximal p-extension of k).

The following theorem provides a sufficient condition for solvability of the embedding problem in the proper sense.

THEOREM 4.2.4. *Let the embedding problem $(K/k, G, \varphi, B)$, where the kernel B is Abelian, have a positive solution (in the sense of algebras). Moreover, if*

$d(k^*) \geq d(G)$, $d(k^*) > d(G/B) + d(\widehat{B}_0)$, then this problem is also solvable in the proper sense.

PROOF. Now we use the additive notation for the elements of the space k^*/k^{*p} and for norm residue symbols. Then we can reformulate the conditions indicated above concerning solvability in the proper sense as follows. There exist vectors y_1, $\dots$, y_l belonging to the space $P = k^*/k^{*p}$ and such that

(a) the vectors $a_1 + y_1$, $\dots$, $a_l + y_l$, where a_i are certain vectors of P, are linearly independent with respect to the subspace $Q = (k^* \cap K^{*p})/k^{*p}$ (here $l = d(G) - d(F)$);

(b) for every χ of the set of generators of $\widehat{B}_0$, we have $\sum_{i=1}^{l}(c(i,\chi), y_i) = 0$; here $(\cdot,\cdot)$ is an alternating bilinear nondegenerate form on P with values in $\mathbb{Z}/p\mathbb{Z}$.□

Let e_1, $\dots$, e_m be a basis of the subspace Q and let e_1, $\dots$, e_m, e_{m+1}, $\dots$, e_n be a basis of the space P (hence $m = d(F)$, $n = d(k^*)$). We denote by f_1, $\dots$, f_n the dual basis of P defined by the relations $(f_i, e_j) = \delta_{ij}$ (Kronecker's symbol). With the elements $c(1,\chi)$, $\dots$, $c(l,\chi)$ we associate the matrix

$$C^{(\chi)} = \left\| \begin{matrix} c_{11}^{\chi} & c_{12}^{\chi} & \cdots & c_{1l}^{\chi} \\ \cdots & \cdots & \cdots & \cdots \\ c_{n1}^{\chi} & c_{n2}^{\chi} & \cdots & c_{nl}^{\chi} \end{matrix} \right\|,$$

whose columns are the coordinates of the vector $c(i,\chi)$ in the basis $\{f_j\}$. The matrix $C^{(\chi)}$ splits into the submatrices $\left\| \begin{smallmatrix} C_1^{(\chi)} \\ C_2^{(\chi)} \end{smallmatrix} \right\|$, where $C_1^{(\chi)}$ is the matrix composed of the first m rows of $C^{(\chi)}$. In this way we construct the matrices $A = \left\| \begin{smallmatrix} A_1 \\ A_2 \end{smallmatrix} \right\|$ and $Y = \left\| \begin{smallmatrix} Y_1 \\ Y_2 \end{smallmatrix} \right\|$ composed of the elements a_1, $\dots$, a_l, and of the unknowns y_1, $\dots$, y_l in the basis $\{e_j\}$, respectively. The relative linear independence of the elements $a_1 + y_1$, $\dots$, $a_l + y_l$ means that the matrix $A_2 + Y_2$ has rank l. The condition $\sum_{i=1}^{l}(c(i,\chi), y_i) = 0$ means that $\sum c_{ji}^{\chi} y_{ji} = 0$ ($i = 1, \dots, l$; $j = 1, \dots, n$) or $\operatorname{Tr}(C^{(\chi)}Y^T) = 0$. Among these equations there are at most $t = d(\widehat{B}_0)$ linearly independent equations. Therefore, the dimension of the space of matrices Y satisfying $\operatorname{Tr}(C^{(\chi)}Y^T) = 0$ is at least $ln - t$. The matrices Y_2 (the lower parts of Y) also form a subspace, and its dimension is at least $l(n-m) - t$.

The condition $\operatorname{rank}(A_2 + Y_2) = l$ implies that $n - m \geq l$, i.e., $d(k^*) \geq d(F) + d(B/B_0) = d(G)$. Thus, the condition $d(k^*) \geq d(G)$ is necessary in order that our problem have a proper solution. (This, however, may be seen directly.)

To complete the proof of the theorem, we need an assertion that will be proved below. In the space of $\mu \times \nu$ matrices over an arbitrary field, let a linear variety be given such that each matrix in this variety has rank not greater than ρ. Then the dimension of this variety does not exceed $\rho \max(\mu, \nu)$.

Now we assume that each matrix $A_2 + Y_2$ is of rank less than l (i.e., not greater than $l-1$). Then the dimension of the linear variety $A_2 + \mathcal{Y}_2$ (where $\mathcal{Y}_2$ is the space of matrices Y_2) is not greater than $(l-1)\max(l, n-m)$. Since $l \leq n-m$, we have $\dim(A_2 + \mathcal{Y}_2) \leq (l-1)(n-m)$. Hence $l(n-m) - t \leq \dim(A_2 + \mathcal{Y}_2) \leq (l-1)(n-m)$, i.e., $n \leq m + t$ or $d(k^*) \leq d(F) + d(\widehat{B}_0)$, which contradicts the assumptions of the theorem. It remains to prove the assertion about the dimension of the variety.

Let M be a given linear variety of matrices, let A be a matrix of maximal rank ρ from M. There exist nonsingular matrices C_1 and C_2 such that $I = C_1 A C_2 =$

$\begin{Vmatrix} E_\rho & 0 \\ 0 & 0 \end{Vmatrix}$, where E_ρ is the identity matrix of order ρ. Clearly, it is sufficient to prove the assertion for the variety $M_1 = C_1 M C_2$; let $P_1 = M_1 - I$ be a subspace corresponding to this variety.

Let θ be a homomorphism that maps each matrix $X \in P_1$ to its upper left square of order ρ. Then $\dim M = \dim M_1 = \dim P_1 = \dim(\operatorname{Im}\theta) + \dim(\operatorname{Ker}\theta) \leq \rho^2 + \dim \operatorname{Ker}\theta$. We estimate the second summand. Let $Y = \begin{Vmatrix} 0 & Y_1 \\ Y_2 & Y_3 \end{Vmatrix} \in \operatorname{Ker}\theta$. Then $I + Y = \begin{Vmatrix} E_\rho & Y_1 \\ Y_2 & Y_3 \end{Vmatrix} \in M_1$ and $\operatorname{rank}(I+Y) \leq \rho$. We multiply the matrix $I+Y$ by the nonsingular square matrix $\begin{Vmatrix} E_\rho & -Y_1 \\ 0 & E_\lambda \end{Vmatrix}$, where $\lambda = \nu - \rho$. Then the rank of the matrix $\begin{Vmatrix} E_\rho & 0 \\ Y_2 & Y_3 - Y_2Y_1 \end{Vmatrix} \leq \rho$, whence $Y_3 = Y_2 Y_1$. For the two matrices $Y, Z \in \operatorname{Ker}\theta$, the sum $Y + Z \in \operatorname{Ker}\theta$, whence we obtain

$$(*) \qquad Y_3 + Z_3 = (Y_2 + Z_2)(Y_1 + Z_1), \qquad Y_2 Z_1 + Z_2 Y_1 = 0.$$

With each matrix $Y \in \operatorname{Ker}\theta$ we associate its right upper corner Y_1. The mapping $Y \longrightarrow Y_1$ is a homomorphism θ_1. For $Y \in \operatorname{Ker}\theta_1$ we have $Y = \begin{Vmatrix} 0 & 0 \\ Y_2 & 0 \end{Vmatrix}$, and for any matrix $Z = \begin{Vmatrix} 0 & Z_1 \\ Z_2 & Z_3 \end{Vmatrix} \in \operatorname{Ker}\theta$, we obtain $Y_2 Z_1 = 0$ by $(*)$. Therefore, each column of the matrix Z_1 is orthogonal to each row of the matrix Y_2. Let $\varkappa$ be the maximal number of linearly independent rows among all the rows of the matrices Y_2, where $Y \in \operatorname{Ker}\theta_1$. Then the number of linearly independent columns of the matrix Z_1 does not exceed $\rho - \varkappa$, whence $\dim(\operatorname{Im}\theta_1) \leq (\rho - \varkappa)(\nu - \rho)$. At the same time, $\dim(\operatorname{Ker}\theta_1) \leq \varkappa(\mu - \rho)$. Therefore, $\dim(\operatorname{Ker}\theta) = \dim(\operatorname{Im}\theta_1) + \dim(\operatorname{Ker}\theta_1) \leq (\rho - \varkappa)(\nu - \rho) + \varkappa(\mu - \rho)$, where $0 \leq \varkappa \leq \rho$. Hence $\dim(\operatorname{Ker}\theta) \leq \rho \max(\nu - \rho, \mu - \rho) = \rho \max(\mu, \nu) - \rho^2$. Thus, $\dim P_1 \leq \rho \max(\mu, \nu)$, and the assertion is proved.

We show that in Theorem 4.2.4 the inequality $d(k^*) > d(F) + d(\widehat{B}_0)$ cannot be weakened in the general case. For example, let $d(k^*) = d(G)$ and $d(\widehat{B}_0) = d(k^*) - d(F)$. It is not difficult to construct a semidirect extension of B by F for which the equations $\prod_i (c(i,\chi), y_i) = 1$ turn into the equations $(a_0, y_i) = 1$ $(i = 1, \dots, l)$, where $a_0 \in k^* \cap K^{*p}$. Moreover, if a_0 is orthogonal to each $z \in k^* \cap K^{*p}$, then y_1, $\dots$, y_l satisfying the conditions $(a_0, y_i) = 1$ cannot be linearly independent over the space $(k^* \cap K^{*p})/k^{*p}$. Thus, the embedding problem has no proper solution.

The problem is considerably simplified if every character $\chi \in \widehat{B}_0$ can be extended to an F-operator character of the group B. In this case, all the elements $c(i,\chi)$ constructed above are equal to unity and, thus, the condition $d(k^*) \geq d(G)$ is necessary and sufficient in order that a solvable embedding problem be solvable in the proper sense.

5°. We have not succeeded in obtaining sufficient solvability conditions from analysis of the system of equations obtained in Theorem 4.2.3 for a non-Abelian kernel (although it is more realistic that Theorem 4.2.4 is valid in the general case). However, we may justify rougher sufficient conditions.

THEOREM 4.2.5. *Let the compatibility condition be fulfilled for the embedding problem $(K/k, G, \varphi, B)$ (where the fields are local, G is a p-group). If $d(k^*) \geq d(G) + d(\widehat{B}_0) + \delta$, then this problem has a proper solution. Here $\widehat{B}_0 = \operatorname{Hom}_G(B_0, K^*)$, where $B_0 = B \cap \Phi(G)$. Moreover, $\delta = 1$ if $p = 2$ and $\sqrt{-1} \notin K$, and $\delta = 0$ in the other cases.*

We proceed by induction on the order of the kernel B. When B is an Abelian group, the compatibility condition guarantees solvability (in the sense of algebras), and the inequality $d(k^*) \geq d(G) + d(\widehat{B}_0)$ is stronger than $d(k^*) > d(F) + d(\widehat{B}_0)$, which guarantees embeddability by Theorem 4.2.4.

First let $(B : B_0) > p$. The embedding problem $(K/k, G/B', B/B')$ is solvable in the proper sense. Let L be a solution of this problem. As before, we can choose a subgroup $B_1 \subset B$ such that $B_0 \subset B_1$, $(B : B_1) = p$, and $L_1 = L^{B_1/B_1'}$ is not a cyclotomic extension of K. Then $\mathrm{Hom}_G(B_1, L_1^*) = \mathrm{Hom}_G(B_1, K^*)$, whence $B' = [B, B_1] \subset \mathrm{Ker}\,\chi$, where $\chi \in \mathrm{Hom}_G(B, K^*)$. The compatibility conditions and the inequality indicated in the theorem are fulfilled for the embedding problem $(L_1/k, G, B_1)$. Therefore, this problem and hence the initial problem are solvable in the proper sense.

It remains to consider the case $(B{:}B_0) = p$. However, in this case $\widetilde{B} = [B, B_0] = B'$. Therefore, if $p \neq 2$, or $p = 2$ but $\sqrt{-1} \in K$, then the problem $(K/k, G, B)$ reduces (by Theorem 4.2.4) to the problem with smaller kernel, which is solvable by the inductive assumption. Thus, we need to investigate only the case $p = 2$, $\sqrt{-1} \notin K$, and $(B : B_0) = p = 2$.

We prove that among the solutions of the associated embedding problem $(K/k, G/B', B/B')$, there exists a proper solution L for which $L^{B_0/B'}$ is not a cyclotomic extension of K, i.e., it does not contain $\sqrt{-1}$. To this end, we turn to the proof of Theorem 4.2.4. Let, as before, P be the space k^*/k^{*2} written additively, let Q be the subspace $(k^* \cap K^{*2})/k^{*2}$. We denote by $\overline{Q}$ the subspace generated by Q and the class $(-1) \bmod k^{*2}$, and let $k(\sqrt{a})$ be a subfield belonging to the group B_0/B' of some proper solution of the problem $(K/k, G/B', B/B')$ (which exists because B/B' is Abelian). The arguments of Theorem 4.2.4 imply that a proper solution L of the indicated Abelian problem, for which $\sqrt{-1} \notin L^{B_0/B'}$, exists if and only if there is an element $y \in P$ such that for any $\chi \in \mathrm{Hom}_{G/B'}(B_0/B', K^*)$, the forms $(c(\chi), y)$ are trivial, and also $\bar{a} + y \notin \overline{Q}$ (where $\bar{a}$ is the class of the element a in P); here $c(\chi) \in P$ and $c(\chi_1\chi_2) = c(\chi_1) + c(\chi_2)$.

Let R be the subspace generated by the elements $c(\chi) \in P$. The conditions $(c(\chi), y) = 0$ mean that $y \in R^\perp$. Clearly, the dimension of R is not greater than $d(\widehat{B}_0)$, whence $\dim R^\perp \geq d(k^*) - d(\widehat{B}_0)$, and, therefore, $\dim R^\perp \geq d(G) + 1 > d(F) + 1 = d(\overline{Q})$. Thus, there exists $y \in R^\perp$ such that $\bar{a} + y \notin \overline{Q}$, and our assertion is proved. Now we are in a position to complete the proof of the theorem.

Let L be the solution of the problem $(K/k, G/B', B/B')$, $L_1 = L^{B_0/B'}$, indicated above. Then for the embedding problem $(L_1/k, G, B_0)$ the assumptions of the theorem are fulfilled (because $\widehat{B}_0 = \mathrm{Hom}_G(B_0, K^*) = \mathrm{Hom}_G(B_0, L_1^*)$), and this problem is solvable in the proper sense. This proves the theorem.

Note that Example 1 in the present section provides an unsolvable (in the proper sense) problem with $d(k^*) = 3$, $d(G) = 2$, $d(\widehat{B}_0) = 1$. Therefore, we cannot eliminate the summand δ in the assumptions of the theorem.

CHAPTER 5

The Embedding Problem with Non-Abelian Kernel for Algebraic Number Fields

The first two sections of this chapter are devoted to the study of the embedding problem with non-Abelian kernel of order p^3 for arbitrary fields, but embedding conditions take a final form only for algebraic number fields. In later sections, for these fields we consider the embedding problem with solvable kernel of odd order and the semidirect problem with nilpotent kernel.

We introduce some notation which will be used in this chapter. For an arbitrary algebraic number field k,

J_k is the idèle group;

C_k is the idèle class group J_k/k^*;

$k_\mathfrak{p}$ is the completion of k at a point $\mathfrak{p}$;

$V_\mathfrak{p}$ is the group of $\mathfrak{p}$-adic units of $k_\mathfrak{p}$;

$V_S = \prod_{\mathfrak{p}\notin S} V_\mathfrak{p}$, $J_{k,S} = J_k/V_S$, $C_{k,S} = C_k/V_S$ for the set S of points of k.

If K is an algebraic extension of the field k, then we also denote by S the set of points of K lying over the points of S. We shall write (m) for the principal divisor corresponding to a number m, A' stands for the group $\mathrm{Hom}(A, \mathbb{Q}/\mathbb{Z})$, where A is an Abelian group.

§1. The embedding problem with non-Abelian kernel of order p^3. I

Groups of order p^3 provide examples of minimal non-Abelian p-groups, thus it is of interest to consider the embedding problems for such kernels as the first step in studying the problems with non-Abelian kernels. In this section we deal with the embedding problems $(K/k, G, \varphi, B)$, where G is also a p-group. Moreover, we assume that the characteristic of k is not equal to p and the field K (and thus k) contains a primitive pth root of 1 (this always occurs for $p = 2$). A specific situation pertaining to the structure of p-extensions of non-Abelian groups of order p^3 allows us to give embedding conditions for such groups in adequate terms. This problem is solved in **[32]** by somewhat different methods.

1°. For each prime number p, there exist exactly two, up to isomorphism, non-Abelian groups of order p^3, and they have different forms for even and odd p. For $p = 2$, these groups are the dihedral group and the quaternion group (generators of them are denoted by α, β):

$\alpha^4 = 1$, $\beta^2 = 1$, $\beta^{-1}\alpha\beta = \alpha^{-1}$ (the dihedral group) (I),

$\alpha^4 = 1$, $\beta^2 = \alpha^2$, $\beta^{-1}\alpha\beta = \alpha^{-1}$ (the quaternion group) (II).

For odd p, these groups are also groups with two generators, and either $\alpha^p = 1$, $\beta^{p^2} = 1$, $\beta^{-1}\alpha\beta = \alpha\beta^p$ (the group of period p^2) (III) or $\alpha^p = \beta^p = [\alpha, \beta]^p = 1$ and $\gamma = [\alpha, \beta]$ is a central element (the group of period p) (IV).

In either case, the commutator subgroup B' of such a group B coincides with the center of B and is the cyclic group of order p.

For the dihedral group the automorphism group is a 2-group (and the requirement that G is a 2-group is unnecessary). In the remaining cases, the full automorphism group $\operatorname{Aut} B$ is not a p-group. However, the automorphisms induced by the elements $g \in G$ form a p-group and are embedded in one of the Sylow p-subgroups of $\operatorname{Aut} B$.

Here we give formulas for the action of the automorphisms of a Sylow p-subgroup of $\operatorname{Aut} B$ on the generators of B. We restrict ourselves to the choice of one of Sylow subgroups because all Sylow subgroups are conjugate and the formulas for the action coincide under a suitable choice of the generators of B.

For the group (I), we have $\alpha^g = \alpha^{\pm 1}$, $\beta^g = \beta\alpha^j$ $(j = 0, 1, 2, 3)$. This is the full automorphism group.

For the group (II), $\alpha^g = \alpha^{\pm 1}$, $\beta^g = \beta\alpha^j$ $(j = 0, 1, 2, 3)$.

For the group (III), $\alpha^g = \alpha\beta^{pm}$, $\beta^g = \alpha^j\beta^{1+pl}$ $(j, l, m = 0, 1, \dots, p-1)$.

For the group (IV), $\alpha^g = \alpha\gamma^m$, $\beta^g = \beta\alpha^j\gamma^l$, $\gamma^g = \gamma$, where $\gamma = [\alpha, \beta]$ $(j, l, m = 0, 1, \dots, p-1)$.

We introduce the following notation:

F is the Galois group of K/k; $F = \operatorname{Im}\varphi$.

G_0 is the subgroup of G, whose elements act on B as inner automorphisms. For $p = 2$, this means that $\beta^g = \beta\alpha^{2j}$ $(j = 0, 1)$, and for $p \neq 2$, $\beta^g = \beta^{1+pl}$ (type III), $\beta^g = \beta\gamma^l$ (type IV). The group G_0 either coincides with G or it has index p in G and thus is normal in G.

The image of G_0 under the homomorphism φ is denoted by F_0.

The subfield of invariant elements in K with respect to the group F_0 is denoted by K_0. If $F_0 = F$, then $K_0 = k$; otherwise, $K_0 = k(\sqrt[p]{c})$, where $c \in k^*$. Moreover, we choose $c \in k^*$ in such a way that (for $F_0 \neq F$) $\sqrt[p]{c^{\varphi(g)}} = \sqrt[p]{c}\zeta^j$ if $\beta^g \equiv \beta\alpha^j$ (mod B'), where ζ is a fixed primitive pth root of 1.

$B' = [B, B]$.

H is the subgroup of G whose elements act on the kernel B as follows: $\alpha^h = \alpha$, $\beta^h = \beta$ or $\alpha^h = \alpha^{-1}$, $\beta^h = \beta\alpha$ for $p = 2$; $\alpha^h = \alpha$, $\beta^h = \beta\alpha^j$ for $p \neq 2$. It is clear that every element $g \in G$ can be mapped to H via multiplication by an element from B, whence $\varphi(H) = F$ and $H \cap B = B'$ (where B' is the commutator subgroup of B). Thus, the extension $1 \to B/B' \to G/B' \to F \to 1$ is semidirect (if $G_0 = G$, this extension is direct).

The subgroup generated by H and by the element $\alpha \in B$ is denoted by $\widetilde{H}$.

The subgroup of B generated by B' and the element α is denoted by A. Thus $A = \widetilde{H} \cap B$, the subgroups $\widetilde{H}/A \approx F$ and B/A commute, and thus the extension

$$1 \to B/A \to G/A \to F \to 1$$

is direct.

We shall solve the embedding problem

$$(K/k, G, \varphi, B) \tag{1}$$

in three stages: at the first stage we solve the direct embedding problem

$$(K/k, G/A, \varphi_2, B/A); \tag{2}$$

at the second stage we try to clarify what solutions L_1 of problem (2) enable us to solve the problem

$$(L_1/k, G/B', \varphi_3, A/B'); \tag{3}$$

at the third stage we find those solutions L_2 of problem (3) that enable us to solve the problem

$$(L_2/k, G, \varphi_4, B'). \tag{4}$$

It is obvious that the solutions of problem (4) (if they exist) also solve problem (1).

2°. Since the group extension determining (2) is direct, it follows that any solution of problem (2) is of the form $L_1 = K(\sqrt[p]{x})$, where $x \in k^*$. Moreover, we may assume that $\sqrt[p]{x}^{\varphi_2(\beta)} = \sqrt[p]{x}\zeta$, where ζ is a fixed pth root of 1, and for $g \in H$ we have $\sqrt[p]{x}^{\varphi_2(g)} = \sqrt[p]{x}$. In what follows we assume that $G_0 \neq G$.

Problem (3) is Brauer, whence the compatibility condition is sufficient for its solvability. In the group algebra $k[A/B']$ we have the decomposition of the unity as a sum of central idempotents $1 = e_0 + e_1 + \cdots + e_{p-1}$ for which $\overline{\alpha}e_i = \zeta^i e_i$, where $\overline{\alpha}$ is the image of α in A/B'.

For $i \neq 0$ we consider the components of the crossed product $((G/B') \times L_1)e_i = D_i$. These simple central algebras contain the subalgebras $(\widetilde{H}/B' \times K)e_i \approx H/B' \times K \approx F \times K$ (with unit factors). Thus, $(\widetilde{H}/B' \times K)e_i$ is the full matrix algebra over k of order $(F:1)$. The centralizer of the algebra $(\widetilde{H}/B' \times K)e_i$ in D_i is generated over k by the elements $\sqrt[p]{x}$ and $\overline{\beta}(\sqrt[p]{c})^{-i}e_i$. In fact, if $\beta^g \equiv \beta\alpha^j \pmod{B'}$, then, by the choice of c, we have $(\sqrt[p]{c})^{\varphi(g)} = \sqrt[p]{c}\zeta^j$, but in the algebra $((\widetilde{H}/B') \times K)e_i$ we then have $\overline{\beta}^g e_i = \overline{\beta}\zeta^{ij}e_i$ so that $(\overline{\beta}(\sqrt[p]{c})^{-i}e_i)^g = \overline{\beta}(\sqrt[p]{c})^{-i}e_i$.

Therefore, the compatibility condition for problem (3) means that the centralizer indicated above is isomorphic to a matrix algebra. At the same time, the elements $\sqrt[p]{x}$ and $\overline{\beta}(\sqrt[p]{c})^{-i}$ generate over k the algebra of generalized quaternions $k[x, c^{-i}]$ and $k[x, c^{-i}] \sim k[x, c]^{-i}$ (here we mean the equivalence in the Brauer group $\operatorname{Br} k$).

Thus, we have the following lemma.

LEMMA 5.1.1. *If $G \neq G_0$, problem* (3) *is solvable if and only if the algebra $k(x, c)$ is trivial. If $G = G_0$, problem* (3) *is solvable for any $x \in k^*$.*

3°. Let the condition $(x, c) \sim 1$ be fulfilled and let L_2^0 be a solution of problem (3). Then L_2^0 is a solution of the semidirect embedding problem $(K/k, G/B', B/B')$. Let $K' \subset L_2^0$ be the subfield (subalgebra) corresponding to the subgroup H/B'. Then K' is not a normal extension of the field k, but it contains the field (Galois algebra) $k(\sqrt[p]{x})$ and is normal over it because H is normal in $\widetilde{H}$. Therefore, $K' = k(\sqrt[p]{x})(\theta)$, where $\theta^p = \xi \in k(\sqrt[p]{x})$, and $L_2^0 = L_1(\theta)$. We choose an element θ such that $\theta^{\overline{\alpha}} = \theta\zeta$, where $\overline{\alpha}$ is the image of α in B/B', and ζ is a fixed pth root of 1. The element θ is invariant with respect to H/B' because $\theta \in K'$. For $\overline{\beta}$ we have $\theta^{\overline{\beta}} = \theta\lambda$, where $\lambda \in L_1 = K(\sqrt[p]{x})$.

We establish some properties of the number λ (one can find λ directly from the representation of x as a norm from the field $k(\sqrt[p]{c})$, but we need not do this). From the relation $\theta\lambda^{\overline{h}} = (\theta\lambda)^{\overline{h}} = \theta^{\overline{\beta}\,\overline{h}} = \theta^{\overline{h}\,\overline{\beta}\overline{\alpha}^j} = \theta^{\overline{\beta}\overline{\alpha}^j} = (\theta\lambda)^{\alpha^j} = \theta\zeta^j\lambda$ (where

$h \in H$) it follows that $\lambda = \lambda_1 \sqrt[p]{c}$, where $\lambda_1 \in k(\sqrt[p]{x})$. Since $\theta^{\overline{\beta}^p} = \theta$, it follows that $\lambda^{1+\overline{\beta}+\cdots+\overline{\beta}^{p-1}} = 1$, whence $\lambda = \mu^{1-\overline{\beta}}$ with $\mu \in k(\sqrt[p]{x}, \sqrt[p]{c})$ (Hilbert's Theorem 90).

The replacement of the element $\xi = \theta^p$ by the element ξy, where y is an arbitrary element of k^*, in the solution L_2^0 yields an arbitrary solution of the Brauer problem (3), as is clear from the method of solution of this problem (Theorem 3.1).

4°. Let L_2 be a solution of problem (3). Consider the Brauer embedding problem

$$(L_2/k, G, \varphi_4, B') \tag{4}$$

(any solution of it also is a solution of problem (1)). The embeddability conditions for this problem are equivalent to the compatibility conditions, i.e., to the splitting of all components of the crossed product $(G \times L_2)E_i$, where $E_i = \frac{1}{p}\sum_{k=0}^{p-1}\gamma^k\zeta^{-ik}$ (where γ is a generator of the commutator subgroup B'). The extension $1 \longrightarrow B' \longrightarrow G \longrightarrow G/B' \longrightarrow 1$ is determined by a cocycle $\gamma_{\sigma_1,\sigma_2} \in Z^2(G/B', B')$, where $\sigma_i \in G/B'$. Since for $\gamma \in B'$ we have $\gamma E_i = \zeta^i E_i$, the algebra $S_i = (G \times L_2)E_i$ is the crossed product of the field L_2 and its Galois group G/B' with factors $\zeta^i_{\sigma_1,\sigma_2}$, where $\zeta_{\sigma_1,\sigma_2}E_1 = \gamma_{\sigma_1,\sigma_2}E_1$. Therefore, all the algebras S_i split if the algebra $S_1 = (G \times L_2)E_1$ splits.

The algebra S_1 is simple and contains the simple central subalgebra $T_1 = (G \times K)E_1$, whence $S_1 = T_1 \otimes V_1$, where V_1 is the centralizer of T_1 in S_1. The centralizer V_1 is generated over k by η_1 and η_2, where $\eta_1 = \alpha\sqrt[p]{x}^{-1}E_1$ for $p \neq 2$, $\eta_1 = \alpha\sqrt{x}\sqrt{c}E_1$ for $p = 2$; $\eta_2 = \beta(\mu + \alpha\mu^h + \cdots + \alpha^{p-1}\mu^{h^{p-1}})\theta E_1$, where $h \in H$ is chosen so that $\beta^h = \beta\alpha$ (and $\sqrt[p]{c}^h = \sqrt[p]{c}\zeta$). It is obvious that η_1 commutes with the elements from T_1. As is easy to see, the element η_2 commutes with the field K, the group H, and the element α; moreover, the summand $\beta\mu\theta$ commutes with β because $\theta^\beta = \theta\mu^{1-\overline{\beta}}$, and for the other summands this follows from $\mu^{1-\overline{\beta}}/\sqrt[p]{c} \in k(\sqrt[p]{x})$.

Since $\eta_2 = \tau\theta$ with $\tau \in (G \times L_1)E_1$, we have $\eta_2\eta_1 = \zeta\eta_1\eta_2$ (because $\eta_1\tau = \tau\eta_1$). Therefore, η_2^p is a central element of S_1 and $\eta_2^p \in k^*$. Moreover, $\eta_1^p = x^{-1}E_1$ for $p \neq 2$ and $\eta_1^2 = -cxE_1$ for $p = 2$. Therefore, V_1 is a cyclic algebra of dimension p^2 over k with the following invariants: (x^{-1}, z) for $p \neq 2$ and $(-cx, z)$ for $p = 2$, where $zE_1 = \tau_2^p E_1$. Note that the element $z \in k^*$ may be arbitrary, since the passage from the fixed solution L_2^0 of problem (3) to an arbitrary solution is given by $z = z^{(0)}y$, where $y \in k^*$ is arbitrary. Thus, we have proved the following lemma.

LEMMA 5.1.2. *The initial embedding problem* (1) *is solvable if and only if there exist elements $x, z \in k^*$ such that*

(1) *the Hilbert symbol (c, x) over k is equal to* 1;

(2) *in the Brauer group* $\mathrm{Br}\, k$ *of the field k, the algebra $T_1 = (G \times K)E_1$ has the following invariants*: (x, z) *for $p \neq 2$ and $(-cx, z)$ for $p = 2$.*

5°. Now let the compatibility condition be fulfilled for the initial embedding problem (1). This means that all the components $T_i = (G \times K)E_i$ are isomorphic to the matrix algebra of order $(F:1)$ over a subalgebra. As before, this is equivalent to such a representation only for the algebra T_1. Therefore, the compatibility condition means that there exists a central algebra P contained in T_1 and such that $T_1 \approx P \otimes (k)_{(F:1)}$. It is clear that the dimension of P over k is equal to p^2.

The algebra T_1, as the matrix algebra over P, is equivalent to P in $\mathrm{Br}\, k$. It follows from Lemma 5.1.2 that for the solvability of problem (1) it is necessary that the algebra P be cyclic, i.e., it must contain a normal (hence radical) splitting field.

Suppose this condition is fulfilled. We examine the hypotheses of Lemma 5.1.2 under this assumption. As a cyclic algebra, P is determined by elements $a, b \in k^*$. The condition $k[a,b] \approx k[x,z]$ (or $k[a,b] = k[-cx,z]$ for $p=2$) means (because z is arbitrary) that $k(\sqrt[p]{x})$ (respectively, $k(\sqrt{-cx})$) must be a splitting field for the algebra P, i.e., such a field is embedded in P. Let p be odd. The condition $(c,x)=1$ means that the element c is the norm of an element $\varkappa \in k(\sqrt[p]{x})$, and thus is the reduced norm of an element from P. For $p=2$, the condition $(c,x)=1$ can be rewritten as $(c,-cx)=1$, and c again is the reduced norm of an element of P. The converse is also true. Namely, if c is the reduced norm of an element $\varkappa' \in P$ and $k(\varkappa')$ is a normal extension of k, then it is radical. We choose $x \in k$ so that $k(\varkappa') = k(\sqrt[p]{x})$ (for $p=2$, $k(\varkappa') = k(\sqrt{-cx})$, respectively). Then $k(\sqrt[p]{x})$ (or $k(\sqrt{-cx})$) is a splitting field of the algebra P, and for some $y \in k^*$ the algebra P and the cyclic algebra $k(x, z^{(0)}y)$ (or $k(-cx, z^{(0)}y)$) are isomorphic. Thus, we have obtained the following theorem.

THEOREM 5.1.1. *For $G_0 \neq G$ the embedding problem* (1) *is solvable if and only if*

(1) *the compatibility condition is fulfilled;*

(2) *the element c and the algebra P defined above are such that c is the reduced norm of an element $\varkappa \in P$ such that the extension $k(\varkappa)$ is normal over k.*

REMARK. The requirement for $k(\varkappa)$ to be normal over k ensures the cyclicity of the algebra P. In general, the cyclicity of the algebra P is obvious for $p=2$ and may be proved for $p=3$. For $p \geq 5$, in the case of an arbitrary field k it is unknown whether or not any algebra of dimension p^2 is cyclic.

Now consider a simpler case of $G_0 = G$, i.e., the elements $g \in G$ act on the kernel B as inner automorphisms. Here, the extension $1 \to B/B' \to G/B' \to F \to 1$ is direct, therefore the field L_2 solving problem (3) is of the form $L_2 = K(\sqrt[p]{x}, \sqrt[p]{y})$, where $x, y \in k^*$. An argument similar to that in the general case (but simpler) allows us to obtain the following theorem.

THEOREM 5.1.2. *For $G_0 = G$, the embedding problem* (1) *is solvable if and only if*

(1) *the compatibility condition is fulfilled;*

(2) *the algebra P subject to condition* (1) *is cyclic.*

6°. We study our problem in the case of local fields or algebraic number fields.

By Theorem 2.9, for such fields it is sufficient to verify the compatibility condition for the associated Abelian problem resulting from factorization by the commutator subgroup of the kernel. But the extension $1 \to B/B' \to G/B' \to F \to 1$ is semidirect, hence for this problem the compatibility condition is trivial. Consequently, it is also valid for the initial embedding problem. The algebra P subject to this condition is either the matrix algebra of order p over k, or a division algebra. In either case, this algebra is cyclic over k (for the matrix algebra this is obvious, for a division algebra over number or local fields, this is a known theorem of Brauer–Hasse–Noether). Thus, for $G_0 = G$, the embedding problem is always solvable.

Now we turn to the case where k is a local field.

THEOREM 5.1.3. *The embedding problem $(K/k, G, \varphi, B)$, where B is a non-Abelian group of order p^3, G is a p-group, k is a local field, is always solvable.*

PROOF. We show that the assumptions of Theorem 5.1.1 are satisfied. The first assumption (the compatibility condition) holds. If an element $c \in k^*$ is a pth power in k^*, then c is the reduced norm (in P) of an element from k^*. Let $c \notin k^{*p}$. As is known, in the case of local fields any irreducible equation of degree n over k has a solution in any central simple algebra of dimension n^2 over k (see [**1**], VI.1). For $p \neq 2$, we consider the equation $t^p - c = 0$. There exists $\varkappa \in P$ such that $\varkappa^p = c$. Therefore, the field $k_1 = k(\varkappa)$ is normal over k and $N_{k_1/k}\varkappa = c$. Thus, c is the reduced norm in the algebra P and the assumptions of Theorem 5.1.1 are fulfilled.

Let $p = 2$, and let $-c$ be a nonsquare in k^*. Then, as an element $\varkappa$, we can take a root of the equation $t^2 + c = 0$. If $p = 2$ and $-c = u^2$ with $u \in k^*$, then we choose $v \in k$ such that $u^2 + v^2$ is a nonsquare in k^* and consider the equation $t^2 - (u^2 + v^2) = 0$. Then we have $c = N_{k_1/k}(v + \varkappa)$, where $\varkappa \in P$ and $\varkappa^2 = u^2 + v^2$; $k_1 = k(\varkappa)$. The theorem is proved. □

We now proceed with consideration of a global number field.

THEOREM 5.1.4. *If k is an algebraic number field, B is a non-Abelian group of order p^3 for odd p, and G is a p-group, then the embedding problem $(K/k, G, \varphi, B)$ is always solvable.*

PROOF. The case $G_0 = G$ was considered above, therefore we assume that $G_0 \neq G$. We find an element $x \in k^*$ for which

1) $(c, x) = 1$;

2) $P \otimes_k k(\sqrt[p]{x})$ is a matrix algebra.

Denote by S the set of points $\mathfrak{p}$ of k at which the algebra P is ramified (i.e., $P_\mathfrak{p}$ is a division algebra). The set S is the union $S_1 \cup S_2$, where for $\mathfrak{p} \in S_1$ c is not a pth power in $k_\mathfrak{p}$, but for $\mathfrak{p} \in S_2$ c is equal to d^p with $d \in k_\mathfrak{p}$.

Now we find a prime element q of k such that q is a pth power in $k_\mathfrak{p}$ for $\mathfrak{p} \in S_1$, q is not a pth power in $k_\mathfrak{p}$ for $\mathfrak{p} \in S_2$, and q is a pth power in $k_\mathfrak{p}$, where $\mathfrak{p} \mid p$, $\mathfrak{p} \notin S$. By Dirichlet's theorem, such a prime element q exists. Now we set $x = cq$. Then $(c, x)_\mathfrak{p} = (c, q)_\mathfrak{p} = 1$ and $P_\mathfrak{p} \otimes_{k_\mathfrak{p}} k_\mathfrak{p}(\sqrt[p]{x})$ is a matrix algebra for all points $\mathfrak{p}$ of the field k (except for, possibly, the point q). By the reciprocity law, we see that x is the required element.

Since the algebra $P(\sqrt[p]{x})$ splits, in the algebra P there exists an element ξ such that $\xi^p = x$ (and $\xi \notin k$ if P is not a matrix algebra). Since $(\zeta\xi)^p = x$, where $\zeta^p = 1$, $\zeta \in k^*$, there exists an $\eta \in P$ such that $\eta^{-1}\xi\eta = \xi\zeta$. Then $P = k(\xi, \eta)$, η^p commutes with η and ξ, i.e., $\eta^p \in k^*$. We set $y = \eta^p \in k^*$. Then P is determined by a pair (x, y), and the elements (x, y) satisfy the assumptions of Lemma 5.1.2. The theorem is proved. □

For the case $p = 2$, the proof of Theorem 5.1.4 does not work, since in addition to the completions by discrete valuations we must consider the case of an Archimedean valuation.

We show with a simple example that Theorem 5.1.4 is false for $p = 2$.

Let B be a dihedral group given by the relations $\alpha^4 = \beta^2 = 1$, $\beta^{-1}\alpha\beta = \alpha^{-1}$. As G, we consider the group with 16 elements that is generated by the group B and an element g such that $g^2 = \alpha^2$, $\alpha^g = \alpha^{-1}$, $\beta^g = \alpha\beta$. Let $K = k(\sqrt{c})$, where $c \in k^*$. The algebra $(G \times K)E$ (for $E = \frac{1-\alpha^2}{2}$) is the tensor product over k of two algebras of generalized quaternions, generated by the pairs of elements $(\beta E, \alpha E)$

and $(\sqrt{c}E, g\beta(1+\alpha)E)$. The first is a matrix algebra because $\beta^2 = 1$, while the second can be defined as $k(c,-2)$. As is shown in Theorem 5.1.1, the solvability of this problem is equivalent to the fact that c can be represented in the form $z_1^2 - cz_2^2 + 2z_3^2 - 2cz_4^2$, where $z_i \in k$ (the general form of the reduced norm for $p = 2$). It is clear that if k is embedded in the field of real numbers $\mathbb{R}$ and $c < 0$, this representation is impossible.

REMARK. If k is a global field such that $\sqrt{-1} \in k$, then Theorem 5.1.4 is also valid for $p = 2$.

§2. The embedding problem with non-Abelian kernel of order p^3. II

In this section we consider the embedding problem $(K/k, G, \varphi, B)$, where B is a non-Abelian group of order p^3 and the group G is not assumed to be a p-group. Unlike the Abelian case where Kochendörffer's theorem is valid, removing the requirement that G is a p-group leads to additional difficulties, and we have not succeeded in resolving them even for the embedding problem over a local field. In order to avoid cumbersome computations, a detailed consideration is performed only for a kernel isomorphic to the quaternion group.

1°. Let B be the quaternion group with generators α, β, where $\alpha^4 = 1$, $\beta^2 = \alpha^2$, $\beta^{-1}\alpha\beta = \alpha^{-1}$, and let G be an arbitrary (finite) extension of the group B.

LEMMA 5.2.1. *The group G/B', where B' is the commutator subgroup of B, is a semidirect extension of the group B/B'.*

As shown in the preceding section, this fact is valid if G is a p-group. Since B/B' is an Abelian group, the lemma now follows from the Gaschütz theorem (see [**1**], IV, 6). However for this case we may give a direct proof of the fact indicated.

We write G_0 for the subgroup of those $g \in G$ that commute with all elements of the kernel. It is obvious that $G_0 \cap B = B'$. If all $g \in G$ act on the elements of the kernel as 2-automorphisms, then the lemma is valid, as shown in the preceding section.

Now we assume that an element $\sigma \in G$ inducing a 3-automorphism on the kernel exists (it is easy to see that the order of the full automorphism group of the quaternion group is equal to 24). By choosing suitable generators $\alpha, \beta \in B$, we may achieve the relations $\alpha^\sigma = \beta$, $\beta^\sigma = \alpha\beta$. It is clear that $\sigma^3 \in G_0$ and $\sigma^{-1}G_0\sigma = G_0$, whence we have $G_1 \cap B = B'$, $(G_1 : G_0) = 3$ for the group G_1 which is generated by G_0 and σ. If $\varphi(G_1) = F = G/B$, where φ is a homomorphism of G onto G/B, the lemma is valid. If $\varphi(G_1) \neq G/B$, then the index of $\varphi(G_1)$ in $\varphi(G)$ equals 2. We take an element $\tau_1 \in G$, $\tau_1 \notin G_1$, $\tau_1^2 \in G_1$. Multiplying τ_1 by a suitable element of the kernel and by σ^i $(i = 0, 1, 2)$, we obtain an element τ, which induces an automorphism on B such that $\alpha^\tau = \beta^{-1}$, $\beta^\tau = \alpha^{-1}$. It is easy to see that $(\tau\sigma)^2 \in G_0$, $\tau^2 \in G_0$, thus $\tau^{-1}\sigma\tau \in \sigma^2 G_0$ and then $\tau^{-1}G_1\tau = G_1$. We have $H \cap B = B'$, $\varphi(H) = G/B$ for the group H generated by G_1 and τ. Therefore, $H/B' \approx G/B$, and the lemma is proved.

2°. We solve the embedding problem $(K/k, G, \varphi, B)$ in two steps. First we consider the semidirect associated problem $(K/k, G/B', \varphi', B/B')$ and find all its solutions. As shown, they are in one-to-one correspondence with the elements of the group of extensions $\operatorname{Ext}_F^1(\widehat{B/B'}, K^*)$, where $F = \operatorname{Gal}(K/k)$ and $\widehat{B/B'}$ is the character group of B/B'. We apply to our case the direct description of solutions of the semidirect problem carried out in §12 of Chapter 3. We assume that there exists an element σ giving rise to a 3-automorphism on the kernel (otherwise the problem is equivalent to the problem considered in the preceding section).

In our case $\widehat{B/B'}$ consists of four elements: χ_0 (the principal character), χ_1, χ_2, $\chi_3 = \chi_1\chi_2$. We assume $\chi_1(\alpha) = 1$, $\chi_1(\beta) = -1$; $\chi_2(\alpha) = -1$, $\chi_2(\beta) = 1$; $\chi_3(\alpha) = -1$, $\chi_3(\beta) = -1$ (here we somewhat abuse the notation; it is more proper to write $\chi(\alpha \bmod B')$ and so on). We have $\chi_i^f = \chi_i$ for $f \in F_0 = \varphi(G_0)$. For the element σ defined above we have $\chi_1^\sigma = \chi_2$, $\chi_2^\sigma = \chi_3$, $\chi_3^\sigma = \chi_1$. For the element τ (if it exists), we have $\chi_1^\tau = \chi_2$, $\chi_2^\tau = \chi_1$, $\chi_3^\tau = \chi_3$.

We show that the solution L of the problem $(K/k, G/B', B/B')$ is determined by a unique element $b = h_1(\chi_1, \chi_2)$ (the notation was introduced in §11 of Chapter 3). Indeed, $h_1(\chi_1, \chi_3) = h_1(\chi_3, \chi_1) = h_1(\chi_1^{\sigma^2}, \chi_2^{\sigma^2}) = h_1(\chi_1, \chi_2)^{\sigma^2}$; similarly, $h_1(\chi_2, \chi_3) = h_1(\chi_1, \chi_2)^\sigma$. The condition that $h_1(\chi_i, \chi_j)$ are cocycles yields

$$h_1(\chi_1, \chi_1)h_1(\chi_1\chi_1, \chi_2) = h_1(\chi_1, \chi_1\chi_2)h_1(\chi_1, \chi_2),$$

whence $h_1(\chi_1, \chi_1) = b^{1+\sigma^2}$ (since $h_1(\chi_1\chi_1, \chi_2) = h_1(1, \chi_2) = 1$). Similarly, $h_1(\chi_2, \chi_2) = b^{1+\sigma}$, $h_1(\chi_3, \chi_3) = b^{\sigma+\sigma^2}$. Moreover, $b^\tau = h_1(\chi_1, \chi_2)^\tau = h_1(\chi_2, \chi_1)$ $= b$ (if τ exists). If $f \in F_0$, then $b^f = b$. Thus, the algebra L is determined by an element b contained in a cubic extension of the field k (if F is generated by the subgroup F_0 and the element σ, then this extension is normal; if τ exists, this is not the case).

Thus, L is a linear space over K possessing a basis $\{1, u_1, u_2, u_3\}$ with (commutative) multiplication table: $u_1u_2 = u_3b$, $u_2u_3 = u_1b^\sigma$, $u_1u_3 = u_2b^{\sigma^2}$, $u_1^2 = b^{1+\sigma^2}$, $u_2^2 = b^{1+\sigma}$, $u_3^2 = b^{\sigma+\sigma^2}$, and with the following actions of automorphisms: $u_i^f = u_i$ $(f \in F_0)$, $u_1^\sigma = u_2$, $u_2^\sigma = u_3$, $u_3^\sigma = u_1$, $u_1^\tau = u_2$, $u_2^\tau = u_1$, $u_3^\tau = u_3$, $u_1^\alpha = u_1$, $u_1^\beta = -u_1$, $u_2^\alpha = -u_2$, $u_2^\beta = u_2$, $u_3^\alpha = -u_3$, $u_3^\beta = -u_3$; here $b \in K^{F_2}$, where F_2 is generated by the subgroup F_0 and the element τ.

3°. Now consider the embedding problem $(L/k, G, B')$, where $L = L(b)$ is the solution indicated above of the semidirect problem. Every solution of this problem, if it exists, is a solution of the original problem (and, conversely, every solution of the original problem is a solution of the problem $(L/k, G, B')$ with a suitable L). This is a problem with Abelian kernel of order 2, consequently it can be reduced to a problem with 2-group. It is more convenient to consider the embedding problem $(L/k_2, G_2, B')$, where $G_2 = \varphi^{-1}(F_2)$, $F_2 = \langle F_0, \tau\rangle$, and $k_2 = K^{F_2}$ (G_2 may contain elements whose order is not a power of two, but the automorphisms induced by them on the kernel are 2-automorphisms; as an intermediate problem between two equivalent problems, one with group G and the other with Sylow 2-subgroup of it, this problem is equivalent to both of them).

The solvability of the problem $(L/k_2, G_2, B')$ is equivalent to the fulfilment of the compatibility condition for this problem (since it is Brauer). The crossed product $G_2 \times L$ is a direct orthogonal sum of two simple algebras $(G_2 \times L)E_0$ and $(G_2 \times L)E_1$, where $E_0 = \frac{1}{2}(1 + \alpha^2)$, $E_1 = \frac{1}{2}(1 - \alpha^2)$. The first component is a full matrix algebra over its center k_2 (since $G_2E_0 \approx G_2/B'$), whence only the second component needs further investigation. The algebra $(G_2 \times L)E_1$ contains the simple central subalgebra $(G_2 \times K)E_1$, therefore $(G_2 \times L)E_1 \approx (G_2 \times K)E_1 \otimes Z$, where Z is the centralizer of $(G_2 \times K)E_1$. Let us find the algebra Z.

If $F = \langle F_0, \sigma\rangle$ (i.e., in G there is no element τ with properties indicated above), then Z is generated by the elements αu_1E_1 and βu_2E_1 over k_2. Then Z is the algebra of generalized quaternions over k_2 with constants $(-b^{1+\sigma^2}, -b^{1+\sigma})$. Now let such a τ exist. Then K^{F_1} ($F_1 = \langle F_0, \sigma\rangle$) is a quadratic extension of k and there exists an element $\theta \in K^{F_1}$, $\theta^2 = c \in k$, such that $\theta^\tau = -\theta$, $\theta^\sigma = \theta$. Then Z is generated by $\alpha\beta u_3\theta E_1$ and $(\alpha u_1 - \beta u_2)E_1$ over k_2 and is isomorphic to the algebra of generalized quaternions $k_2(-cb^{\sigma+\sigma^2}, -(b^{1+\sigma} + b^{1+\sigma^2}))$.

Now we assume that the compatibility condition is fulfilled for the original embedding problem $(K/k, G, \varphi, B)$. As shown in Theorem 2.9, this condition is always satisfied in the case of local and global number fields. Then the algebra $(G \times K)E_1$ is isomorphic to the matrix algebra of order $(F:1)$ over a subalgebra P. The algebra P is a simple central algebra and its dimension over k equals 4. Therefore, P is an algebra of generalized quaternions $k(a_1, a_2)$, where $a_1, a_2 \in k^*$. Hence, the algebra $(G_2 \times K)E_1$ is isomorphic to the matrix algebra of order $(F_2:1)$ over the subalgebra $P \otimes_k k_2 = k_2(a_1, a_2)$. Therefore, the compatibility condition for the embedding problem $(L/k_2, G_2, B')$ means that $P \otimes k_2 \approx Z$, and the following theorem is proved.

THEOREM 5.2.1. *Let the compatibility condition be fulfilled for the embedding problem $(K/k, G, \varphi, B)$, and let a_1, a_2, c be the elements of k constructed above. Then the solvability of this problem is equivalent to the existence of an element $y \in k_2$, for which the following algebras are isomorphic: $k_2(a_1, a_2)$ and $k_2(-y^{1+\sigma^2}, -y^{1+\sigma})$ if $F = F_1 = \langle F_0, \sigma\rangle$; $k_2(a_1, a_2)$ and $k_2(-cy^{\sigma+\sigma^2}, -(y^{1+\sigma} + y^{1+\sigma^2}))$ if $F \neq F_1 = \langle F_0, \sigma\rangle$.*

Note that in the second case the element $y^{\sigma+\sigma^2}$ is a norm from $k_2(\sqrt{c})$ to k_2. Indeed, let $y^\sigma = z_1 + z_2\sqrt{c}$, where $z_1, z_2 \in k_2$. Then $y^{\sigma+\sigma^2} = (y^\sigma)^{1+\tau} = z_1^2 - cz_2^2$. It follows from the isomorphism $k_2(-cy^{\sigma+\sigma^2}, -(y^{1+\sigma} + y^{1+\sigma^2})) \approx k_2(a_1, a_2)$ that $-cy^{\sigma+\sigma^2}$ is a square in $k_2(a_1, a_2)$, i.e., there exist $x_1, x_2, x_3 \in k_2$ such that $-c(z_1^2 - cz_2^2) = -cy^{\sigma+\sigma^2} = a_1x_1^2 + a_2x_2^2 - a_1a_2x_3^2$. Hence, c is a reduced norm in the algebra $k_2(a_1, a_2)$. This necessary condition for the solvability of the original embedding problem is (as shown in the preceding section) a criterion for the solvability of the Sylow associated problem.

4°. Now consider a specific example which is of interest from the standpoint of distinctions between the Abelian and the non-Abelian embedding problem (see also [**33**]).

Let $B = \langle\alpha, \beta\rangle$ be the quaternion group, and let the group G be generated over B by an element σ, for which $\sigma^6 = \alpha^2$, $\alpha^\sigma = \beta$, $\beta^\sigma = \beta\alpha$. The group $F = G/B$ is a cyclic group of order 6. The element σ^3 is central, and the Sylow 2-subgroup of G generated by the kernel and the element σ^3 is a semidirect extension of the kernel because $(\sigma^3\alpha)^2 = 1$. At the same time, the group G itself is not a semidirect extension of the kernel because the order of all elements of the coset σB is equal to 12.

Consider the embedding problem $(K/k, G, \varphi, B)$, where $F = \mathrm{Gal}(K/k)$ is a cyclic group of order 6 and $\varphi(\sigma)$ is a generator of this group. Since $\mathbb{Z}/6\mathbb{Z} \approx \mathbb{Z}/2\mathbb{Z} \times \mathbb{Z}/3\mathbb{Z}$, the field K is the composite of a quadratic extension k_1 of k and a normal cubic extension k_2 of k. Let $k_1 = k(\sqrt{d})$ with $d \in k^*$ and $\sqrt{d}^{\varphi(\sigma^3)} = -\sqrt{d}$. The Galois group of the extension k_2/k is generated by $\varphi(\sigma^2)$.

Consider the crossed product $G \times K$. By Theorem 2.4, the compatibility condition is fulfilled for our problem, therefore $(G \times K)E_1$ (where $E_1 = \frac{1}{2}(1 - \alpha^2)$) is isomorphic to the algebra of matrices of order 6 over the subalgebra P. Then $(G_2 \times K)E_1$ is a matrix algebra over $P \otimes k_2$ (of order 2). We have $(G_2 \times K)E_1 \approx$

$k_2(\alpha\sigma^3 E_1, \sqrt{d})\otimes_k k_2(\alpha E_1, \beta\sqrt{d}E_1)$. Therefore, (since $(\alpha\sigma^3)^2 = 1$) the algebra $P\otimes k_2$ is the algebra of generalized quaternions $k_2(-1,-d)$. Since $d \in k_2$, the algebra P is generated (over k) by anticommuting elements, whose squares are -1 and $-d$. These elements are a_1 and a_2 specified in our case.

According to the theorem proved above (in the previous subsection), the solvability of the problem in our case means the existence of an element $y \in k_2$ such that the algebra of generalized quaternions over k_2 with the constants $-y^{1+\sigma}$, $-y^{1+\sigma^2}$ is isomorphic to $k_2(-1,-d)$. Thus, our problem reduces to the existence of a nontrivial solution of the following system of equations over k_2:

$$(*)\qquad \begin{cases} x_1^2 + dx_2^2 + dx_3^2 = y^{1+\sigma}x_4^2, \\ x_5^2 + dx_6^2 + dx_7^2 = y^{1+\sigma^2}x_8^2, \\ x_1x_5 + dx_2x_6 + dx_3x_7 = 0. \end{cases}$$

We show that this system is not always consistent even for local fields. Let $p \equiv -1 \pmod 4$ be a prime number, k the p-adic number field, $d = p$, and let k_2 be a normal cubic extension of the field $\mathbb{Q}_p$, in which p is unramified. We prove that system $(*)$ has no nontrivial solutions in this case.

Suppose the contrary. Denote the valuation in k_2 by the letter ω. Since $k_2/\mathbb{Q}_p$ is normal, we have $\omega(y) = \omega(y^\sigma) = \omega(y^{\sigma^2})$, and thus the number $\omega(y^{1+\sigma}) = \omega(y^{1+\sigma^2}) = \omega(y^{\sigma+\sigma^2})$ is even. Without loss of generality, we may assume that $\omega(y^{1+\sigma}) = 0$ and then $\omega(y) = 0$. We may also assume that the numbers x_i are integral numbers in k_2 and at least one of them is not divisible by p. In this case $\omega(x_1) = \omega(x_4) = \omega(x_5) = \omega(x_8) = 0$. Indeed, if, for example, $\omega(x_1) > 0$, $\omega(x_4) > 0$, then $\omega(x_2) = \omega(x_3) = 0$ and $x_2^2 + x_3^2 \equiv 0 \pmod p$, but, by assumption, $(\frac{-1}{p}) = -1$. If $\omega(x_1) = \omega(x_5) = 0$, then the third equation of the system is not solvable in integral elements x_i. Thus, our problem is not solvable.

This example shows that Kochendörffer's theorem cannot be applied to the case of a non-Abelian kernel. This is also an example of the embedding problem for local fields, where compatibility does not imply embeddability.

5°. Now we outline the embedding problem $(K/k, G, \varphi, B)$, where B is a noncommutative group of order p^3 with odd p. It follows from the Gaschütz theorem (see above) that G/B' is a semidirect extension of B/B' by F. Let L be an arbitrary solution of the semidirect problem $(K/k, G/B', B/B')$. As above, the solvability of the original problem is equivalent to the existence of an algebra L such that the compatibility condition is fulfilled for the embedding problem $(L/k_p, G_p, B')$, where G_p is a Sylow p-subgroup of G, $k_p = K^{F_p}$. In other words, the component of $(G_p \times L)E_1$ is isomorphic to a matrix algebra over k_p (here E_1, as above, is the idempotent $\frac{1}{p}\sum_{i=0}^{p-1}\gamma^i\zeta^{-i}$, where γ is a generator of B'). We have $(G_p \times L)E_1 \approx (G_p \times K)E_1 \otimes Z$, where Z is the centralizer of the subalgebra $(G_p \times K)E_1$. Clearly, Z is a central simple algebra over k_p of dimension p^2. An explicit computation of the structure constants of $\mathbb{Z}$ is difficult, but we can claim that this algebra contains a maximal commutative subfield which is a cyclic extension of k_p. We shall show this.

Let λ be an inclusion of F in G/B', for which the composite mapping $F \to G/B' \to F$ is identical. For G_p we choose the full preimage of the subgroup generated in G/B' by the subgroups B/B' and $\lambda(F_p)$, where F_p is a Sylow p-subgroup of F. In the kernel B there exists a noncentral element α such that $\alpha^g \equiv \alpha \pmod{B'}$ for every $g \in G_p$. We denote by β an element of B such that $\beta^{-1}\alpha\beta = \alpha\gamma$, where γ is the element of B' mentioned above. Thus, $\beta^{-1}\alpha\beta E_1 = \alpha\zeta E_1$. Denote by $\widetilde{G}_p$ the preimage of $\lambda(F_p)$ in G_p and by $\widetilde{G}_0$ the subgroup of the elements of $\widetilde{G}_p$ that act trivially on α. Clearly, either $\widetilde{G}_0 = \widetilde{G}_p$, or the index of $\widetilde{G}_0$ in $\widetilde{G}_p$ equals p. In the second case we assume that $\widetilde{k} = K^{\varphi(G_0)}$. The field $\widetilde{k}$ is a normal extension of k_p of degree p, whence $\widetilde{k} = k_p(\sqrt[p]{c})$ with $c \in k_p^*$. If $\alpha^{\overline{f}} = \alpha\gamma^l$, where $f \in F_p$ and $\overline{f}$ is its preimage in $\widetilde{G}_p$, we choose $\sqrt[p]{c}$ in such a way that $\sqrt[p]{c}^f = \sqrt[p]{c}\zeta^l$.

Now let $\chi_1 \in \operatorname{Hom}(B/B', K^*)$ be a character such that $\chi_1(\alpha) = 1$ and $\chi_1(\beta) = \zeta$, and let $u_1 \in L$ be the element corresponding to χ_1. Then $u_1^\alpha = u_1$, $u_1^\beta = u_1\zeta^{-1}$, $u_1^{\lambda(f)} = u_1$. Let $\xi = \alpha(\sqrt[p]{c})^{-1}u_1E_1$ or let $\xi = \alpha u_1 E_1$ if $\widetilde{G}_0 = \widetilde{G}_p$. The element ξ is invariant with respect to $\widetilde{G}_p$, as well as to α and β, i.e., ξ is invariant with respect to the action of G. Moreover, it commutes with the elements of K. Therefore, $\xi \in Z$. In addition, we have $\xi^p = \alpha^p E_1 c^{-1} u_1^p$ (or $\xi^p = \alpha^p E_1 u_1^p$). The element u_1^p is contained in the center of the algebra $(G_p \times L)E_1$, i.e., it belongs to k_p, and $\alpha^p E = \zeta^i E$ for some i. Thus, $\xi^p \in k_p$. Hence, $k(\xi)$ is a normal extension of k_p and our assertion is proved. Therefore, the algebra Z is a crossed product over k_p with structure constants $y_1, y_2 \in k_p$ determined by the solution L_1.

Now let the compatibility condition be fulfilled for the original embedding problem $(K/k, Q, \varphi, B)$. Then $(G \times K)E_1$ is a matrix algebra over its simple subalgebra P of rank p over k and $(G_p \times K)E_1$ is a matrix algebra over $P \otimes_k k_p$. Thus, the complete splitting of $(G_p \times L)E_1$ means that the algebra Z is opposite to the algebra $P \otimes_k k_p$. The elements y_1 and y_2 that determine Z are not arbitrary; they may be connected by some relations, however, there is freedom in choosing them because to each element in $\operatorname{Ext}_F^1(\widehat{B/B'}, K^*)$ there correspond y_1 and y_2 of its own.

A detailed analysis of the obtained embedding condition is beyond the scope of this book.

§3. The Shafarevich lemma

In this section we show that in a Galois extension, there exist algebraic numbers with prescribed local properties at finitely many points.

Let L/k be a finite normal extension with Galois group G, K a normal intermediate subfield with group F, L_0 a maximal Abelian extension of period p of K, contained in L, where p is prime. Suppose K contains all pth roots of 1.

Let T be a finite set of prime divisors of K such that the following conditions are fulfilled:

1) T does not contain divisors of p, divisors of the discriminant of K/k, and for $p = 2$ infinite divisors;

2) if $\mathfrak{P} \in T$, then $\mathfrak{P}^f \in T$ for $f \in F$.

Suppose for every $\mathfrak{P} \in T$ a $\mathfrak{P}$-adic unit $\overline{m}_\mathfrak{P}$ is given and let $m_\mathfrak{P}$ be its image in the group $K_\mathfrak{P}^*/K_\mathfrak{P}^{*p}$. We associate with $\prod_{\mathfrak{P}\in T} m_\mathfrak{P}$ an element m_T of the group J_K/J_K^p by setting

$$m_{T,\mathfrak{P}} = \begin{cases} m_\mathfrak{P} & \text{if } \mathfrak{P} \in T, \\ 1 & \text{if } \mathfrak{P} \notin T, \end{cases}$$

where $m_{T,\mathfrak{P}}$ is the projection of m_T on the $\mathfrak{P}$-component.

In an algebraic number field we may define the pth power residue symbol $\left(\frac{m}{\mathfrak{P}}\right)$ (the Legendre symbol) for a prime divisor $\mathfrak{P}$ and a number m such that $\mathfrak{P}$ does not divide the divisor (m) (the definition of the symbol and its property may be found in Exercises in [**1**]). It immediately follows from the definition that the value of the symbol $(\frac{m}{\mathfrak{P}})$ is completely determined by the image of the number m in the group $K^*_{\mathfrak{P}}/K^{*p}_{\mathfrak{P}}$.

We denote by θ_0 the Artin homomorphism of the group J_k onto the Galois group F_0 of the extension L_0/K (for the definition and properties of the Artin mapping, see Chapter VII of [**1**]). The period of F_0 is equal to p; so, θ_0 annihilates the subgroup J_k^p and we may assume that θ_0 maps J_k/J_k^p onto F_0.

LEMMA 5.3.1 (I. R. Shafarevich). *If $\theta_0(m_T) = 1$, then there exists a number $\overline{m}$ in K with the following properties:*

1) *the decomposition of the divisor $(\overline{m})$ is of the form $\mathfrak{P}_1\mathfrak{P}_2\mathfrak{a}^p$ for odd p and $\mathfrak{P}_1\mathfrak{P}_2\mathfrak{P}_3\mathfrak{a}^2$ for $p = 2$, where the $\mathfrak{P}_i$ are prime divisors of $\mathfrak{p}_i$ from k, the $\mathfrak{p}_i$ completely decompose in L, the element $\mathfrak{a}$ is an arbitrary divisor of K, relatively prime to $\mathfrak{P}_i$; moreover, $\mathfrak{P}_i^f$ does not divide $(\overline{m})$ for $f \in F$ and $f \neq 1$, and the symbols $(\frac{\overline{m}}{\mathfrak{P}_i^f})$ are equal to* 1;

2) *the image of $\overline{m}$ in the group $K^*_{\mathfrak{P}}/K^{*p}_{\mathfrak{P}}$ coincides with $m_{\mathfrak{P}}$ for $\mathfrak{P} \in T$;*

3) *$\overline{m}$ is a pth power in the completions of K that correspond to the divisors of p and infinite divisors.*

PROOF. First, let p be an odd prime number. We construct recursively a sequence $\{m_i\}$ of numbers in k with the following properties:

(1) $(m_i) = \mathfrak{P}_i\mathfrak{a}_i^p$, where $\mathfrak{P}_i$ is a prime divisor of $\mathfrak{p}_i$ from k, and $\mathfrak{p}_i$ completely decomposes in the field L; in addition, $\mathfrak{P}_i \notin T$, the divisors $\mathfrak{p}_i$ are distinct and relatively prime to $\mathfrak{a}_j$, where $i, j = 1, 2, \ldots$.

(2) $(\frac{m_i}{\mathfrak{P}_j^f}) = (\frac{m_j}{\mathfrak{P}_j^f})^{-1}$ for $j = 1, 2, \ldots, i-1$; $f \in F$ and $f \neq 1$.

(3) For $\mathfrak{P} \in T$ the image of the number m_i in $K^*_{\mathfrak{P}}/K^{*p}_{\mathfrak{P}}$ equals $m_{\mathfrak{P}}^{1/2}$ (the element $m_{\mathfrak{P}}^{1/2}$ is well defined because the number p is odd).

(4) m_i is a pth power in the completions of the field K that correspond to the divisors of p.

Suppose that the numbers $m_1, \ldots, m_{t-1}$ have been constructed. We denote by T_{t-1} the set of prime divisors of K consisting of the set T, the divisors of p, and the divisors $\mathfrak{P}_j^f$ for $j = 1, \ldots, t-1$; $f \in F$. Let L_{t-1} be a maximal Abelian extension of K of period p, which is unramified outside the set T_{t-1}. Hence, L_{t-1} is the composite of all Abelian extensions of K in which only the divisors from the set T_{t-1} ramify, and the Galois group of these extensions is an elementary Abelian p-group. Clearly, L_{t-1} is a finite normal extension over k with Galois group H, and the subgroup $V_{T_{t-1}}K^*J_K^p$ of the idèle group J_K corresponds to the extension L_{t-1}/K, according to class field theory.

Let $m_{\mathfrak{P}_j^f}$ be the image of m_j in $K^*_{\mathfrak{P}_j^f}/K^{*p}_{\mathfrak{P}_j^f}$, where $j = 1, \ldots, t-1$; $f \in F$. Consider the element $m_{T_{t-1}}$ in the group J_K/J_K^p, the projections $m_{T_{t-1},\mathfrak{P}}$ of which are equal to $m_{\mathfrak{P}}^{-1/2}$ if $\mathfrak{P} \in T$, to $m_{\mathfrak{P}_j^f}$ if $\mathfrak{P} = \mathfrak{P}_j^f$, $f \neq 1$, and to the unity element in the other components. Since $\mathfrak{P}_j^f$ are unramified in L_0/K, we have $\theta_0(m_{T_{t-1}}) = \theta_0(m_T)^{-1/2} = 1$.

Let K_1 be the composite of the extensions L and L_{t-1}, K_2 the intersection of these fields, G_1 and G_2 the Galois groups of K_1/k and K_2/k, respectively. Then G_1 is isomorphic to the direct product of H and G with amalgamated quotient group G_2. We denote by θ and θ_2 the Artin homomorphisms of J_K/J_K^p onto the Galois groups of the extensions L_{t-1}/K and K_2/K, respectively. The Galois group of the extension K_2/K is a homomorphic image of F_0; consequently, $\theta_0(m_{T_{t-1}}) = 1$ implies that $\theta_2(m_{T_{t-1}}) = 1$. We set $h = \theta(m_{T_{t-1}})$; then h may be regarded as an element of H. The relation $\theta_2(m_{T_{t-1}}) = 1$ implies that h induces the trivial automorphism in the extension K_2/k and in G_1 there exists an element $(h, 1)$, i.e., an automorphism of the extension K_1/k, that induces the automorphism h in L_{t-1}/k and the trivial automorphism in L/k. We denote by $[(h, 1)]$ the class of the elements of G_1 conjugate with $(h, 1)$.

By the Chebotarev Density Theorem, there exist infinitely many divisors of k belonging to the class $[(h, 1)]$. Therefore, among them we may choose a divisor $\mathfrak{p}_t$ different from $\mathfrak{p}_1, \dots, \mathfrak{p}_{t-1}$ and relatively prime to $\mathfrak{a}_1, \dots, \mathfrak{a}_{t-1}$. Let $\mathfrak{P}_t$ be a prime divisor of $\mathfrak{p}_t$ in K, belonging to the automorphism $(h, 1)$ in the extension K_1/K and to the automorphism h in the extension L_{t-1}/K. This means that $\widetilde{\pi}_{\mathfrak{P}_t} m_{T_{t-1}}^{-1} \bmod V_{T_{t-1}} K^* J_K^p = 1$, where $\widetilde{\pi}_{\mathfrak{P}_t}$ is an idèle such that its component corresponding to the divisor $\mathfrak{P}_t$ equals a prime element $\pi_{\mathfrak{P}_t}$ of $K_{\mathfrak{P}_t}$ and the other components are trivial. Thus, there exists a number $\overline{m}_t$, the decomposition of the principal divisor $(\overline{m}_t)$ of which is of the form $\mathfrak{P}_t\, \overline{\mathfrak{a}}_t^p$, and the image of the number $\overline{m}_t$ in $\prod_{\mathfrak{P} \in T_{t-1}} K_{\mathfrak{P}}^* / K_{\mathfrak{P}}^{*p}$ equals the projection of $m_{T_{t-1}}^{-1}$ on the components from the set T_{t-1}. Since every class of divisors contains infinitely many divisors, we may choose a divisor $\mathfrak{a}_t$ in the class with representative $\overline{\mathfrak{a}}_t$ such that it is not contained in the set T_{t-1} and is relatively prime to $\mathfrak{p}_1, \dots, \mathfrak{p}_t$. Then the divisor $\mathfrak{a}_t \overline{\mathfrak{a}}_t^{-1}$ is principal, i.e., there exists a number x in K such that $(x) = \mathfrak{a}_t \overline{\mathfrak{a}}_t^{-1}$. We put $m_t = \overline{m}_t x^p$. Conditions (1)–(4) that determine the sequence $\{m_i\}$ are immediately verified.

Suppose the sequence $\{m_i\}$ contains sufficiently many terms, namely, greater than $p^{[F:1]-1}$. The symbols $\left(\frac{m_i}{\mathfrak{P}_i^f}\right)$ with $i = 1, 2, \dots$; $f \in F$ and $f \neq 1$ may take only finitely many values, namely, no more than $p^{[F:1]-1}$. Therefore, there exist numbers m_{i_1} and m_{i_2} such that for all elements f of F different from 1, the equalities $\left(\frac{m_{i_1}}{\mathfrak{P}_{i_1}^f}\right) = \left(\frac{m_{i_2}}{\mathfrak{P}_{i_2}^f}\right)$ hold. Put $\overline{m} = m_{i1} m_{i2}$ and show that $\overline{m}$ is the required number. For definiteness, let $i_1 < i_2$. Clearly, conditions (2) and (3) are fulfilled. We check that the corresponding Legendre symbols are equal to unity:

$$\left(\frac{\overline{m}}{\mathfrak{P}_{i_1}^f}\right) = \left(\frac{m_{i_1}}{\mathfrak{P}_{i_1}^f}\right)\left(\frac{m_{i_2}}{\mathfrak{P}_{i_1}^f}\right) = \left(\frac{m_{i_1}}{\mathfrak{P}_{i_1}^f}\right)\left(\frac{m_{i_1}}{\mathfrak{P}_{i_1}^f}\right)^{-1} = 1.$$

Next, when computing the symbols we shall use the reciprocity law for pairs m_{i_1}, $m_{i_1}^f$ and m_{i_2}, $m_{i_2}^f$:

$$\begin{aligned}\left(\frac{\overline{m}}{\mathfrak{P}_{i_2}^f}\right) &= \left(\frac{m_{i_1}}{\mathfrak{P}_{i_2}^f}\right)\left(\frac{m_{i_2}}{\mathfrak{P}_{i_2}^f}\right) = \left(\frac{m_{i_2}^f}{\mathfrak{P}_{i_1}}\right)\left(\frac{m_{i_1}}{\mathfrak{P}_{i_1}^f}\right) = \left(\frac{m_{i_2}}{\mathfrak{P}_{i_1}^{f^{-1}}}\right)^f\left(\frac{m_{i_1}}{\mathfrak{P}_{i_1}^f}\right) \\ &= \left(\frac{m_{i_1}}{\mathfrak{P}_{i_1}^{f^{-1}}}\right)^{-f}\left(\frac{m_{i_1}}{\mathfrak{P}_{i_1}^f}\right) = \left(\frac{m_{i_1}^f}{\mathfrak{P}_{i_1}}\right)^{-1}\left(\frac{m_{i_1}}{\mathfrak{P}_{i_1}^f}\right) = \left(\frac{m_{i_1}}{\mathfrak{P}_{i_1}^f}\right)^{-1}\left(\frac{m_{i_1}}{\mathfrak{P}_{i_1}^f}\right) = 1.\end{aligned}$$

Now consider the case $p = 2$. Denote by $\overline{T}$ the set of the prime divisors of the field K that divide the discriminant of the extension K/k.

Let $\{F_1, F_2, F_3\}$ be a partition of the set of all elements of F different from the unit element such that

(1) F_1 consists of all elements of order 2.

(2) If $f \in F_2$, then $f^{-1} \in F_3$, and, conversely, if $f \in F_3$, then $f^{-1} \in F_2$.

We construct recurrently the sequence $\{m_i\}$ of numbers in K that satisfy the following conditions:

(1) $(m_i) = \mathfrak{P}_i \mathfrak{a}_i^2$, where $\mathfrak{P}_i$ is a prime divisor of $\mathfrak{p}_i$ that completely decomposes in the field L, and $\mathfrak{P}_i \notin T$; the divisors $\mathfrak{P}_i$ are distinct and relatively prime with $\mathfrak{a}_j$, where $i, j = 1, 2 \ldots$; $\mathfrak{a}_i$ is an integral divisor.

(2) $(\frac{m_i}{\mathfrak{P}}) = 1$ for $\mathfrak{P} \in \overline{T}$.

(3) The image of the number m_i in the group $K_{\mathfrak{P}}^*/K_{\mathfrak{P}}^{*2}$ is equal to $m_{\mathfrak{P}}$ for $\mathfrak{P} \in T$.

(4) The m_i is a square in the completions of K corresponding to infinite divisors and divisors of 2.

We shall require that the numbers m_i satisfy two more conditions, which we state below.

Suppose that the set M_{t-1} consisting of the numbers $m_1, \ldots, m_{t-1}$ has been constructed. We split M_{t-1} into subsets formed by the numbers for which the symbols $(\frac{m_i}{\mathfrak{P}_i^f})$ have the same values for all i, where $f \in F$, $f \neq 1$. Then the number of these subsets does not exceed $2^{[F:1]-1}$. The set M_{t-1} is ordered in a natural way, namely, $m_{i_1} < m_{i_2}$ if $i_1 < i_2$. This order on M_{t-1} induces orders on the subsets. We construct the sequence $\{m_i\}$ until some subset consists of three elements. Find a number m_t satisfying conditions (1)–(4) and the following two additional conditions:

(5) $(\frac{m_t}{\mathfrak{P}_j^f}) = 1$ for $j = 1, \ldots, t-1$; $f \in F_1$.

(6) If m_j is a maximal element in its subset, with $1 \leq j \leq t-1$, then

$$\left(\frac{m_t}{\mathfrak{P}_j^f}\right) = \begin{cases} 1 & \text{if } f \in F_2, \\ \left(\dfrac{m_j}{\mathfrak{P}_j^{f-1}}\right) & \text{if } f \in F_3; \end{cases}$$

if m_j is not maximal, then

$$\left(\frac{m_t}{\mathfrak{P}_j^f}\right) = \begin{cases} \left(\dfrac{m_j}{\mathfrak{P}_j^{f}}\right) & \text{if } f \in F_2, \\ 1 & \text{if } f \in F_3. \end{cases}$$

The existence of a number m_t satisfying conditions (1)–(6) is proved similar to the case of odd p.

Let the sequence $\{m_i\}$ have sufficiently many elements; then some subset will consist of three elements m_{i_1}, m_{i_2}, m_{i_3}. For definiteness, we assume that $i_1 < i_2 < i_3$. Put $\overline{m} = m_{i_1} m_{i_2} m_{i_3}$ and show that $\overline{m}$ is the required number. Compute the symbols $(\frac{m_{i_j}}{\mathfrak{P}_{i_j}^f})$, where $j = 1, 2, 3$; $f \in F_1$. We denote by K_f the subfield of K belonging to the cyclic subgroup of order 2 with generator f. We may assume that $K = K_f(w)$, where $w \in K$ and w^2 is an integer in K_f. The divisor (w) is the product of a divisor $\mathfrak{b}_1$ of the field K_f and a divisor $\mathfrak{b}_2$ of K, and the decomposition of $\mathfrak{b}_2$ into prime factors contains only divisors of the discriminant of the extension K/K_f, i.e., the divisors that occur in the set $\overline{T}$. Let $m_{i_j} = c_1 + c_2 w$, where $c_1, c_2 \in K_f$. Then $m_{i_j} - m_{i_j}^f = 2c_2 w$, and $2c_2 w$ is integral in the field K. We have

$$\left(\frac{m_{i_j}}{\mathfrak{P}_{i_j}^f}\right) = \left(\frac{m_{i_j} + 2c_2 w}{\mathfrak{P}_{i_j}^f}\right) = \left(\frac{2c_2 w}{\mathfrak{P}_{i_j}^f}\right).$$

By the reciprocity law,

$$\left(\frac{2c_2w}{\mathfrak{P}_{i_j}^f}\right) = \prod_{\mathfrak{P}}\left(\frac{m_{i_j}^f}{\mathfrak{P}}\right),$$

where $\mathfrak{P}$ ranges over all prime divisors of $(2c_2w)$ (this form of the reciprocity law is valid because the number m_{i_j} satisfies condition (4)). If $\mathfrak{P}$ is a divisor of 2 and is contained in $\overline{T}$, then $(\frac{m_{i_j}^f}{\mathfrak{P}}) = 1$ by construction. Otherwise,

$$\left(\frac{m_{i_j}^f}{\mathfrak{P}}\right) = \left(\frac{c_1 - c_2w}{\mathfrak{P}}\right) = \left(\frac{c_1}{\mathfrak{P}}\right),$$

and either $\mathfrak{P}$ is a divisor from the field K_f or $\mathfrak{P}^f$ also divides $(2c_2w)$. Since $c_1 \in K_f$, it follows that either $(\frac{c_1}{\mathfrak{P}}) = 1$ ($\mathfrak{P}$ is a divisor from the field K_f) or $(\frac{c_1}{\mathfrak{P}}) = (\frac{c_1}{\mathfrak{P}^f})$. Therefore, in either case $\prod_{\mathfrak{P}}(\frac{m_{i_j}^f}{\mathfrak{P}}) = 1$.

Now we compute the symbols $(\frac{\overline{m}}{\mathfrak{P}_{i_j}^f})$ for $f \in F_1$:

$$\left(\frac{\overline{m}}{\mathfrak{P}_{i_1}^f}\right) = \left(\frac{m_{i_1}}{\mathfrak{P}_{i_1}^f}\right)\left(\frac{m_{i_2}}{\mathfrak{P}_{i_1}^f}\right)\left(\frac{m_{i_3}}{\mathfrak{P}_{i_1}^f}\right) = 1,$$

$$\left(\frac{\overline{m}}{\mathfrak{P}_{i_2}^f}\right) = \left(\frac{m_{i_1}}{\mathfrak{P}_{i_2}^f}\right)\left(\frac{m_{i_2}}{\mathfrak{P}_{i_2}^f}\right)\left(\frac{m_{i_3}}{\mathfrak{P}_{i_2}^f}\right) = \left(\frac{m_{i_1}}{\mathfrak{P}_{i_2}^f}\right).$$

By the reciprocity law, $(\frac{m_{i_1}}{\mathfrak{P}_{i_2}}) = (\frac{m_{i_2}^f}{\mathfrak{P}_{i_1}})$. Furthermore,

$$\left(\frac{m_{i_2}^f}{\mathfrak{P}_{i_1}}\right) = \left(\frac{m_{i_2}}{\mathfrak{P}_{i_1}^f}\right) = 1, \quad \text{i.e.,} \quad \left(\frac{\overline{m}}{\mathfrak{P}_{i_2}^f}\right) = 1;$$

$$\left(\frac{\overline{m}}{\mathfrak{P}_{i_3}^f}\right) = \left(\frac{m_{i_1}}{\mathfrak{P}_{i_3}^f}\right)\left(\frac{m_{i_2}}{\mathfrak{P}_{i_3}^f}\right)\left(\frac{m_{i_3}}{\mathfrak{P}_{i_3}^f}\right) = \left(\frac{m_{i_3}^f}{\mathfrak{P}_{i_1}}\right)\left(\frac{m_{i_3}^f}{\mathfrak{P}_{i_2}}\right) = \left(\frac{m_{i_3}}{\mathfrak{P}_{i_1}^f}\right)\left(\frac{m_{i_3}}{\mathfrak{P}_{i_2}^f}\right) = 1.$$

Now let $f \in F_2$; then we have

$$\left(\frac{\overline{m}}{\mathfrak{P}_{i_1}^f}\right) = \left(\frac{m_{i_1}}{\mathfrak{P}_{i_1}^f}\right)\left(\frac{m_{i_2}}{\mathfrak{P}_{i_1}^f}\right)\left(\frac{m_{i_3}}{\mathfrak{P}_{i_1}^f}\right) = \left(\frac{m_{i_1}}{\mathfrak{P}_{i_1}^f}\right)\left(\frac{m_{i_1}}{\mathfrak{P}_{i_1}^f}\right) = 1,$$

$$\begin{aligned}\left(\frac{\overline{m}}{\mathfrak{P}_{i_2}^f}\right) &= \left(\frac{m_{i_1}}{\mathfrak{P}_{i_2}^f}\right)\left(\frac{m_{i_2}}{\mathfrak{P}_{i_2}^f}\right)\left(\frac{m_{i_3}}{\mathfrak{P}_{i_2}^f}\right) = \left(\frac{m_{i_2}^f}{\mathfrak{P}_{i_1}}\right)\left(\frac{m_{i_2}}{\mathfrak{P}_{i_2}^f}\right)\\ &= \left(\frac{m_{i_2}}{\mathfrak{P}_{i_1}^{f^{-1}}}\right)\left(\frac{m_{i_2}}{\mathfrak{P}_{i_2}^f}\right) = \left(\frac{m_{i_1}}{\mathfrak{P}_{i_1}^f}\right)\left(\frac{m_{i_2}}{\mathfrak{P}_{i_2}^f}\right) = 1,\end{aligned}$$

$$\begin{aligned}\left(\frac{\overline{m}}{\mathfrak{P}_{i_3}^f}\right) &= \left(\frac{m_{i_1}}{\mathfrak{P}_{i_3}^f}\right)\left(\frac{m_{i_2}}{\mathfrak{P}_{i_3}^f}\right)\left(\frac{m_{i_3}}{\mathfrak{P}_{i_3}^f}\right) = \left(\frac{m_{i_3}^f}{\mathfrak{P}_{i_1}}\right)\left(\frac{m_{i_3}^f}{\mathfrak{P}_{i_2}}\right)\left(\frac{m_{i_3}}{\mathfrak{P}_{i_3}^f}\right)\\ &= \left(\frac{m_{i_3}}{\mathfrak{P}_{i_1}^{f^{-1}}}\right)\left(\frac{m_{i_3}}{\mathfrak{P}_{i_2}^{f^{-1}}}\right)\left(\frac{m_{i_3}}{\mathfrak{P}_{i_3}^f}\right) = \left(\frac{m_{i_2}}{\mathfrak{P}_{i_2}^f}\right)\left(\frac{m_{i_3}}{\mathfrak{P}_{i_3}^f}\right) = 1.\end{aligned}$$

The case $f \in F_3$ is handled similarly. Clearly, the number $\overline{m}$ satisfies conditions (2) and (3) of the lemma. □

We prove another auxiliary statement.

LEMMA 5.3.2. *Let k be a local field (a finite extension of the field of p-adic numbers), K a finite l-extension of k, where l is a prime number different from p. Then the Galois group F of the extension K/k has at most two generators.*

PROOF. We may assume that k contains all of the lth roots of 1. If K_0 is the fixed field for the Frattini subgroup of F, then it suffices to prove the statement for an extension K_0/k with an elementary Abelian l-group F_0.

The number of generators of F_0 does not exceed the dimension of the space k^*/k^{*l} over the finite field $\mathbb{Z}/l\mathbb{Z}$. The group k^* is the direct product of the group V of units and the infinite cyclic group generated by a prime element π.

Let V_1 be the subgroup of principal units of V, i.e., the units v with $v \equiv 1 \pmod{\pi}$. The quotient group V/V_1 is isomorphic to the multiplicative group of the residue field of the ring of integral elements of k modulo π ([**1**, Chapter 1, Proposition 1.4]). By Hensel's lemma ([**4**, Chapter IV, §3, Theorem 2]), every principal unit is an lth power and, thus, V/V^l is a group of order l and k^*/k^{*l} has dimension 2 over $\mathbb{Z}/l\mathbb{Z}$. □

§4. The Neukirch theorem

In the present section we consider the embedding problem with solvable kernel of odd order in the case where the field to be embedded does not contain a primitive pth root of unity, where p is a prime number dividing the order of the kernel. In this special case the solvability of the global embedding problem is equivalent to the solvability of associated local problems.

1°. Let $(K/k, G, \varphi, B)$ be the embedding problem for algebraic number fields which is related to the exact sequence of groups $1 \longrightarrow B \longrightarrow G \longrightarrow F = \mathrm{Gal}(K/k) \longrightarrow 1$. We shall assume the following:

1. B is a solvable group of odd order and exponent n.
2. If p is a prime number dividing n, then K does not contain a primitive pth root of unity.

Let S be a finite set of prime divisors of k containing the divisors of n and the divisors of the discriminant of the extension K/k. For an arbitrary point $\mathfrak{p}$ of k, we denote by $P_\mathfrak{p}$ the set of solutions of the associated embedding problem $(K_\mathfrak{p}/k_\mathfrak{p}, G, \varphi, B)$. If P is the set of solutions of the global problem, then there is a natural mapping ψ of P to $\prod_{\mathfrak{p}\in S} P_\mathfrak{p}$ taking $L \in P$ to the set of local solutions $L_\mathfrak{p} = L \otimes_k k_\mathfrak{p}$. We call $\psi(L)_\mathfrak{p}$ *the local behavior of the solution L at the point $\mathfrak{p}$*.

Since the solvability of the local embedding problems is a necessary condition for the solvability of the global problem, the condition $P_\mathfrak{p} = \varnothing$ (the local problem is unsolvable) implies that $P = \varnothing$; consequently, in the sequel we shall consider the global problems for which the set $P_\mathfrak{p}$ is not empty for all $\mathfrak{p} \in S$.

THEOREM 5.4 (J. Neukirch). *The mapping ψ is surjective. Moreover, one may choose a proper solution of the global problem with given local behavior at the points of the set S.*

The theorem not only reduces the solution of the global embedding problem to the solution of finitely many local problems, but it enables us to construct a solution of the global problem with given local behavior at finitely many points.

The proof of the theorem proceeds in several stages. First let the kernel B be an Abelian group that is a simple F-module, i.e., a module not containing proper submodules different from zero. In this case we shall denote the kernel by A. It is clear that A is an elementary Abelian p-group and, as an F-module, it has a single generator.

Denote by K_0 the field that is obtained by adjoining to K all of the nth roots of 1. Let K_1 be a maximal Abelian extension of K_0 of period p, unramified outside S. It is clear that K_1 is a finite extension of K_0 and is normal over k. We denote by F_1 and F_0 the Galois groups of K_1/k and K_1/K_0, respectively.

Let θ_0 be the Artin homomorphism of the idèle group J_{K_0} onto the group F_0. Note that θ_0 annihilates the subgroup $V_{K_0,S}$. We show the existence of a finite set $\overline{S}_1$ of divisors of the field k, which completely decompose in the field K_0 and such that the image $\theta_0(V_{K_0,S}\prod_{\mathfrak{p}\in\overline{S}_1} K_{0\mathfrak{p}}^*)$ generates the group F_0. Indeed, let a be an arbitrary element of the group F_0 and let $[a]$ be the class of conjugates to a in F_1. By the Chebotarev Density Theorem, there exists a prime divisor $\mathfrak{p}_a$ of the field k that belongs to the class $[a]$. It is clear that if a set of elements a is such that the union $\bigcup_a [a]$ coincides with the group F_0, then the prime divisors $\mathfrak{p}_a$ form the set $\overline{S}_1$. We put $S_1 = S \cup \overline{S}_1$.

Let $\overline{k}$ be an algebraic closure of the field k with Galois group $\overline{F}$, and let Ω be the maximal algebraic extension of the field k, unramified outside S_1, with Galois group F_{S_1}, i.e., the composite of the finite normal field extensions of k, in which only the divisors from S_1 are ramified.

We assume that composition law on the set of solutions of the embedding problems $(K/k, G_h, \varphi_h, A)$ is given; here h ranges over all elements of the group $H^2(F, A)$ (see §15, Chapter III). In the semidirect extension G_0 the system of representatives $\{\overline{f}_0\}_{f\in F}$ of the cosets of A forms a subgroup isomorphic to F. Thus, an inclusion map β of F in G_0 is given.

It is clear that K is a subfield of Ω. Therefore, the lifting homomorphism λ of $H^2(F, A)$ to $H^2(F_{S_1}, A)$ is well defined. Let h be the cohomology class in $H^2(F, A)$ that corresponds to the group extension G.

LEMMA 5.4.1. *The existence of a solution of $(K/k, G, \varphi, A)$, unramified outside S_1, is equivalent to the fulfilment of the relation $\lambda(h) = 0$.*

The proof repeats almost literally the argument of Theorem 3.13.2. The distinction between the arguments is in the necessity of replacing the Galois group of the algebraic closure by the Galois group F_{S_1} of Ω/k.

PROPOSITION 5.4.1. *If A is a simple module, then there exists a solution of $(K/k, G, \varphi, A)$, unramified outside S_1.*

PROOF. In accordance with the previous lemma, it suffices to deduce that $\lambda(h) = 0$. Consider the commutative diagram

$$\begin{array}{ccc} H^2(F_{S_1}, A) & \xrightarrow{\psi_1} & \sum\limits_{\mathfrak{p}\in S_1}^{\oplus} H^2(\overline{F}_{\mathfrak{p}}, A) \\ \uparrow{\scriptstyle\lambda} & & \uparrow{\scriptstyle\lambda_{S_1}} \\ H^2(F, A) & \xrightarrow{\psi_2} & \sum\limits_{\mathfrak{p}\in S_1}^{\oplus} H^2(F_{\mathfrak{p}}, A) \end{array}$$

where $\overline{F}_{\mathfrak{p}}$ is the Galois group of the algebraic closure of the local field $k_{\mathfrak{p}}$, ψ_2 is a collection of restriction homomorphisms onto the decomposition groups of the divisors from S_1, λ_{S_1} is a collection of local lifting homomorphisms, ψ_1 is a natural mapping whose description may be found in §3 of the Appendix. We note that for $\mathfrak{p} \in S_1$ the associated local embedding problems are solvable. For $\mathfrak{p} \in \overline{S}_1$ this follows from Lemma 3.14, and for $\mathfrak{p} \in S$ they are solvable by assumption. Thus, $\lambda_{S_1}\psi_2(h) = 0$ or $\psi_1\lambda(h) = 0$ by the commutativity of the diagram.

Now we show that ψ_1 is an isomorphism, whence the proposition follows. We make use of the Tate duality. We have the commutative diagram

$$\begin{array}{ccc} H^2(F_{S_1}, A) & \xrightarrow{\psi_1} & \sum\limits_{\mathfrak{p}\in S_1}^{\oplus} H^2(\overline{F}_{\mathfrak{p}}, A) \\ \downarrow & & \downarrow \\ \operatorname{Hom}_F(A, C_{K,S_1})' & \xrightarrow{\overline{\psi}_4} & \operatorname{Hom}_F(A, J_{K,S_1})' \end{array}$$

where $\overline{\psi}_4$ is the mapping dual to the natural homomorphism $\psi_4\colon \operatorname{Hom}_F(A, J_{K,S_1}) \longrightarrow \operatorname{Hom}_F(A, C_{K,S_1})$ induced by the mapping ψ_3 of J_{K,S_1} onto C_{K,S_1}. Clearly, it suffices to show that ψ_4 is an isomorphism.

The exact sequence $1 \longrightarrow K^* \longrightarrow J_{K,S_1} \xrightarrow{\psi_3} C_{K,S_1} \longrightarrow 1$ induces the exact sequence $1 \longrightarrow \operatorname{Hom}(A, K^*) \longrightarrow \operatorname{Hom}(A, J_{K,S_1}) \xrightarrow{\psi_5} \operatorname{Hom}(A, C_{K,S_1})$. The group K^* contains only the trivial pth root of unity; so, the group $\operatorname{Hom}(A, K^*)$ is trivial. Now we show that ψ_5 is an isomorphism. Let x be an element of order p in C_{K,S_1}. Then we have $x_1^p \in K^*$ for a preimage $x_1 = \psi_3^{-1}(x)$. We put $x_2 = x_1^p$. The extension $K_0(\sqrt[p]{x_2})$ is contained in the field K_1. Since x_2 is a pth power in all of the completions $K_{0\mathfrak{P}}$ with $\mathfrak{P} \in S_1$, the element x_2 is a pth power in K_0 and there is an element $x_3 \in K_0^{*p}$ such that $x_3^p = x_2$ (here we use the fact that the divisors from $\overline{S}_1$ generate the Galois group F_0 of K_1/K_0).

The field K_0 is obtained by adjoining to K all of the nth roots of unity. There is a subfield K_2 in K_0 such that it does not contain nontrivial pth roots of unity and K_0 is obtained by adjoining to K_2 a primitive pth root ζ of unity. Let γ be a generator of the Galois group of K_0/K_2, and let F_2 be the Galois group of K_2/K. The action of γ on the pth roots of unity is defined by the relation $\zeta^\gamma = \zeta^j$, where j is an integer, $1 < j < p$. We have $x_3^{p\gamma} = x_2$ or $x_3^{p(\gamma-1)} = 1$, i.e., $x_3^{\gamma-1} = \zeta^t$ for some integer t. Since $j \not\equiv 1 \pmod p$, there exists an integer $\overline{j}$ with the property $\overline{j}(j-1) \equiv 1 \pmod p$. Then $\zeta^{\overline{j}t(\gamma-1)} = \zeta^t$ is fulfilled. We put $x_4 = x_3\zeta^{-\overline{j}t}$, then we have $x_4^p = x_2$ and $x_4^{\gamma-1} = 1$, i.e., $x_4 \in K_2$. Further, let f be an arbitrary element of F_2. Then $x_4^{(f-1)p} = 1$. The field K_2 contains only the trivial pth root of unity, consequently, $x_4^{f-1} = 1$ and $x_4 \in K^*$. By setting $x_5 = x_1x_4^{-1}$, we have $x_5^p = 1$ and $\psi_3(x_5) = x$. Thus, the groups $\operatorname{Hom}(A, J_{K,S_1})$ and $\operatorname{Hom}(A, C_{K,S_1})$ are isomorphic, whence ψ_4 is an isomorphism.

2°. Let μ be the homomorphism of the group $\overline{F}$ onto F corresponding to the inclusion map of K in $\overline{k}$. Consider the Galois group $\overline{F}_{\mathfrak{p}}$ of the algebraic closure $\overline{k}_{\mathfrak{p}}$ of the local field $k_{\mathfrak{p}}$ and the composition $\mu_{\mathfrak{p}}\colon \overline{F}_{\mathfrak{p}} \longrightarrow F$ of the homomorphism from $\overline{F}_{\mathfrak{p}}$ onto the Galois group $F_{\mathfrak{P}}$ of an extension $K_{\mathfrak{P}}/k_{\mathfrak{p}}$ (of the field-kernel of the algebra $K_{\mathfrak{p}} = K \otimes k_{\mathfrak{p}}$) and the inclusion map $F_{\mathfrak{P}} \longrightarrow F$. The homomorphism $\mu_{\mathfrak{p}}$ that we assign to $K_{\mathfrak{p}}$ clearly depends on the choice of a prime point $\mathfrak{P}$ of the field K that is lying over $\mathfrak{p}$ and on the choice of an inclusion map $K_{\mathfrak{P}} \longrightarrow \overline{k}_{\mathfrak{p}}$. Now note that there is a homomorphism $\sigma_{\mathfrak{p}} \colon \overline{F}_{\mathfrak{p}} \longrightarrow \overline{F}$ such that $\mu\sigma_{\mathfrak{p}} = \mu_{\mathfrak{p}}$. (We give a description of similar mappings and induced mappings of the cohomology group $H^i(\overline{F}, A) \longrightarrow H^i(\overline{F}_{\mathfrak{p}}, A)$ in the Appendix, §3.) □

The inclusion map of Ω in $\overline{k}$ induces a homomorphism $\overline{\mu}_{S_1}$ of $\overline{F}$ onto F_{S_1}. The field K is contained in Ω, whence the homomorphism μ_{S_1} of F_{S_1} onto F is well defined. We define the mapping $\sigma_{1\mathfrak{p}}$ of $\overline{F}_{\mathfrak{p}}$ to F_{S_1} by the commutativity of the diagram

$$\begin{array}{ccc} \overline{F}_{\mathfrak{p}} & \xrightarrow{\sigma_{\mathfrak{p}}} & \overline{F} \\ & \sigma_{1\mathfrak{p}} \searrow & \downarrow \overline{\mu}_{S_1} \\ & & F_{S_1}. \end{array}$$

Note that if a homomorphism $\nu \colon F_{S_1} \longrightarrow G$ corresponds to a solution L of $(K/k, G, \varphi, A)$, unramified outside S_1, then the homomorphism $\nu\sigma_{1\mathfrak{p}}$ corresponds to the local solution $L_{\mathfrak{p}} = L \otimes_k k_{\mathfrak{p}}$.

PROPOSITION 5.4.2. *Let A be a simple F-module and let the natural mapping $H^1(F_{S_1}, A) \longrightarrow \prod_{\mathfrak{p}\in S} H^1(\overline{F}_{\mathfrak{p}}, A)$ be surjective. Then the Neukirch theorem is valid for the embedding problem with kernel A.*

PROOF. Given an element q_S in $\prod_{\mathfrak{p}\in S} P_{\mathfrak{p}}$, we may relate a homomorphism $\nu_{\mathfrak{p}} \colon \overline{F}_{\mathfrak{p}} \longrightarrow G$ to every solution $L_{\mathfrak{p}}$ of the local problem. We note that $\nu_{\mathfrak{p}}$ is defined up to coboundary from $B^1(\overline{F}_{\mathfrak{p}}, A)$ (see Chapter 1, §14).

We must show the existence of a homomorphism $\nu \colon F_{S_1} \longrightarrow G$ such that the following conditions are fulfilled:

(1) $\varphi\nu = \mu_{S_1}$ (ν corresponds to some solution L of $(K/k, G, \varphi, A)$, unramified outside S_1);

(2) $\nu\sigma_{1\mathfrak{p}}(\overline{f}_{\mathfrak{p}}) = \nu_{\mathfrak{p}}(\overline{f}_{\mathfrak{p}}) a_{\mathfrak{p}}^{\overline{f}\mathfrak{p}-1}$, where $\overline{f} \in \overline{F}_{\mathfrak{p}}$, $a_{\mathfrak{p}} \in A$ (the local behavior $L \otimes_k k_{\mathfrak{p}}$ of the solution L coincides with the given local solution $L_{\mathfrak{p}}$);

(3) ν is a surjective homomorphism (L is a field).

By Proposition 5.4.1, there is a solution L_1 of $(K/k, G, \varphi, A)$, unramified outside S_1. Let a homomorphism $\nu_1 : F_{S_1} \longrightarrow G$ correspond to L_1. As mentioned above, the local behavior $L_1 \otimes_k k_{\mathfrak{p}}$ corresponds to the homomorphism $\nu_1\sigma_{1\mathfrak{p}} : \overline{F}_{\mathfrak{p}} \longrightarrow G$. Then $\nu_{\mathfrak{p}}(\overline{f}_{\mathfrak{p}}) = \nu_1\sigma_{1\mathfrak{p}}(\overline{f}_{\mathfrak{p}})\overline{h}_{\mathfrak{p}}(\overline{f}_{\mathfrak{p}})$, where $\overline{f}_{\mathfrak{p}} \in \overline{F}_{\mathfrak{p}}$ and $\overline{h}_{\mathfrak{p}}$ is a one-dimensional cocycle of the group $Z^1(\overline{F}_{\mathfrak{p}}, A)$ (see Theorem 1.15.2). □

By assumption, in $Z^1(F_{S_1}, A)$ there is a cocycle $\overline{h}$ such that its images in $Z^1(\overline{F}_{\mathfrak{p}}, A)$ are cohomologous to the cocycles $\overline{h}_{\mathfrak{p}}$ for $\mathfrak{p} \in S$. Let $\nu(\overline{f}) = \nu_1(\overline{f})\overline{h}(\overline{f})$, where $\overline{f} \in F_{S_1}$. Since $\overline{h}$ is a cocycle, $\nu \colon F_{S_1} \longrightarrow G$ is a group homomorphism. Clearly, $\varphi\nu = \mu_{S_1}$.

Further, $\nu\sigma_{1\mathfrak{p}}(\overline{f}_{\mathfrak{p}}) = \nu_1\sigma_{1\mathfrak{p}}(\overline{f}_{\mathfrak{p}})\overline{h}(\sigma_{1\mathfrak{p}}(\overline{f}_{\mathfrak{p}})) = \nu_{\mathfrak{p}}(\overline{f}_{\mathfrak{p}}) a_{\mathfrak{p}}^{1-\overline{f}_{\mathfrak{p}}}$ for $a_{\mathfrak{p}} \in A$ (here we use the fact that $\overline{h}_{\mathfrak{p}}$ is cohomologous to the image of $\overline{h}$).

Generally speaking, it may occur that the Galois algebra L corresponding to ν is not a field. We show that in this case L splits completely. Let L_0 be the field-kernel of L and H the Galois group of L_0/k. Then, by the simplicity of the F-module A, the kernel A_0 of the natural mapping of H onto F is trivial (A_0 is a proper submodule of A). It is clear that in this case the embedding problem $(K/k, G, \varphi, A)$ is semidirect.

Now we choose a solution L_2 so that $L_3 = L \circ L_2$ is a field and the local behavior of L_3 coincides with the algebras $L_\mathfrak{p}$ for $\mathfrak{p} \in S$.

Let $\mathbb{Z}/p\mathbb{Z}[F]$ be a group ring of F over the field with p elements and D a free $\mathbb{Z}/p\mathbb{Z}[F]$-module with one generator, i.e., a set of elements of the form $\prod_{f\in F} d_f^{n_f}$, where $n_f \in \mathbb{Z}/p\mathbb{Z}$.

Since A is a simple F-module, there is a mapping δ of D onto A (δ takes the generator d_1 to a generator of A). The sequence

$$1 \longrightarrow D_0 \longrightarrow D \xrightarrow{\delta} A \longrightarrow 1$$

is exact.

The mapping δ induces the homomorphism $\overline{\delta}$ of a semidirect extension of D by F onto G_0, namely, $\overline{\delta}(f \cdot d) = \beta(f) \cdot \delta(d)$, where $f \in F$ and $d \in D$ (F acts on the elements $d \in D$ in a natural way, namely, $(\prod_{f\in F} d_f^{n_f})^{f'} = \prod_{f\in F} d_{ff'}^{n_f}$, β is the inclusion $F \longrightarrow G$). Let K_3 be the field obtained by adjoining to K all the pth roots of unity, F_3 the Galois group of K_3/k. Let γ_1 be a generator of the Galois group Γ_1 of K_3/K, and let c be the order of γ_1. There is a number j such that $\zeta^{\gamma_1} = \zeta^j$. Note that $e_t = \frac{1}{c}\sum_{i=0}^{c-1} j^{-it}\gamma_1^i$ for $t = 0, \ldots, c-1$ are pairwise orthogonal idempotents in the group ring $\mathbb{Z}/p\mathbb{Z}[\Gamma_1]$. We may speak about the action of the ring $\mathbb{Z}/p\mathbb{Z}[\Gamma_1]$ on the group K_3^*/K_3^{*p}. Then the module K_3^*/K_3^{*p} is a direct product of the modules $(K_3^*/K_3^{*p})^{e_t}$, and the action of γ_1 on $(K_3^*/K_3^{*p})^{e_t}$ is of the form $z^{\gamma_1} = z^{j^t}$. The converse is also valid, i.e., if $z \in K_3^*/K_3^{*p}$ and $z^{\gamma_1} = z^{j^t}$, then $z \in (K_3^*/K_3^{*p})^{e_t}$.

LEMMA 5.4.2. *All the solutions of the semidirect embedding problem $(K/k, F \cdot D, D)$ are in one-to-one correspondence with the elements of the group $(K_3^*/K_3^{*p})^{e_1}$.*

PROOF. The semidirect extension $F \cdot D$ is the wreath product of the cyclic group $\mathbb{Z}/p\mathbb{Z}$ and F. Therefore, the tensor product $L = \bigotimes_k L_f$ of the conjugate algebras L_f (with $f \in F$) over the field K with group $\mathbb{Z}/p\mathbb{Z}$ (see Chapter 3, §6) is a solution L of the problem under consideration.

It remains to make clear the structure of a cyclic algebra of order p over K. Let $m^{e_1} \in (K_3^*/K_3^{*p})^{e_1}$, and let $\overline{m}$ be a preimage of m^{e_1} in the group K_3^*. Consider the algebra $\widetilde{K}_3 = K_3(\sqrt[p]{\overline{m}})$. Let $g \in \mathrm{Gal}(\widetilde{K}_3/K_3)$ act on $\sqrt[p]{\overline{m}}$ by the rule $\sqrt[p]{\overline{m}}^{\,g-1} = \zeta$. Let $\widetilde{\gamma}_1$ be an extension of γ such that $\sqrt[p]{\overline{m}}^{\,\widetilde{\gamma}_1} = \sqrt[p]{\overline{m}}^{\,j} m_0$, where $m_0 \in K_3^*$. We have $(\sqrt[p]{\overline{m}}^{\,g})^{\widetilde{\gamma}_1} = (\sqrt[p]{\overline{m}}\zeta)^{\widetilde{\gamma}_1} = \sqrt[p]{\overline{m}}^{\,j} m_0\zeta^j$. On the other hand, $(\sqrt[p]{\overline{m}}^{\,\widetilde{\gamma}_1})^g = (\sqrt[p]{\overline{m}}^{\,j} m_0)^g = \sqrt[p]{\overline{m}}^{\,j}\zeta^j m_0$. Thus, $\widetilde{K}_3$ is a Galois algebra over the field K with group $\mathbb{Z}/p\mathbb{Z} \times \Gamma_1$. The subalgebra L_1 belonging to the cyclic subgroup $\{\widetilde{\gamma}_1\}$ is a Galois algebra over the field K with group $\mathbb{Z}/p\mathbb{Z}$.

Conversely, let L_1 be a Galois algebra over K with group $\mathbb{Z}/p\mathbb{Z}$. The algebra $\widetilde{K}_3 = L_1 \otimes_K K_3$ has group $\mathbb{Z}/p\mathbb{Z} \times \Gamma_1$ over K. Observe that $\widetilde{K}_3 = K_3(\sqrt[p]{\overline{m}})$, where $\overline{m} \in K_3^*$.

Let g and $\widetilde{\gamma}_1$ be generators of $\mathrm{Gal}(\widetilde{K}_3/K)$ such that g leaves the elements of K_3 fixed, and $\widetilde{\gamma}_1$ leaves the elements of L_1 fixed. Note that $\sqrt[p]{\overline{m}}^{\,\widetilde{\gamma}_1} = \sqrt[p]{\overline{m}}^{\,v} m_0$, where $v \in \mathbb{Z}$ and $m_0 \in K_3$. If $\sqrt[p]{\overline{m}}^{\,g} = \sqrt[p]{\overline{m}}\zeta^u$, then $\sqrt[p]{\overline{m}}^{\,g\widetilde{\gamma}_1} = \sqrt[p]{\overline{m}}^{\,v} m_0\zeta^{uj}$. Since $\sqrt[p]{\overline{m}}^{\,g\widetilde{\gamma}_1} = \sqrt[p]{\overline{m}}^{\,\widetilde{\gamma}_1 g}$, we have $\sqrt[p]{\overline{m}}^{\,v} m_0\zeta^{uj} = \sqrt[p]{\overline{m}}^{\,v} m_0\zeta^{uv}$, i.e., $j \equiv v \pmod p$. Therefore, $\overline{m}^{\widetilde{\gamma}_1}\overline{m}^{-j} \in K_3^{*p}$. The lemma is proved. □

By the Chebotarev density theorem, there is a divisor $\mathfrak{p}_0$ of the field k that completely decomposes in K_1. Let $\mathfrak{P}_0$ be a prime divisor of $\mathfrak{p}_0$ in K_0, and let π be a prime element in the field $k_{\mathfrak{p}_0}$. Consider the idèle y_1 from the group J_{K_0} whose component at the point $\mathfrak{P}_0$ is equal to π and the other components are equal to 1. By class field theory, the Galois group of K_1/K_0 is isomorphic to $J_{K_0}/J_{K_0}^p K_0^* V_{K_0,S}$; so, $y_1 \in J_{K_0}^p K_0^* V_{K_0,S}$, i.e., there exists $y_2 \in K_0$ such that the divisor (y_2) is of the form $\mathfrak{P}_0 \mathfrak{a}^p$, where $\mathfrak{a}$ is a divisor of K_0. Moreover, y_2 is a pth power in the completions of K_0 corresponding to the divisors from S.

Let $\overline{m}$ be the norm of y_2 from K_0 to K_3. Clearly, $(\overline{m}) = \mathfrak{P}_3 \mathfrak{a}_3^p$, where $\mathfrak{P}_3$ is a prime divisor of $\mathfrak{p}_0$ in K_3, and $\mathfrak{a}_3$ is a divisor of K_3. Moreover, $\overline{m}$ is a pth power in the completions of K_3 corresponding to the divisors from S.

By Lemma 5.4.2, m^{e_1} determines a solution L_4 of the embedding problem $(K/k, F \cdot D, D)$, where m is the image of $\overline{m}$ in K_3^*/K_3^{*p}.

We show that L_4 is a field. It suffices to prove that $\widetilde{L}_4 = L_4 \otimes_K K_3$ is a field. Let $\{\overline{f}\}_{f \in F}$ be representatives of cosets of Γ_1 in F_3. In K_3^* we choose preimages $\overline{m}_f$ of the elements $m^{e,\overline{f}}$. Then we may assume that $\widetilde{L}_4$ is obtained by adjoining to K_3 the radicals $\sqrt[p]{\overline{m}_f}$. We note that $\widetilde{L}_4$ is a field if and only if the fact that $\prod_{f \in F} \overline{m}_f^{z_f}$ is a pth power in K_3 implies that p divides all of the integers z_f.

Consider the divisor $\prod_{f \in F} (\overline{m}_f)^{z_f}$ and its image $\prod_{f \in F} (\overline{m}_f)^{z_f} \bmod \mathfrak{M}^p$ in the quotient group $\mathfrak{M}/\mathfrak{M}^p$, where $\mathfrak{M}$ is the group of divisors of K_3. Then the condition $\prod_{f \in F} (\overline{m}_f)^{z_f} \bmod \mathfrak{M}^p = 1$, or $\prod_{f \in F} (\mathfrak{P}_3 \bmod \mathfrak{M}^p)^{z_f \overline{f} e_1} = 1$, implies that p divides all of the numbers z_f.

Let ν_4 be the homomorphism of $\overline{F}$ onto $F \cdot D$ corresponding to the field L_4. Then the homomorphism $\nu_4 \sigma_{\mathfrak{p}}$ corresponds to the local algebra $L_{4\mathfrak{p}} = L_4 \otimes_k k_{\mathfrak{p}}$. Since the numbers $\overline{m}_f$ are pth powers in $K_{3\mathfrak{P}}$ for $\mathfrak{P} \in S$, $L_{4\mathfrak{p}}$ for $\mathfrak{p} \in S$ completely splits. Therefore, the image of $\nu_4 \sigma_{\mathfrak{p}}$ is isomorphic to the decomposition group of a prime divisor $\mathfrak{P}$ of $\mathfrak{p}$ in the extension K/k, i.e., $\nu_4 \sigma_{\mathfrak{p}}(\overline{f}_{\mathfrak{p}}) = \mu_{\mathfrak{p}}(\overline{f}_{\mathfrak{p}}) \overline{h}_{\mathfrak{p}}(\overline{f}_{\mathfrak{p}})$, where $\overline{f}_{\mathfrak{p}} \in \overline{F}_{\mathfrak{p}}$ and $\overline{h}_{\mathfrak{p}}$ is a cocycle in $Z^1(\overline{F}_{\mathfrak{p}}, D)$.

Let $\overline{F}_{0\mathfrak{p}}$ be the subgroup of $\overline{F}_{\mathfrak{p}}$ belonging to the subfield $K_{\mathfrak{P}}$. Now if $\overline{f}_{\mathfrak{p}} \in \overline{F}_{0\mathfrak{p}}$, then $\nu_4 \sigma_{\mathfrak{p}}(\overline{f}_{\mathfrak{p}}) = 1$ (this follows from the fact that the image of $\nu_4 \sigma_{\mathfrak{p}}$ is isomorphic to the Galois group $F_{\mathfrak{P}}$ of $K_{\mathfrak{P}}/k_{\mathfrak{p}}$, and, by assumption, $\varphi \nu_4 \sigma_{\mathfrak{p}} = \mu_{\mathfrak{p}}$). Therefore, the values of $\nu_4 \sigma_{\mathfrak{p}}$ depend on cosets with respect to $\overline{F}_{0\mathfrak{p}}$, i.e., the cocycle $\overline{h}_{\mathfrak{p}}$ is the lifting of a cocycle $\overline{h}_{1\mathfrak{p}} \in Z^1(F_{\mathfrak{P}}, D)$.

We show that $H^1(F_{\mathfrak{P}}, D)$ is trivial. Note that D is a p-group. Then, by Corollary 3 to Proposition 6.3. of [**1**, Chapter IV], it suffices to prove the triviality of the group $H^1(F_{\mathfrak{P},p}, D)$, where $F_{\mathfrak{P},p}$ is a Sylow p-subgroup of $F_{\mathfrak{P}}$. Now we note that D is a free $\mathbb{Z}/p\mathbb{Z}[F_{\mathfrak{P},p}]$-module. Therefore, by Theorem 9.1 of [**1**, Chapter IV], $H^1(F_{\mathfrak{P},p}, D) = 0$.

Thus, the cocycle $\overline{h}_{1\mathfrak{p}}$ splits; so, there is a $\widetilde{d} \in D$ such that

$$\nu_4 \sigma_{\mathfrak{p}}(\overline{f}_{\mathfrak{p}}) = \mu_{\mathfrak{p}}(\overline{f}_{\mathfrak{p}}) \widetilde{d}^{\overline{f}_{\mathfrak{p}} - 1}.$$

Let L_2 be the subfield of L_4 belonging to the subgroup D_0 (L_2 is a solution of $(K/k, G, \varphi, A)$). Then the homomorphism $\nu_2 = \overline{\delta} \nu_4$ corresponds to the field L_2. The homomorphism $\overline{\delta} \nu_4 \sigma_{\mathfrak{p}}$ defined by $\overline{\delta} \nu_4 \sigma_{\mathfrak{p}}(\overline{f}_{\mathfrak{p}}) = \beta \mu_{\mathfrak{p}}(\overline{f}_{\mathfrak{p}}) a^{\overline{f}_{\mathfrak{p}} - 1}$, where $a = \delta(\widetilde{d})$ corresponds to the local behavior $L_{2\mathfrak{p}} = L_2 \otimes_k k_{\mathfrak{p}}$ at the point $\mathfrak{p} \in S$.

Show that $L_3 = L \circ L_2$ is the required solution. We note that a homomorphism $\overline{\nu}\colon \overline{F} \longrightarrow G$ extending ν corresponds to the solution L. The homomorphism $\nu_3(\overline{f}) =$

$(\overline{\nu}(\overline{f}), \nu_2(\overline{f}))$ mod $\widetilde{A}$, where $\overline{f} \in \overline{F}$ corresponds to the algebra L_3. Since the image of $\overline{\nu}$ is isomorphic to F, ν_3 is surjective.

The local behavior $L_{3\mathfrak{p}} = L_3 \otimes k_{\mathfrak{p}}$ corresponds to the homomorphism $\nu_3\sigma_{\mathfrak{p}}(\overline{f}_{\mathfrak{p}})$ $= (\nu\sigma_{\mathfrak{p}}(\overline{f}_{\mathfrak{p}}), \nu_2\sigma_{\mathfrak{p}}(f_{\mathfrak{p}}))$ mod $\widetilde{A}$. We have $\nu_3\sigma_{\mathfrak{p}}(\overline{f}_{\mathfrak{p}}) = (\nu\sigma_{\mathfrak{p}}(\overline{f}_{\mathfrak{p}}), \beta\mu_{\mathfrak{p}}(\overline{f}_{\mathfrak{p}})a_{\mathfrak{p}}^{\overline{f}_{\mathfrak{p}}-1}) = \nu\sigma_{\mathfrak{p}}(\overline{f}_{\mathfrak{p}})a_{\mathfrak{p}}^{\overline{f}_{\mathfrak{p}}-1}$, i.e., the solutions $L_3 \otimes_k k_{\mathfrak{p}}$ and $L \otimes_k k_{\mathfrak{p}}$ coincide for $\mathfrak{p} \in S$. Proposition 5.4.2 is proved.

Thus, to complete the consideration of the case where the kernel of the embedding problem is a simple F-module A, the following proposition must be proved.

PROPOSITION 5.4.3. *If A is a simple F-module, then the natural mapping $H^1(F_{S_1}, A) \to \prod_{\mathfrak{p}\in S} H^1(\overline{F}_{\mathfrak{p}}, A)$ is surjective.*

PROOF. Consider again the exact sequence $1 \to D_0 \to D \to A \to 1$. We note that all the modules in this sequence are naturally F_{S_1}- and $\overline{F}_{\mathfrak{p}}$-modules.

Consider the following commutative diagram with natural homomorphisms:

$$\begin{array}{ccccccc}
H^1(F_{S_1}, D) & \xrightarrow{\tau_1} & H^1(F_{S_1}, A) & \xrightarrow{\tau_2} & H^2(F_{S_1}, D_0) & \xrightarrow{\tau_3} & H^2(F_{S_1}, D) \\
\downarrow{\scriptstyle \tau_4} & & \downarrow{\scriptstyle \tau_5} & & \downarrow{\scriptstyle \tau_6} & & \downarrow{\scriptstyle \tau_7} \\
\prod_{\mathfrak{p}\in S_1} H^1(\overline{F}_{\mathfrak{p}}, D) & \xrightarrow{\tau_8} & \prod_{\mathfrak{p}\in S_1} H^1(\overline{F}_{\mathfrak{p}}, A) & \xrightarrow{\tau_9} & \prod_{\mathfrak{p}\in S_1} H^2(\overline{F}_{\mathfrak{p}}, D_0) & \xrightarrow{\tau_{10}} & \prod_{\mathfrak{p}\in S_1} H^2(\overline{F}_{\mathfrak{p}}, D)
\end{array}$$

The diagram makes it sufficient to prove the proposition for D. Indeed, let $r_S \in \prod_{\mathfrak{p}\in S} H^1(\overline{F}_{\mathfrak{p}}, A)$ and $r_{\overline{S}_1} \in \prod_{\mathfrak{p}\in\overline{S}_1} H^1(\overline{F}_{\mathfrak{p}}, A)$. We put $r_{S_1} = (r_S, r_{\overline{S}_1})$. In proving Proposition 5.4.1 we have shown that the mapping $H^2(F_{S_1}, A) \to \prod_{\mathfrak{p}\in S_1} H^2(\overline{F}_{\mathfrak{p}}, A)$ is an isomorphism for any finite F-module A. Therefore, τ_6 and τ_7 are isomorphisms. Then there is $x_1 \in H^2(F_{S_1}, D_0)$ with $\tau_6(x_1) = \tau_9(r_{S_1})$. Note that $\tau_7\tau_3(x_1) = \tau_{10}\tau_6(x_1) = \tau_{10}\tau_9(r_{S_1}) = 0$; so, $\tau_3(x_1) = 0$, and there exists an element $x_2 \in H^1(F_{S_1}, A)$ such that $\tau_2(x_2) = x_1$. Then $\tau_9(\tau_5(x_2) - r_{S_1}) = \tau_6\tau_2(x_2) - \tau_6(x_1) = 0$, and, thus, there exists an element $\overline{r}_{S_1} \in \prod_{\mathfrak{p}\in S_1} H^1(\overline{F}_{\mathfrak{p}}, D)$ such that $\tau_8(\overline{r}_{S_1}) = \tau_5(x_2) - r_{S_1}$. □

Suppose the composition of τ_4 and the projection on the S-components is surjective. Then there is $x_3 \in H^1(F_{S_1}, D)$ with $\mathrm{pr}_S\,\tau_4(x_3) = \mathrm{pr}_S\,\overline{r}_{S_1}$. Set $x_4 = x_2 - \tau_1(x_3)$. Then we have $\tau_5(x_4) = \tau_5(x_2) - \tau_5\tau_1(x_3) = \tau_8(\overline{r}_{S_1}) + r_{S_1} - \tau_8\tau_4(x_3)$ and $\mathrm{pr}_S\,\tau_5(x_4) = \mathrm{pr}_S\,\tau_8(\overline{r}_{S_1} - \tau_4(x_3)) + r_S = r_S$. Thus, we have reduced the statement to the case of the module D.

Let $\overline{F}_0$ be the subgroup of F_{S_1} corresponding to the subfield K, and let $\mathbb{Z}[F_{S_1}/\overline{F}_0]$ be a free Abelian group with generators u_f, where u_f are in one-to-one correspondence with the cosets of $\overline{F}_0$ in F_{S_1} (with the elements of F). There is an operator isomorphism η: $\mathrm{Hom}(\mathbb{Z}[F_{S_1}/\overline{F}_0], \mathbb{Z}/p\mathbb{Z}) \to D$ given by $\eta(\chi) = \prod_f d_f^{\chi(u_f)}$.

Consider the commutative diagram with natural homomorphisms

$$\begin{array}{ccc}
H^1(F_{S_1}, D) & \xrightarrow{\tau_4} & \prod_{\mathfrak{p}\in S_1} H^1(\overline{F}_{\mathfrak{p}}, D) \\
\downarrow{\scriptstyle \tau_{11}} & & \downarrow{\scriptstyle \tau_{12}} \\
H^1(\overline{F}_0, \mathbb{Z}/p\mathbb{Z}) & \xrightarrow{\tau_{13}} & \prod_{\mathfrak{P}\in S_1} H^1(\overline{F}_{0\mathfrak{P}}, \mathbb{Z}/p\mathbb{Z}),
\end{array}$$

where the product $\prod_{\mathfrak{p}\in S_1} H^1(\overline{F}_\mathfrak{p}, D)$ is taken over the divisors of k, and the product $\prod_{\mathfrak{P}\in S_1} H^1(\overline{F}_{0\mathfrak{P}}, \mathbb{Z}/p\mathbb{Z})$ is taken over the divisors of K. It follows from Lemma A.3.9 that τ_{11} and τ_{12} are isomorphisms.

It suffices to show that the composition of τ_{13} and the projection on the S-components is surjective. Note that $H^1(\overline{F}_0, \mathbb{Z}/p\mathbb{Z})$ is dual to $C_{K,S_1}/C^p_{K,S_1}$, and $\prod_{\mathfrak{P}\in S} H^1(F_{0\mathfrak{P}}, \mathbb{Z}/p\mathbb{Z})$ is dual to $\prod_{\mathfrak{P}\in S} K^*_\mathfrak{P}/K^{*p}_\mathfrak{P}$.

Consider the exact sequence

$$1 \longrightarrow K^*/K^{*p} \xrightarrow{\tau_{14}} J_{K,S_1}/J^p_{K,S_1} \xrightarrow{\tau_{15}} C_{K,S_1}/C^p_{K,S_1} \longrightarrow 1$$

(as we saw, τ_{14} is injective). The surjectivity of the composition of τ_{13} and the projection on the S-components is equivalent to the following assertion. If $x_5 \in J_{K,S_1}/J^p_{K,S_1}$, $\tau_{15}(x_5) = 1$, and the projection of x_5 on the components complementary to S is trivial, then $x_5 = 1$. Let $\tau_{15}(x_5) = 1$. Then there is $x_6 \in K^*/K^{*p}$ with $\tau_{14}(x_6) = x_5$. Let x_7 denote a preimage of x_6 in K^*. Then x_7 is a pth power in the completions $K_\mathfrak{P}$ of K for $\mathfrak{P} \in \overline{S}_1$. Moreover, the extension $K_0(\sqrt[p]{x_7})/K_0$ is unramified outside S, whence $K_0(\sqrt[p]{x_7}) \subset K_1$. Since x_7 is a pth power in $K_\mathfrak{P}$ for $\mathfrak{P} \in \overline{S}_1$, x_7 is a pth power in K. Thus, Proposition 5.4.3 is proved.

3°. Now consider the general case. We use induction on the order of the kernel B. Note that there is a subgroup B_1 of B such that B_1 is normal in B and the quotient group $A = B/B_1$ is a simple F-module.

Let φ_1 be the natural homomorphism from $G_1 = G/B_1$ onto F induced by φ. Then the embedding problem $(K/k, G_1, \varphi_1, A)$ is associated with the original problem. For $\mathfrak{p} \in S$ let $L_{1\mathfrak{p}}$ be the subalgebra of a given algebra $L_\mathfrak{p}$, corresponding to the subgroup B_1. As shown above, there is a solution L_3 of the global problem $(K/k, G_1, \varphi_1, A)$ whose local behavior coincides with given solutions $L_{1\mathfrak{p}}$ for $\mathfrak{p} \in S$.

We denote by S_2 the set of ramification points of the extension L_3/k. Then only $\mathfrak{p}_0$ and the divisors from S_1 may belong to S_2. Note that, by choice, the divisors from $S_2 \setminus S$ completely decompose in K_0, i.e., the local field $K_\mathfrak{P}$ contains all the nth roots of unity for $\mathfrak{P} \in S_2 \setminus S$.

Generally speaking, it may be impossible to embed the solution L_3 in an extension with group G, since the embedding conditions may break down for the local problems at the points from $S_2 \setminus S_1$.

Let a homomorphism ν_3 from $\overline{F}$ onto G_1 correspond to the field L_3. Then the homomorphism $\nu_3\sigma_\mathfrak{p}$ corresponds to the local behavior $L_{3\mathfrak{p}} = L_3 \otimes_k k_\mathfrak{p}$. If $\mathfrak{p} \in S_2 \setminus S$, then $\mathfrak{p}$ completely decomposes in K. Therefore, the image of $\nu_3\sigma_\mathfrak{p}$ is a subgroup of A. By Lemma 5.3.2, $\nu_3\sigma_\mathfrak{p}$ induces the mapping $\overline{\nu_3\sigma_\mathfrak{p}}$ of the elementary Abelian p-group $H_\mathfrak{p}$ with two generators $f_{1\mathfrak{p}}$ and $f_{2\mathfrak{p}}$ to the group A (since A is elementary Abelian p-group, the subgroup of $\overline{F}_\mathfrak{p}$ generated by the commutator subgroup and pth powers is contained in the kernel of $\nu_3\sigma_\mathfrak{p}$). We assume that the generator $f_{1\mathfrak{p}}$ is chosen so that the automorphism $f_{1\mathfrak{p}}$ acts trivially on the unramified extension of $k_\mathfrak{p}$ of degree p.

Let $\overline{\nu_3\sigma_\mathfrak{p}}(f_{1\mathfrak{p}}) = a_{1\mathfrak{p}}$ and $\overline{\nu_3\sigma_\mathfrak{p}}(f_{2\mathfrak{p}}) = a_{2\mathfrak{p}}$. We define the homomorphism $\nu_{5\mathfrak{p}} \colon H_\mathfrak{p} \longrightarrow G_0$ (G_0 is a semidirect extension of A by F) by setting

$$\nu_{5\mathfrak{p}}(f_{1\mathfrak{p}}) = 1, \qquad \nu_{5\mathfrak{p}}(f_{2\mathfrak{p}}) = a_{2\mathfrak{p}}^{-1}.$$

A solution $L_{5\mathfrak{p}}$ of the semidirect embedding problem $(K_\mathfrak{p}/k_\mathfrak{p}, G_0, A)$ corresponding to $\nu_{5\mathfrak{p}}$. For $\mathfrak{p} \in S_2 \setminus S$ let $\overline{d}_\mathfrak{p}$ denote a preimage of $a_{2\mathfrak{p}}$ in D under the mapping δ. Let $\overline{d}_\mathfrak{p}^{-1} = \prod_{f\in F} d_f^{n_f}$.

The homomorphism $\nu_{6\mathfrak{p}}\colon H_\mathfrak{p} \longrightarrow F \cdot D$ annihilating $f_{1\mathfrak{p}}$ and taking $f_{2\mathfrak{p}}$ to $\overline{d}_\mathfrak{p}^{-1}$ corresponds to a solution $L_{6\mathfrak{p}}$ of the embedding problem $(K_\mathfrak{p}/k_\mathfrak{p}, F\cdot D, D)$ (a homomorphism of $\overline{F}_\mathfrak{p}$ to $F\cdot D$ naturally corresponds to $\nu_{6\mathfrak{p}}$). Clearly, the subalgebra of $L_{6\mathfrak{p}}$ belonging to the subgroup D_0 coincides with $L_{5\mathfrak{p}}$.

As before, let $\{\overline{f}\}_{f\in F}$ be representatives of the cosets of Γ_1 in F_3. We fix a primitive pth root of unity $\zeta \in K_3$. Let $z_{\overline{f}}$ be an integer with $\zeta^{\overline{f}} = \zeta^{z_{\overline{f}}}$.

We note that the mapping $\sigma_\mathfrak{p}$ from $\overline{F}_\mathfrak{p}$ to $\overline{F}$ depends on the choice of a point $\overline{\mathfrak{P}}$ in $\overline{k}$ lying over the point $\mathfrak{p}$ of k. Let $\mathfrak{P}$ be a point of K_3 over which $\overline{\mathfrak{P}}$ lies.

Denote by $w_\mathfrak{p}$ a $\mathfrak{p}$-adic unit in $k_\mathfrak{p}$ with $\sqrt[p]{w_\mathfrak{p}}^{f_{2\mathfrak{p}}-1} = \zeta$. For a prime divisor $\mathfrak{b}$ of K_3 we set

$$m_\mathfrak{b} = \begin{cases} 1 & \text{for } \mathfrak{b} \notin S_2 \setminus S, \\ (w_\mathfrak{p} \bmod K^{*p}_{\mathfrak{P}^{\overline{f}^{-1}\gamma_1^{-i}}})^{n_f z_{\overline{f}} j^i} & \text{for } \mathfrak{b} = \mathfrak{P}^{\overline{f}^{-1}\gamma_1^{-i}}. \end{cases}$$

Note that we have $m_{S_2}^{e_1} = m_{S_2}$ for the element $m_{S_2} = \prod_{\mathfrak{b}\in\mathfrak{M}} m_\mathfrak{b}$.

Let θ be the Artin mapping from $J_{K_3}/J_{K_3}^p$ onto the Galois group A of the extension $L_3 \otimes_K K_3/K_3$. We show that $\theta(m_{S_2}) = 1$. By Proposition 4.3 in Chapter VII of [**1**], we have $\theta(x^{\gamma_1}) = \theta(x)$ for every $x \in J_{K_3}/J_{K_3}^p$. Then $\theta(m_{S_2}) = \prod \theta(w_\mathfrak{p} \bmod K^{*p}_{3,\mathfrak{P}^{\overline{f}^{-1}}})^{n_f z_{\overline{f}} \sum_{i=0}^{c-1} j^i} = 1$, since in $\mathbb{Z}/p\mathbb{Z}$ the element $\sum_{i=0}^{c-1} j^i$ equals 0 (here $w_\mathfrak{p} \bmod K^{*p}_{3,\mathfrak{P}^{\overline{f}^{-1}}}$ stands for an element in $J_{K_3}/J_{K_3}^p$ whose components except for $\mathfrak{P}^{\overline{f}^{-1}}$ are trivial).

Now we apply Lemma 5.3.1, using $S\cup S_2$ as T and $L_3 \otimes_K K_3$ as L. Then there is a number $\overline{m} \in K_3$ satisfying the following conditions:

1) $(\overline{m}) = \mathfrak{P}_1\mathfrak{P}_2\mathfrak{a}^p$, where $\mathfrak{P}_1$ and $\mathfrak{P}_2$ are prime divisors in K_3 of $\mathfrak{p}_1$ and $\mathfrak{p}_2$ from k, respectively. Moreover, $\mathfrak{p}_1$ and $\mathfrak{p}_2$ completely decompose in $L_3 \otimes_K K_3$ and the Legendre symbols $(\frac{\overline{m}}{\mathfrak{P}_i^f})$ are equal to 1, where $i = 1, 2$, $f \in F_3$, and $f \neq 1$;

2) the image of $\overline{m}$ in $K_{3\mathfrak{b}}^*/K_{3\mathfrak{b}}^{*p}$ is equal to $m_\mathfrak{b}$ for $\mathfrak{b} \in S\cup S_2$.

Let m be the image of $\overline{m}$ in K_3^*/K_3^{*p}. Since e_1 is an idempotent of the ring $\mathbb{Z}/p\mathbb{Z}[\Gamma_1]$, the image of m^{e_1} in $K_{3\mathfrak{b}}^*/K_{3\mathfrak{b}}^{*p}$ equals $m_\mathfrak{b}$ for $\mathfrak{b} \in S\cup S_2$. By Lemma 5.4.2, the element m^{e_1} yields a solution L_6 of the embedding problem $(K/k, F\cdot D, D)$. More precisely, if $\overline{m^{e_1}} \in K_3^*$ is a preimage of m^{e_1}, then the algebra L_7 generated over K_3 by the radicals $\sqrt[p]{\overline{m^{e_1}}^{\overline{f}}}$, where $\overline{f}$ ranges over all representatives of the cosets of Γ_1 in F_3, is a Galois algebra over k with group $F\cdot D \times_F F_3$. Then the subalgebra L_6 corresponding to Γ_1 is a solution of the problem $(K/k, F\cdot D, D)$.

By construction, the local behavior $L_6 \otimes_K k_\mathfrak{p}$ of L_6 coincides with $L_{6\mathfrak{p}}$ for $\mathfrak{p} \in S_2\setminus S$. Indeed, for $\mathfrak{p} \in S_2\setminus S$ the local behavior $L_7 \otimes_k k_\mathfrak{p}$ of L_7 corresponds to the homomorphism $\nu_{7\mathfrak{p}}\colon H_\mathfrak{p} \longrightarrow F\cdot D\times_F F_3$ defined as follows: $\nu_{7\mathfrak{p}}(f_{1\mathfrak{p}}) = 1$, $\nu_{7\mathfrak{p}}(f_{2\mathfrak{p}}) = (\overline{d}_\mathfrak{p}^{-1}, 1)$. (The divisor $\mathfrak{P}$ completely decomposes in the subfield of the field L_7 consisting of the elements that are invariant with respect to the automorphism

$(\overline{d}_{\mathfrak{p}}^{-1}, 1)$.) Moreover, in the completion $K_{3\mathfrak{P}}$ we have

$$\sqrt[p]{\overline{m^{e_1 f}}}^{(\overline{d}_{\mathfrak{p}}^{-1},1)} = \sqrt[p]{\overline{m^{e_1 f}}}\, \zeta^{n_f z_f}$$

and, on the other hand, we have

$$\sqrt[p]{w_{\mathfrak{p}}^{n_f z_f} y_{\mathfrak{p}}^{p}}^{\,f_{2\mathfrak{p}}} = \sqrt[p]{w_{\mathfrak{p}}^{n_f z_f} y_{\mathfrak{p}}^{p}}\, \zeta^{n_f z_f}.$$

Now we note that the subalgebra $L_6 \otimes_k k_{\mathfrak{p}}$ corresponds to the homomorphism $\nu_{6\mathfrak{p}}$ which takes $f_{1\mathfrak{p}}$ to 1 and $f_{2\mathfrak{p}}$ to $\overline{d}_{\mathfrak{p}}^{-1}$.

Now let L_5 be the subfield of the field L_6 belonging to the subgroup D_0. We observe that L_5 is a solution of the semidirect embedding problem $(K/k, G_0, A)$. Then the algebra $L_8 = L_3 \circ L_5$ is a solution of the problem $(K/k, G_1, \varphi_1, A)$. Show that L_8 is a field. Suppose the contrary. Then L_8 is a completely splitting Galois algebra (A is a simple F-module), and the extension G_1 coincides with G_0. Consider the local behavior $L_8 \otimes_k k_{\mathfrak{p}_1}$ of L_8 at the point $\mathfrak{p}_1$, where $\mathfrak{p}_1$ is a divisor in k whose prime divisor $\mathfrak{P}_1$ divides the principal divisor $(\overline{m})$. As above, let $H_{\mathfrak{p}_1}$ be an Abelian Galois group of period p of the field $k_{\mathfrak{p}_1}$ with generators $f_{1\mathfrak{p}_1}$ and $f_{2\mathfrak{p}_1}$. Then the algebra $L_3 \otimes_k k_{\mathfrak{p}_1}$ corresponds to the trivial homomorphism $\nu_{3\mathfrak{p}_1} : H_{\mathfrak{p}_1} \longrightarrow G_0$ ($\mathfrak{p}_1$ completely decomposes in the extension L_3/k), and the algebra $L_5 \otimes_k k_{\mathfrak{p}_1}$ corresponds to the homomorphism $\nu_{5\mathfrak{p}_1}$ whose kernel and image are groups of order p. Let $f_{\mathfrak{p}_1}$ be an element in $H_{\mathfrak{p}_1}$ that does not belong to the kernel of $\nu_{5\mathfrak{p}_1}$. The homomorphism $\nu_{8\mathfrak{p}_1} : H_{\mathfrak{p}_1} \longrightarrow G_0$ corresponding to $L_8 \otimes_k k_{\mathfrak{p}_1}$ is given by the relations

$$\nu_{8\mathfrak{p}_1}(f_{i\mathfrak{p}_1}) = (\nu_{3\mathfrak{p}_1}(f_{i\mathfrak{p}_1}), \nu_{5\mathfrak{p}_1}(f_{i\mathfrak{p}_1})) \bmod \widetilde{A}, \quad i = 1, 2$$

(see §15 of Chapter 3). Then $\nu_{8\mathfrak{p}_1}(f_{\mathfrak{p}_1}) = \nu_{5\mathfrak{p}_1}(f_{\mathfrak{p}_1}) \neq 1$, which means that the local algebra $L_8 \otimes_k k_{\mathfrak{p}_1}$ does not split completely. Therefore, L_8 is a field.

Let $h \in H^2(F, A)$ be a cohomology class corresponding to the group extension G_1. Note that the local behavior of L_3 and L_8 at the points $\mathfrak{p} \in S$ coincide (a similar proof was presented in 2°). The divisors from S_2, as well as $\mathfrak{p}_1$ and $\mathfrak{p}_2$, are critical divisors of the extension L_8/k. The divisors $\mathfrak{p}_1$ and $\mathfrak{p}_2$ completely decompose in L_3 and decompose into divisors of order one in L_5. Therefore, in the composite of L_3 and L_5 they decompose into divisors of order one. Note that L_8 is a subfield of $L_5 \otimes_K L_3$, whence in L_8/k the divisors $\mathfrak{p}_1$ and $\mathfrak{p}_2$ decompose into divisors of order one.

Now consider the local behavior $L_{8\mathfrak{p}} = L_8 \otimes_k k_{\mathfrak{p}}$ for $\mathfrak{p} \in S_2 \setminus S$. The algebra $L_{8\mathfrak{p}}$ corresponds to the homomorphism $\nu_{8\mathfrak{p}} : H_{\mathfrak{p}} \longrightarrow G_1$ defined by the relations

$$\nu_{8\mathfrak{p}}(f_{i\mathfrak{p}}) = (\nu_{3\mathfrak{p}}(f_{i\mathfrak{p}}), \nu_{5\mathfrak{p}}(f_{i\mathfrak{p}})) \bmod \widetilde{A}, \quad i = 1, 2$$

(see §15 of Chapter 3). We have $\nu_{8\mathfrak{p}}(f_{1\mathfrak{p}}) = (a_{1\mathfrak{p}}, 1) \bmod \widetilde{A} = a_{1\mathfrak{p}}$, $\nu_{8\mathfrak{p}}(f_{2\mathfrak{p}}) = (a_{2\mathfrak{p}}, a_{2\mathfrak{p}}^{-1}) \bmod \widetilde{A} = 1$.

Thus, in L_8/k the divisors from $S_2 \setminus S$ also have a cyclic decomposition group.

Denote by S_3 the set of critical divisors for L_8/k, not contained in S. Now consider the embedding problem $(L_8 \otimes_k k_{\mathfrak{p}}/k_{\mathfrak{p}}, G, \varphi_2, B_1)$, where φ_2 is the natural homomorphism from G onto G_1. We show that for $\mathfrak{p} \in S_3$ the problem is solvable.

Our assertion is equivalent to the existence of a homomorphism $\nu_{9\mathfrak{p}}\colon \overline{F}_{\mathfrak{p}} \longrightarrow G$ such that the following diagram commutes:

$$\begin{array}{ccc} & & \overline{F}_{\mathfrak{p}} \\ & {\scriptstyle \nu_{9\mathfrak{p}}} \swarrow & \big\downarrow {\scriptstyle \nu_{8\mathfrak{p}}} \\ G & \xrightarrow[\varphi_2]{} & G_1 \end{array}$$

Note that the image of $\nu_{8\mathfrak{p}}$ is a cyclic subgroup of G_1 with generator $a_{\mathfrak{p}}$ ($a_{\mathfrak{p}} \in A$ because $\mathfrak{p}$ completely decomposes in K). Let t be the period of $a_{\mathfrak{p}}$, $k_{\mathfrak{p}}(\sqrt[t]{\pi_{\mathfrak{p}}})$ the field-kernel of the algebra $L_8 \otimes_k k_{\mathfrak{p}}$, $\overline{a}_{\mathfrak{p}} = \varphi_2^{-1}(a_{\mathfrak{p}})$ the preimage of $a_{\mathfrak{p}}$ in the group B, and q the period of $\overline{a}_{\mathfrak{p}}$. Also note that q divides the exponent n of the group B, the field $k_{\mathfrak{p}}$ contains all of the nth roots of unity, consequently the field $k_{\mathfrak{p}}(\sqrt[q]{\pi_{\mathfrak{p}}})$ is well defined, and there is a homomorphism $\nu_{9\mathfrak{p}}$ of $\overline{F}_{\mathfrak{p}}$ to the cyclic subgroup $\{\overline{a}_{\mathfrak{p}}\}$ generated by $\overline{a}_{\mathfrak{p}}$, such that $\nu_{8\mathfrak{p}} = \varphi_2\nu_{9\mathfrak{p}}$. Denote by $L_{9\mathfrak{p}}$ the solution (corresponding to $\nu_{9\mathfrak{p}}$) of the local problem $(L_{8\mathfrak{p}} \otimes_k k_{\mathfrak{p}}/k_{\mathfrak{p}}, G, \varphi_2, B_1)$.

By the induction hypothesis, there is a solution L_9 of the global embedding problem $(L_8/k, G, \varphi_2, B_1)$ with the properties:

1) the local behavior $L_9 \otimes_k k_{\mathfrak{p}}$ for $\mathfrak{p} \in S$ coincides with a given solution $L_{\mathfrak{p}}$, and for $\mathfrak{p} \in S_3$ it coincides with $L_{9\mathfrak{p}}$;

2) L_9 is a field.

Clearly, L_9 is the required solution. Theorem 5.4 is proved.

§5. The semidirect embedding problem with nilpotent kernel

This section is devoted to proving the existence of a proper solution of the semidirect embedding problem with nilpotent kernel for algebraic number fields.

1°. It is more convenient to consider the embedding problem for extensions of algebraic number fields such that the compatibility condition is fulfilled for a sufficiently broad class of group extensions. Scholzian extensions (or universally compatible extensions with respect to groups of fixed exponent) represent a class of such field extensions.

DEFINITION 5.5.1. We call an extension K/k *Scholzian* with respect to an integer n if for all critical, i.e., ramified in K/k, divisors $\mathfrak{p}$ of k, the extension $K_{\mathfrak{P}}/k_{\mathfrak{p}}$ is a cyclic totally ramified extension and the field $k_{\mathfrak{p}}$ contains all of the nth roots of unity, where $\mathfrak{P}$ is a prime divisor of $\mathfrak{p}$ in K.

At the same time field extensions with arbitrary local structure are frequently dealt with. In this connection we consider a larger class of fields.

DEFINITION 5.5.2. We call an extension K/k *quasi-Scholzian* with respect to a natural number n and a finite set S_0 of prime divisors of k if for all critical divisors of k outside S_0 the extension $K_{\mathfrak{P}}/k_{\mathfrak{p}}$ is a cyclic totally ramified extension and $k_{\mathfrak{p}}$ contains all the nth roots of unity.

Let $(K/k, G, \varphi, B)$ be a semidirect embedding problem with nilpotent kernel and F the Galois group of the extension K/k. Then an inclusion of F in G exists and we have the action of F on B. Since a finite nilpotent group is a direct product of its Sylow p-groups and each of them is invariant under the action of F, by Theorem 1.12, it suffices to consider the case where the kernel is a p-group.

Let the kernel B of the embedding problem be a p-group with generators $b_1, \dots, b_d$, and let $B_1 = B$, $B_2 = [B_1, B_1]B_1^p, \dots, B_{i+1} = [B_i, B_1]B_i^p$ be the decreasing p-central series in B. Then there exists c such that $B_c \neq 1$, $B_{c+1} = 1$, and the group B is said to be of class c. In the sequel we assume that K contains all p^{c+1}th roots of unity (otherwise we may adjoin them to K in the ordinary way).

Let U_δ be a free group with generators $u_{i,f}$, where $i = 1, \dots, \delta$; $f \in F$. We denote by $B_{\delta,c}$ the quotient group of U_δ by the $(c+1)$th term $U_{\delta,c+1}$ of the decreasing p-central series. The group F acts on U_δ if we put $u_{i,f}^{f'} = u_{i,ff'}$. In the sequel we shall consider the groups $A_{\delta,c} = U_{\delta,c-1}/U_{\delta,c+1}$, $Z_{\delta,c} = U_{\delta,c}/U_{\delta,c+1}$, and $D_{\delta,c} = B_{\delta,1} \times (B_{\delta,1} \otimes B_{\delta,1}) \times \dots \times (B_{\delta,1} \otimes \dots \otimes B_{\delta,1})$, where the number of tensorial factors increases from 1 to c. Note that the groups $A_{\delta,c}$ are Abelian for $c > 2$. Clearly, the action of F on U_δ induces an action of F on each of the newly introduced groups.

PROPOSITION 5.5.1. *Let B be a group of class c with d generators. Then B is an operator-homomorphic image of $B_{d,c}$.*

PROOF. Let b_i be generators of B. Consider the mapping ψ of U_d defined on the generators by the relations $\psi(u_{i,f}) = b_i^f$ and extended to the whole of U_d by multiplicativity. It suffices to show that ψ maps the subgroups $U_{d,i}$ to B_i for all i. For $i = 1$ the assertion is trivial. Let $\psi(U_{d,i}) = B_i$; then $\psi(U_{d,i+1}) = \psi([U_{d,i}, U_{d,1}]U_{d,i}^p) = [B_i, B_1]B_i^p = B_{i+1}$. Consequently, ψ induces a homomorphism of $B_{d,c}$ onto B. The fact that this homomorphism is an operator homomorphism can be verified directly. □

The proposition we have proved allows us to reduce the embedding problem with kernel of class c with d generators to the embedding problem with kernel $B_{d,c}$. The latter will be considered in what follows. The image of $u_{i,f}$ in $B_{d,c}$ will be denoted by $b_{i,f}$.

Let S_0 be a finite set of prime divisors in k, containing divisors of p, all of the critical divisors for the extension K/k, and infinite divisors. We seek a solution of the semidirect embedding problem $(K/k, F \cdot B_{d,c}, \varphi, B_{d,c})$ in the class of quasi-Scholzian extensions with respect to S_0 and p^{c+1}. Note that if L is a quasi-Scholzian solution, $\mathfrak{p}$ is a critical divisor of L/k, and $\mathfrak{p} \notin S_0$, then $\mathfrak{p}$ completely decomposes in K. In the sequel we shall omit the homomorphism φ in the notation of the semidirect embedding problem.

2°. Let $\mathbb{Z}/p\mathbb{Z}[F]$ be the group ring of F over the field with p elements. Denote by D a free $\mathbb{Z}/p\mathbb{Z}[F]$-module of rank 1, i.e., D consists of elements of the form $\prod_{f\in F} d_f^{n_f}$ $(n_f \in \mathbb{Z}/p\mathbb{Z})$. The module D is naturally an F-module, namely,

$$\Big(\prod_{f\in F} d_f^{n_f}\Big)^{f'} = \prod_{f\in F} d_{ff'}^{n_f}.$$

LEMMA 5.5.1. *There is a one-to-one correspondence between the solutions of the semidirect embedding problem $(K/k, F \cdot D, D)$ and the elements of K^*/K^{*p}.*

PROOF. Note that the group $F \cdot D$ is the wreath product of the cyclic group $\mathbb{Z}/p\mathbb{Z}$ of order p by F. In §6 of Chapter 3 we have established that the tensor product $\bigotimes_K L_f$ of the Galois algebras L_f with $f \in F$ over the field K with group $\mathbb{Z}/p\mathbb{Z}$ is a solution L of this embedding problem. It remains to observe that the algebra L is of the form $K(\sqrt[p]{m})$, where $m \in K^*$, and if $m' = mx^p$, then the algebras $K(\sqrt[p]{m})$ and $K(\sqrt[p]{m'})$ are isomorphic. □

THEOREM 5.5.1. *For any δ there is a quasi-Scholzian solution of the embedding problem $(K/k, F\cdot B_{\delta,1}, B_{\delta,1})$.*

PROOF. We note that $B_{\delta,1}$ is the direct product of δ copies of D. Let $\delta = 1$. By Lemma 5.5.1, the solution $K_{1,1}$ of the embedding problem $(K/k, F\cdot D, D)$ is determined by an element $m \in K^*/K^{*p}$. Choose $\overline{m} \in K^*$ in accordance with Lemma 5.3.1 assuming that T is the empty set and L is equal to K. Let m be the coset in K^*/K^{*p} with representative $\overline{m}$. We show that $K_{1,1}$ is a quasi-Scholzian solution. Clearly, if $\mathfrak{p}$ is a divisor of k critical for $K_{1,1}/k$ and $\mathfrak{p} \notin S_0$, then the prime divisors $\mathfrak{P}$ of $\mathfrak{p}$ occur in the decomposition of the principal divisor $(\overline{m})$. Since $\mathfrak{p}$ completely decomposes in K, the decomposition of $\mathfrak{p}$ into divisors of order one is equivalent to the relations $(\frac{\overline{m}^f}{\mathfrak{P}}) = 1$ for $f \in F$ and $f \neq 1$. We note that $(\frac{\overline{m}^f}{\mathfrak{P}}) = (\frac{\overline{m}}{\mathfrak{P}^{f^{-1}}})^f$ and, by construction, $(\frac{\overline{m}}{\mathfrak{P}^{f^{-1}}}) = 1$. Clearly, $K_{1,1}$ is a field, for if $(\overline{m})^{\sum_f t_f f}$ is a pth power in the group of divisors, then every coefficient t_f is divisible by p.

Let a quasi-Scholzian solution $K_{\delta-1,1}$ of the embedding problem $(K/k, F\cdot B_{\delta-1,1}, B_{\delta-1,1})$ be constructed, let the elements m_i of K^*/K^{*p} with $i = 1,\dots,\delta-1$ completely determine the field $K_{\delta-1,1}$, and let $\overline{m}_i$ (representatives of m_i in K^*) be chosen in accordance with Lemma 5.3.1. Then $\overline{m}_i = m_{i_1}m_{i_2}$ for odd p and $\overline{m}_i = m_{i_1}m_{i_2}m_{i_3}$ for $p = 2$. Recall that the principal divisors (m_{i_j}) are of the form $\mathfrak{P}_{ij}\mathfrak{a}_{ij}^p$.

Denote by $L_{\delta-1}$ the field obtained by adjoining to K all of the radicals $\sqrt[p]{m_{i_j}^f}$, where f ranges over all the elements of F. We again use Lemma 5.3.1 by taking the empty set as T and the field $L_{\delta-1}$ as L. We denote by $\overline{m}_\delta$ the number from Lemma 5.3.1 and by m_δ its image in K^*/K^{*p}. Then m_δ determines the field $\widetilde{K}_{1,1}$ which is a solution of the embedding problem $(K/k, F\cdot D, D)$. We put $K_{\delta,1} = K_{\delta-1,1}\otimes\widetilde{K}_{1,1}$ and we show that $K_{\delta,1}$ is a quasi-Scholzian solution of the problem $(K/k, F\cdot B_{\delta,1}, B_{\delta,1})$. It is clear that if $\overline{m}_\delta^{\sum_f t_f f}$ is a pth power in $K_{\delta-1,1}$, then every coefficient t_f is divisible by p, i.e., $K_{\delta,1}$ is a field. The critical divisors of $K_{\delta,1}/k$ are, in addition to the divisors from S_0, the divisors $\mathfrak{p}_{ij}$ whose prime divisors $\mathfrak{P}_{ij}$ divide the corresponding principal divisors $(\overline{m}_i)$. Since the divisors $\mathfrak{p}_{ij}$ completely decompose in K, the decomposition of $\mathfrak{p}_{ij}$ in $K_{\delta,1}$ into divisors of order one is equivalent to the relations $(\frac{\overline{m}_e^f}{\mathfrak{P}_{ij}}) = 1$ for $e = 1,\dots,\delta$, $i = 1,\dots,\delta$, $f \in F$, and $f \neq 1$, and to the relations $(\frac{\overline{m}_e^f}{\mathfrak{P}_{ij}}) = 1$ for $i \neq e$ and $f \in F$. By construction, the divisors $\mathfrak{P}_{\delta j}$ completely decompose in $L_{\delta-1}$; consequently, for $e < \delta$ and $f \in F$, the relations $(\frac{\overline{m}_{ej'}^f}{\mathfrak{P}_{\delta j}}) = 1$ are fulfilled, where j and j' independently take the values 1 and 2 for odd p and 1, 2, and 3 for $p = 2$. Consequently, we have $(\frac{\overline{m}_e^f}{\mathfrak{P}_{\delta j}}) = 1$ for $e < \delta$ and $f \in F$. Since $K_{\delta-1,1}$ is a quasi-Scholzian solution, the relations $(\frac{\overline{m}_e^f}{\mathfrak{P}_{ij}}) = 1$ are fulfilled for $e < \delta$ and $i < \delta$, where $f \in F$, $f \neq 1$, and $e \neq i$. It remains to consider the symbols $(\frac{\overline{m}_\delta^f}{\mathfrak{P}_{ij}})$ with $f \in F$ and $i < \delta$. By the reciprocity law, $(\frac{\overline{m}_{\delta j'}^f}{\mathfrak{P}_{ij}}) = (\frac{\overline{m}_{ij}}{\mathfrak{P}_{\delta j'}^f})$.

By construction, the symbols on the right-hand side in the last relations are equal to 1, which completes the proof of the theorem.

3°. In the sequel we construct a solution of the problem by induction on c, the class of the group $B_{d,c}$. Let the existence of a quasi-Scholzian solution $K_{\delta,c-1}$ of the embedding problem $(K/k, F \cdot B_{\delta,c-1}, B_{\delta,c-1})$ be proved for every positive integer δ. Generally speaking, the solution $K_{\delta,c-1}$ cannot be embedded in a quasi-Scholzian extension with group $F \cdot B_{\delta,c}$. In this subsection we find a sufficient condition for such an embedding. First we prove two auxiliary facts. □

LEMMA 5.5.2. *The group $Z_{d,c}$ is an operator-homomorphic image of $D_{d,c}$.*

PROOF. In §4 of Appendix we show that the inclusion of U_d in the Magnus algebra with variables $x_{i,f}$, where $i = 1, \dots, d$; $f \in F$, induces an operator isomorphism ψ_1 of $Z_{d,c}$ onto $\sum^{\oplus c}_{j=1} L_j/pL_j$, where L_j is the group of homogeneous Lie polynomials of degree j. We note that there is a natural operator epimorphism $\psi_2 \colon D_{d,c} \longrightarrow \sum^{\oplus c}_{j=1} L_j/pL_j$ given by the relation

$$\begin{aligned}\psi_2(b^{n_1}_{i_{11},f_{11}}, (b_{i_{21},f_{21}} \otimes b_{i_{22},f_{22}})^{n_2}, \dots, (b_{i_{c1},f_{c1}} \otimes b_{i_{cc},f_{cc}})^{n_c}) \\ = n_1 x_{i_{11},f_{11}} \bmod pL_1 \oplus n_2[x_{i_{21},f_{21}}, x_{i_{22},f_{22}}] \bmod pL_2 \\ \oplus \cdots \oplus n_c[x_{i_{c1},f_{c1}}, \dots, x_{i_{cc},f_{cc}}] \bmod pL_c.\end{aligned}$$

Clearly, $\psi_1^{-1}\psi_2$ is the required homomorphism.

The kernel of the mapping of $D_{d,c}$ onto $Z_{d,c}$ will be denoted by $E_{d,c}$. □

LEMMA 5.5.3. *Let u_1 and u_2 be elements of U_d, and let $u_1 \notin U_{d,2}$, $u_2 \in U_{d,c}$, and $u_2 \notin U_{d,c+1}$. Moreover, we assume that $u_2 u_1^{np^{c-1}} \notin U_{d,c+1}$ for every positive integer n. Then the commutator $[u_2, u_1] \notin U_{d,c+2}$.*

PROOF. We again use Theorem A.4. Denote by H the ideal of the Magnus algebra generated by p and the variables $x_{i,f}$ where $i = 1, \dots, d$, $f \in F$. In this algebra, an element $u_1' = 1 + l + P_2$, where l is a linear Lie polynomial, i.e., $l = c_1 x_{i_1,f_1} + \dots + c_m x_{i_m,f_m}$ and $P_2 \equiv 0 \pmod{H^2}$, corresponds to u_1. Since $u_1 \notin U_{d,2}$, we may assume that at least one of the coefficients, say c_1, is not divisible by p. An element $u_2' = 1 + p^{c-1}l_1 + \dots + l_c + P_{c+1}$ for odd p and $u_2' = 1 + 2^{c-1}l_1 + 2^{c-2}(l_1^2 + l_2) + \dots + l_c + P_{c+1}$ for $p = 2$, where l_j are Lie polynomials of degree j and $P_{c+1} \equiv 0 \pmod{H^{c+1}}$, corresponds to u_2. Since $u_2 \notin U_{d,c+1}$, among $l_1, \dots, l_c$ there is a polynomial l_{j_0} such that $l_{j_0} \not\equiv 0 \pmod p$.

Consider the case $p = 2$. In this algebra an element $1 + 2^{c-1}(l_1 l - l l_1) + 2^{c-1}(l_1^2 l - l l_1^2 + l_2 l - l l_2) + \dots + (l_c l - l l_c) + P_{c+2}$, where $P_{c+2} \equiv 0 \pmod{H^{c+2}}$ corresponds to the commutator $[u_2, u_1]$.

Suppose that our assertion is not true. Then the congruences $l_1 l - l l_1 \equiv 0 \pmod 2$, $l_1^2 l - l l_1^2 + l_2 l - l l_2 \equiv 0 \pmod 2$, … , $l_c l - l l_c \equiv 0 \pmod 2$ hold. If $j_0 = 1$, i.e., $l_1 \not\equiv 0 \pmod 2$ and $l_j \equiv 0 \pmod 2$ for $j > 1$, then the congruence $l_1 l - l l_1 \equiv 0 \pmod 2$ implies $l_1 \equiv l \pmod 2$ and $u_2 \equiv u_1^{2^{c-1}} \pmod{U_{d,c+1}}$, which contradicts the assumption. Consequently, we may assume that $j_0 > 1$. Clearly, the congruence $l_1 l - l l_1 \equiv 0 \pmod 2$ implies $l_1^2 l - l l_1^2 \equiv 0 \pmod 2$. Thus, it remains to show that the congruence $l_{j_0} l - l l_{j_0} \equiv 0 \pmod 2$ for $j_0 > 1$ contradicts the condition $l_{j_0} \not\equiv 0 \pmod 2$. Indeed, since $c_1 \not\equiv 0 \pmod 2$, among the summands that occur in l_{j_0} there are monomials of the form $c_0 x_{i,f} \dots x^g_{i_1,f_1}$, where c_0 is not divisible by 2. Let g be a maximal number for such monomials. Then the monomial $c_1 c_0 x_{i,f} \dots x^{g+1}_{i_1,f_1}$ must be among the summands of $(c_1 x_{i_1,f_1} + \dots + c_m x_{i_m,f_m}) l_{j_0}$, which is impossible. Consequently, the congruence $l_{j_0} l - l l_{j_0} \equiv 0 \pmod 2$ is not valid. The case where p is odd is treated similarly. □

COROLLARY. *Let b_1 and b_2 be elements in $B_{d,c+1}$, $b_1 \notin B_{d,c+1,2}$, $b_2 \in A_{d,c+1}$, and $b_2 \notin Z_{d,c+1}$. Moreover, for every positive integer n let $b_1^{np^{c-1}} b_2 \notin Z_{d,c+1}$. Then the commutator $[b_2, b_1]$ is different from 1, where $B_{d,c+1,2}$ is the second term of the decreasing p-central series of $B_{d,c+1}$.*

Now we prove the main result of this subsection.

THEOREM 5.5.2. *Let the embedding problem $(K_{d,c-1}/k, F \cdot B_{d,c+1}, \varphi_{d,c+1}, A_{d,c+1})$ be solvable, where $\varphi_{d,c+1}$ is the natural homomorphism of the group $F \cdot B_{d,c+1}$ onto $F \cdot B_{d,c-1}$, and let there exist a solution unramified outside the set S of prime divisors consisting of S_0 and the divisors of k that decompose in $K_{d,c-1}$ into divisors of order one. Then there exists a quasi-Scholzian solution of the embedding problem $(K_{d,c-1}/k, F \cdot B_{d,c}, \varphi_{d,c}, Z_{d,c})$, where $\varphi_{d,c}$ is the natural homomorphism of the group $F \cdot B_{d,c}$ onto $F \cdot B_{d,c-1}$.*

PROOF. Let $L_{d,c+1}$ be a solution that exists by assumption and let $L_{d,c}$ be the subfield corresponding to the subgroup $Z_{d,c+1}$. A subfield $K_{d,1}$ corresponds to the subgroup $B_{d,c+1,2}$. We denote by S_1 the set of divisors in k that completely decompose in K and ramify in the extension $K_{d,1}/k$ and by S_2 the set of critical divisors for $L_{d,c}/k$ that completely decompose in $K_{d,1}$. The sets of their divisors in K will be denoted similarly.

Let $\mathfrak{p}$ be a divisor from $S_1 \cup S_2$, $\mathfrak{P}$ its prime divisor in K, $\mathfrak{P}_{c-1}$ a prime divisor of $\mathfrak{P}$ in $K_{d,c-1}$, $\mathfrak{P}_c$ and $\mathfrak{P}_{c+1}$ the corresponding divisors in $L_{d,c}$ and $L_{d,c+1}$. We denote by $G_{c-1,\mathfrak{p}}$, $G_{c,\mathfrak{p}}$, and $G_{c+1,\mathfrak{p}}$ the decomposition groups of the divisors $\mathfrak{P}_{c-1}$, $\mathfrak{P}_c$, and $\mathfrak{P}_{c+1}$ in the corresponding extensions. We observe that the extension $K_{d,c-1,\mathfrak{P}_{c-1}}/k_{\mathfrak{p}}$ either is a totally ramified extension and its Galois group is a subgroup of $B_{d,c-1}$ or is trivial if $\mathfrak{p}$ completely decomposes in $K_{d,c-1}$.

Let the order of $G_{c-1,\mathfrak{p}}$ be equal to p^i. Then $G_{c,\mathfrak{p}}$ is Abelian as a central extension of an Abelian group of period p by the cyclic group $G_{c-1,\mathfrak{p}}$. From the structure of the center of $B_{d,c}$ it follows that the order of the preimage of a generator of $G_{c-1,\mathfrak{p}}$ in $G_{c,\mathfrak{p}}$ increases p-fold. Lemma 5.3.2 implies that $G_{c,\mathfrak{p}}$ has at most two generators. Thus, $G_{c,\mathfrak{p}}$ is either a cyclic group of order p^{i+1} or a direct product of cyclic groups of orders p^{i+1} and p. In the first case, the extension $L_{d,c,\mathfrak{P}_c}$ is of the form $k_{\mathfrak{p}}(\sqrt[p^{i+1}]{\pi_{\mathfrak{p}}})$, in the second it is of the form $k_{\mathfrak{p}}(\sqrt[p^{i+1}]{\pi_{\mathfrak{p}}}, \sqrt[p]{v_{\mathfrak{p}}})$, where $\pi_{\mathfrak{p}}$ is a prime element of $k_{\mathfrak{p}}$ and $v_{\mathfrak{p}}$ is a $\mathfrak{p}$-adic unit. □

We fix a primitive p^{c+1}th root ζ_{c+1} of unity in K. Let $\zeta = \zeta_{c+1}^{p^c}$. Denote by $b_{1\mathfrak{p}}$ and $b_{2\mathfrak{p}}$ the automorphisms from $G_{c,\mathfrak{p}}$ that act on $L_{d,c,\mathfrak{P}_c}$ as follows:

$$\sqrt[p^{i+1}]{\pi_{\mathfrak{p}}}^{b_{1\mathfrak{p}}-1} = \zeta_{c+1}^{p^{c-i}}, \qquad \sqrt[p]{v_{\mathfrak{p}}}^{b_{1\mathfrak{p}}-1} = 1,$$
$$\sqrt[p^{i+1}]{\pi_{\mathfrak{p}}}^{b_{2\mathfrak{p}}-1} = 1, \qquad \sqrt[p]{v_{\mathfrak{p}}}^{b_{2\mathfrak{p}}-1} = \zeta.$$

Now we show that if $\mathfrak{p} \in S_i$, then $G_{c,\mathfrak{p}}$ is cyclic. Note that in this case $i = c-1$. Suppose the contrary, i.e., $G_{c,\mathfrak{p}}$ is the direct product of the cyclic groups of orders p^c and p with generators $b_{1\mathfrak{p}}$ and $b_{2\mathfrak{p}}$. Let $\bar{b}_{1\mathfrak{p}}$ and $\bar{b}_{2\mathfrak{p}}$ be preimages of $b_{1\mathfrak{p}}$ and $b_{2\mathfrak{p}}$ in $G_{c+1,\mathfrak{p}}$. Then $\bar{b}_{1\mathfrak{p}} \in B_{d,c+1}$ and $\bar{b}_{1\mathfrak{p}} \notin B_{d,c+1,2}$, $\bar{b}_{2\mathfrak{p}} \in A_{d,c+1}$ and $\bar{b}_{2\mathfrak{p}} \notin Z_{d,c+1}$. By Corollary to Lemma 5.5.3, the commutator $[\bar{b}_{2\mathfrak{p}}, \bar{b}_{1\mathfrak{p}}]$ is different from the unit element, whence $G_{c+1,\mathfrak{p}}$ is not Abelian. The extension $L_{d,c,\mathfrak{P}_c}$ can be embedded in the Abelian extension $L_{\mathfrak{p}}$ with Galois group $\mathbb{Z}/p^{c+1}\mathbb{Z} \times \mathbb{Z}/p^2\mathbb{Z}$ over the field $k_{\mathfrak{p}}$. Indeed, we can take the field $k_{\mathfrak{p}}(\sqrt[p^{c+1}]{\pi_{\mathfrak{p}}}, \sqrt[p^2]{v_{\mathfrak{p}}})$ as $L_{\mathfrak{p}}$. The extension $L_{d,c+1,\mathfrak{P}_{c+1}}$

over $L_{d,c,\mathfrak{P}_c}$ has an Abelian group of period p, but it is not contained in the field $L_{\mathfrak{p}}$. Then the Galois group of the composite of the fields $L_{\mathfrak{p}}$ and $L_{d,c+1,\mathfrak{P}_{c+1}}$ over the field $L_{d,c,\mathfrak{P}_c}$ is an Abelian group of period p and has more than two generators, which is impossible. Thus, $b_{2\mathfrak{p}}$ is a trivial automorphism, and the assumption that the extension $L_{d,c}/k$ is quasi-Scholzian at the points of S_1 is valid. The field $L_{d,c}$ is a solution of the embedding problem $(K/k, F \cdot B_{d,c}, \varphi_{d,c}, Z_{d,c})$. Let $m_\chi \in L_{d,c}$ be the Lagrange resolvents corresponding to the characters $\chi \in \widehat{Z}_{d,c}$, i.e., the elements possessing the properties $m_\chi^z = m_\chi \chi(z)$, where $z \in Z_{d,c}$.

Let $\mathfrak{p} \in S_2$. We observe that $m_\chi^p \in K_{d,c-1}$, and in the completion $K_{d,c-1,\mathfrak{P}_{c-1}}$ the element m_χ^p is of the form $\sqrt[p^i]{\pi_\mathfrak{p}}^e v_\mathfrak{p}^g y_{\chi,\mathfrak{p}}^p$ if $\chi(b_{1\mathfrak{p}}^{p^i}) = \zeta^e$ and $\chi(b_{2\mathfrak{p}}) = \zeta^g$.

Now let L_1 be a solution of the semidirect embedding problem $(K/k, F \cdot Z_{d,c}, Z_{d,c})$ generated over K by the elements $m_{0\chi}$, where $\chi \in \widehat{Z}_{d,c}$. Suppose the elements $y_\chi = m_{0\chi}^p \in K^*$ are chosen so that if $\chi(b_{2\mathfrak{p}}) = \zeta^g$, then $y_\chi = v_\mathfrak{p}^{-g} y_{0,\chi,\mathfrak{p}}^p$ in the completion $K_\mathfrak{P}$.

We put $L_2 = K_{d,c-1} \otimes_K L_1$; then L_2/k is an extension with group $(F \cdot B_{d,c-1}) \times_F (F \cdot Z_{d,c})$. Let G_0 be a semidirect extension of the group $Z_{d,c}$ by $F \cdot B_{d,c-1}$. Clearly, the groups G_0 and $(F \cdot B_{d,c-1}) \times_F (F \cdot Z_{d,c})$ are isomorphic, and we shall identify them in what follows. We define the embedding β_0 of the group $F \cdot B_{d,c-1}$ in G_0 by setting $\beta_0(f \cdot b) = (f \cdot b, f)$, where $f \in F$, $b \in B_{d,c-1}$.

We assume that on the set of solutions of the embedding problems $(K_{d,c-1}/k, G_h, \varphi_h, Z_{d,c})$, where h ranges over the elements of the group $H^2(F \cdot B_{d,c-1}, Z_{d,c})$, a composition law is given (see §15 of Chapter 3), and the set of representatives of the cosets of $Z_{d,c}$ in G_0 is chosen as the image with respect to β_0.

The algebra $K_{d,c} = L_{d,c} \circ L_2$ is a solution of the embedding problem $(K_{d,c-1}/k, F \cdot B_{d,c}, \varphi_{d,c}, Z_{d,c})$ and is generated over $K_{d,c-1}$ by the elements $m_\chi \otimes m_{0\chi}$, where $\chi \in \widehat{Z}_{d,c}$. Clearly, $K_{d,c}$ is a field (this immediately follows from Corollary 5 to Theorem 1.6 because the subgroup $Z_{d,c}$ is contained in the Frattini subgroup) and the condition that $K_{d,c}/k$ is quasi-Scholzian is fulfilled at the points $\mathfrak{p} \in S_2$. Thus, the problem consists of choosing a suitable solution L_1 of the semidirect problem $(K/k, F \cdot Z_{d,c}, Z_{d,c})$.

We observe that $D_{d,c}$ is a free $\mathbb{Z}/p\mathbb{Z}[F]$-module and that $D_{d,c}$ is a direct product of t copies of D for some integer t. For $\mathfrak{p} \in S_2$ let $d_\mathfrak{p} \in D_{d,c}$ be a preimage of $b_{2\mathfrak{p}}$ under the natural homomorphism from $D_{d,c}$ onto $Z_{d,c}$; $d_\mathfrak{p} = \prod_{\substack{f \in F \\ j=1,\dots,t}} d_{j,f}^{n_{j,f}}$, where $d_{j,1}$ is a generator of the jth component of $D_{d,c}$. We denote by w_f an integer with $\zeta^f = \zeta^{w_f}$.

Let $T_1 = S_1 \cup S_2$ and let θ_1 be the Artin homomorphism from the group J_k/J_k^p onto the Galois group $B_{d,1}$ of $K_{d,1}/K$. For $m_{1,\mathfrak{b}} \in K_\mathfrak{b}^*/K_\mathfrak{b}^{*p}$ we put

$$m_{1,\mathfrak{b}} = \begin{cases} 1 & \text{if } \mathfrak{b} \notin S_2, \\ v_\mathfrak{p}^{-n_{1,f} w_f} \bmod K_{\mathfrak{P}^{f^{-1}}}^{*p} & \text{if } \mathfrak{b} = \mathfrak{P}^{f^{-1}},\ \mathfrak{P} \in S_2,\ f \in F \end{cases}$$

and $m_{1,T_1} = \prod_\mathfrak{b} m_{1,\mathfrak{b}}$. Since the divisors of S_2 are unramified in $K_{d,1}/K$, we have $\theta_1(m_{1,T_1}) = 1$.

Now we use Lemma 5.3.1, taking the set T_1 as T and the field $L_{d,c}$ as L. Let $\overline{m}_1 \in K$ be a number from this lemma; then $\overline{m}_1 = m_{11}m_{12}$ for odd p, $\overline{m}_1 = m_{11}m_{12}m_{13}$ for $p = 2$.

Further, let θ_2 be the Artin homomorphism from J_K/J_K^p onto the Galois group of the extension L_3/K obtained by adjoining to $K_{d,1}$ the radicals $\sqrt[p]{m_{1j}^f}$, where $f \in F$, $j = 1, 2$ for odd p and $j = 1, 2, 3$ for $p = 2$. We denote by T_2 the set of prime divisors of K consisting of T_1 and the divisors $\mathfrak{P}_{1j}^f$ that occur in the decomposition of the principal divisors $(m_{1j}^f) = \mathfrak{P}_{1j}^f \mathfrak{a}_{1j}^{pf}$. For $m_{2,\mathfrak{b}} \in K_\mathfrak{b}^*/K_\mathfrak{b}^{*p}$ we set

$$m_{2,\mathfrak{b}} = \begin{cases} 1 & \text{if } \mathfrak{b} \in S_1, \\ v_\mathfrak{p}^{-n_{2,f} w_f} \bmod K_{\mathfrak{P}^{f^{-1}}}^{*p} & \text{if } \mathfrak{b} = \mathfrak{P}^{f^{-1}},\ \mathfrak{P} \in S_2. \end{cases}$$

Let $m_{2,T_2} = \prod_\mathfrak{b} m_{2,\mathfrak{b}}$. Since the divisors from S_2 are unramified in L_3/K, we have $\theta_2(m_{2,T_2}) = 1$. We again use Lemma 5.3.1, taking the set T_2 as T and the composite of the fields L_3 and $L_{d,c}$ as L. We denote by $\overline{m}_2$ the number obtained from this lemma.

Continuing this process, we find the numbers $\overline{m}_1, \dots, \overline{m}_t$ in K with the following properties:

1) the decomposition of the principal divisor $(\overline{m}_j)$ is of the form $\mathfrak{P}_{j1}\mathfrak{P}_{j2}\mathfrak{a}_j^p$ for odd p and $\mathfrak{P}_{j1}\mathfrak{P}_{j2}\mathfrak{P}_{j3}\mathfrak{a}_j^2$ for $p = 2$;

2) $\overline{m}_j$ is a pth power in the completions of K corresponding to the infinite divisors and the divisors of p;

3) the image of $\overline{m}_j$ in the group $K_\mathfrak{b}^*/K_\mathfrak{b}^{*p}$ coincides with $m_{j,\mathfrak{b}}$ for $\mathfrak{b} \in T_j$.

In accordance with Lemma 5.5.1, each number $\overline{m}_j$ determines a solution K_j of the semidirect embedding problem $(K/k, F \cdot D, D)$, namely, K_j is generated over K by the radicals $\sqrt[p]{\overline{m}_j^f}$ with $f \in F$. By the first Kochendörffer theorem, the algebra $L_4 = \bigotimes_K K_j$ (K_j runs over the algebras $K_1, \dots, K_t$) is a solution of the embedding problem $(K/k, F \cdot D_{d,c}, D_{d,c})$. We note that the action of the group $D_{d,c}$ on the elements of L_4 is defined by the relations

$$\sqrt[p]{\overline{m}_j^g}^{\prod_{j,f} d_{j,f}^{n_{j,f}}} = \sqrt[p]{\overline{m}_j^g}\,\zeta^{w_g n_{j,g}}, \qquad j = 1, \dots, t; \quad f, g \in F.$$

Let $\chi_{j,f} \in \widehat{D}_{d,c}$ be the characters of the group $D_{d,c}$ defined by the relations $\chi_{j,f}(\prod_{j,f} d_{j,f}^{n_{j,f}}) = \zeta^{w_f n_{j,f}}$. Clearly, the elements $\chi_{j,f}$, where $f \in F$ and $j = 1, \dots, t$, generate the group $\widehat{D}_{d,c}$. For $\chi = \prod_{j,f} \chi_{j,f}^{c_{j,f}} \in \widehat{D}_{d,c}$ we put $\overline{m}_{0\chi} = \prod_{j,f} \sqrt[p]{\overline{m}_j^f}^{c_{j,f}}$. Then we have $\overline{m}_{0\chi}^{d_{j,f}} = \overline{m}_{0\chi}\zeta^{w_f c_{j,f}} = \overline{m}_{0\chi}\chi(d_{j,f})$, i.e., the elements $\overline{m}_{0\chi}$ with $\chi \in D_{d,c}$ generate the algebra L_4 over K and satisfy the condition $\overline{m}_{0\chi}^{\overline{d}} = \overline{m}_{0\chi}\chi(\overline{d})$, where $\overline{d} \in D_{d,c}$. It can be immediately verified that for $\mathfrak{P} \in S_2$ the number $\overline{m}_{0\chi}^p$ is of the form $v_\mathfrak{p}^{-g}\overline{y}_{\chi,\mathfrak{p}}^p$ in the completion $K_\mathfrak{P}$ if $\chi(d_\mathfrak{p}) = \zeta^g$.

Now we note that L_4 is a field and every divisor that is critical for L_4/k and is not contained in S_0 decomposes into divisors of order one (the proof is similar to that of Theorem 5.5.1). Let L_1 be the subfield of L_4 belonging to the subgroup $E_{d,c}$ (the kernel of the mapping from $D_{d,c}$ onto $Z_{d,c}$). Then L_1 is a solution of the semidirect embedding problem $(K/k, F \cdot Z_{d,c}, Z_{d,c})$. It is directly verified that if $m_{0\chi}$ with $\chi \in \widehat{Z}_{d,c}$ generate L_1 over K, then for $\mathfrak{P} \in S_2$ the elements $m_{0\chi}^p$ are of the form $v_\mathfrak{p}^{-g} z_{\chi,\mathfrak{p}}^p$ in the completions $K_\mathfrak{P}$, provided that $\chi(b_{2\mathfrak{p}}) = \zeta^g$.

We consider the field $K_{d,c} = L_{d,c} \circ L_2$, where $L_2 = K_{d,c-1} \otimes_K L_1$. Note that $K_{d,c}$ is a subfield of the composite of the fields $L_{d,c}$ and L_2. The divisors from

S_1 completely decompose in L_2 and decompose into divisors of order one in $L_{d,c}$. Consequently, in $K_{d,c}$ these divisors also can be decomposed into divisors of order one. By the choice of L_1, and also of L_2, in $K_{d,c}$ the divisors from S_2 decompose into divisors of order one. It remains to consider the divisors of k that are critical for L_1/k. By construction, these divisors completely decompose in $L_{d,c}$, and each of them decomposes into divisors of order one in L_1/k. Thus, $K_{d,c}$ is a quasi-Scholzian solution of the embedding problem $(K_{d,c-1}/k, F \cdot B_{d,c}, Z_{d,c})$.

4°. Thus, it remains to find a solution $K_{d,c-1}$ of the embedding problem $(K/k, F \cdot B_{d,c-1}, B_{d,c-1})$ such that it can be embedded in the extension $K_{d,c+1}$ with Galois group $F \cdot B_{d,c+1}$ and is unramified outside the set S consisting of S_0 and the divisors of k that decompose into divisors of order one in $K_{d,c-1}$.

If we take a quasi-Scholzian solution $K_{\delta,c-1}$ of the embedding problem $(K/k, F \cdot B_{\delta,c-1}, B_{\delta,c-1})$ with sufficiently large δ, namely, δ must be greater than a certain constant depending only on c and d, then we shall always have a chance of choosing a subfield $K_{d,c-1}$ with Galois group $F \cdot B_{d,c-1}$ over k that can be embedded in an extension with group $F \cdot B_{d,c+1}$ and with necessary bounded ramification. For every δ we denote by S_δ the set of prime divisors of k, consisting of S_0, the divisors critical for $K_{\delta,c-1}/k$, and those divisors of k whose prime divisors in $K_{\delta,c-1}$ generate the p-component of the divisor class group of this field.

Now we consider a homomorphism ε of a free F-operator group U_δ onto $U_{\delta'}$ with $\delta' < \delta$ that takes the generators $u_{i,f}$ either to $u_{i',f}$ or to 1, where $i = 1, \dots, \delta$; $i' = 1, \dots, \delta'$. Such homomorphisms will be called *canonical*. Clearly, the canonical homomorphisms are F-operator homomorphisms and induce operator homomorphisms of $B_{\delta,c}$, $D_{\delta,c}$, $A_{\delta,c}$, and $Z_{\delta,c}$ onto the corresponding groups with subscripts δ' and c. These homomorphisms will also be called *canonical*. The homomorphism ε from $B_{\delta,c+1}$ onto $B_{d,c+1}$ puts in correspondence to the embedding problem $(K_{\delta,c-1}/k, F \cdot B_{\delta,c+1}, \varphi_{\delta,c+1}, A_{\delta,c+1})$ the associated problem $(K_{d,c-1}/k, F \cdot B_{d,c+1}, \varphi_{d,c+1}, A_{d,c+1})$, where $K_{d,c-1}$ is the subfield belonging to the kernel of ε, and $\varphi_{\delta,c+1} \colon F \cdot B_{\delta,c+1} \longrightarrow F \cdot B_{\delta,c-1}$ is the natural epimorphism. We must choose ε in such a way that the associated embedding problem has a solution unramified outside S_d. We prove several auxiliary assertions.

PROPOSITION 5.5.2. *For any natural c, d, and e, there is a natural number $N_1 = N_1(c, d, e)$ such that for $\delta \geq N_1$ and for arbitrary elements $x_1, \dots, x_e$ in $D_{\delta,c}$, a canonical epimorphism $\varepsilon \colon D_{\delta,c} \longrightarrow D_{d,c}$ annihilating $x_1, \dots, x_e$ exists.*

We prove the proposition by induction on d. The case $d = 1$ and the induction passage will be considered simultaneously. For brevity of notation, we shall omit the subscript c. We put $N_{11} = N_1(c, d-1, e)$ and $N_{11} = 0$ for $d = 1$. We take $\delta > N_{11}$. By the induction hypothesis, among the canonical epimorphisms $D_\delta \longrightarrow D_{d-1}$ annihilating the generators $b_{i,f}$ with $i > N_{11}$ there is a mapping ε_0 such that $\varepsilon_0(x_1) = \varepsilon_0(x_2) = \cdots = \varepsilon_0(x_e) = 1$. Let $\delta_1 = \delta - N_{11} + d - 1$. Consider the epimorphism $\varepsilon_1 \colon D_\delta \longrightarrow D_{\delta_1}$ defined by the formula

$$\varepsilon_1(b_{i,f}) = \begin{cases} \varepsilon_0(b_{i,f}) & \text{if } i \leq N_{11}, \\ b_{i-N_{11}+d-1,f} & \text{if } i > N_{11}. \end{cases}$$

Clearly, if $\varepsilon_2 \colon D_{\delta_1} \longrightarrow D_{d-1}$ is the epimorphism leaving fixed the generators whose first subscript is less than d and annihilating the others, then

$$\varepsilon_2\varepsilon_1(x_1) = \varepsilon_2\varepsilon_1(x_2) = \cdots = \varepsilon_2\varepsilon_1(x_e) = 1.$$

We denote $\varepsilon_1(x_i)$ by x_{1i} for $i = 1, \ldots, e$. In the sequel, unless otherwise stated, for natural δ' and δ'' with $\delta' > \delta''$ we shall consider canonical epimorphisms $\varepsilon\colon D_{\delta'} \longrightarrow D_{\delta''}$ such that their restriction to the subgroup $D_{\delta''-1}$ is identical, and the generators $b_{i,f}$ with $i \geq \delta''$ may be mapped only to $b_{\delta'',f}$ or to 1. The generators $b_{i,f}$ with $i \geq \delta''$ will be called *independent* with respect to the epimorphisms just mentioned.

Let x be an element of $D_{\delta'}$. Then x consists of factors of the form $b_{i,f}\otimes\ldots$. Such factors will be called *monomials*. The maximal number of independent generators with distinct first subscripts that occur in a monomial is called *the degree of x with respect to the epimorphism $D_{\delta'} \longrightarrow D_{\delta''}$*.

For $j \geq d$ we consider the epimorphisms $\varepsilon_{1j}\colon D_{\delta_1} \longrightarrow D_d$ defined by the relations

$$\varepsilon_{1j}(b_{i,f}) = \begin{cases} b_{i,f} & \text{if } i \leq d-1, \\ b_{d,f} & \text{if } i = j, \\ 1 & \text{if } i > d-1,\ i \neq j. \end{cases}$$

Since d is a fixed number, for sufficiently large δ_1 the values of ε_{1j} on the elements $x_{11}, \ldots, x_{1e}$ coincide. More precisely, there are $j_1, \ldots, j_p$ with $\varepsilon_{1j_1}(x_{1i}) = \varepsilon_{1j_2}(x_{1i}) = \cdots = \varepsilon_{1j_p}(x_{1i})$ for $i = 1, \ldots, e$. For definiteness, let $j_1 < j_2 < \cdots < j_p$. We consider the epimorphism $\varepsilon_3\colon D_{\delta_1} \longrightarrow D_{\delta_1+1-p}$ defined as follows:

$$\varepsilon_3(b_{i,f}) = \begin{cases} b_{i,f} & \text{if } i < j_2,\ i \neq d,\ i \neq j_1, \\ b_{j_1,f} & \text{if } i = d, \\ b_{d,f} & \text{if } i = j_1, j_2, \ldots, j_p, \\ b_{i-1,f} & \text{if } j_2 < i < j_3,\ i \neq d, \\ \ldots & \\ b_{i-p+1,f} & \text{if } j_p < i,\ i \neq d. \end{cases}$$

We continue the proof by induction on the degrees of $x_{11}, \ldots, x_{1e}$ with respect to the epimorphism $D_{\delta_1} \longrightarrow D_d$. Obviously, the degrees of the elements do not exceed c. Let these degrees be at most 1. Then, by setting

$$\varepsilon_4(b_{i,f}) = \begin{cases} b_{i,f} & \text{if } i \leq d, \\ 1 & \text{if } d < i \leq \delta_1 - 1 + p, \end{cases}$$

we have $\varepsilon_4\varepsilon_3(x_{1i}) = 1$ for $i = 1, \ldots, e$. Suppose that for the degrees not exceeding n the value $N_1(c, d, e)$ is equal to N_{12}, where $n < c$. Consider the case where the degrees of the elements x_{1i} do not exceed $n+1$.

For $j \geq d+1$ we introduce the epimorphisms $\varepsilon_{2j}\colon D_{\delta_1-p+1} \longrightarrow D_{d+1}$ by setting

$$\varepsilon_{2j}(b_{i,f}) = \begin{cases} b_{i,f} & \text{if } i \leq d, \\ b_{d+1,f} & \text{if } i = j, \\ 1 & \text{if } i > d,\ i \neq j. \end{cases}$$

Then for δ_1 large enough there are epimorphisms $\varepsilon_{2j_1}, \ldots, \varepsilon_{2j_p}$ with $\varepsilon_{2j_1}\varepsilon_3(x_{1i}) = \varepsilon_{2j_2}\varepsilon_3(x_{1i}) = \cdots = \varepsilon_{2j_p}\varepsilon_3(x_{1i})$ for $i = 1, \ldots, e$. We may assume that $j_1 < j_2 < \cdots < j_p$.

Introduce the epimorphism $\varepsilon_5\colon D_{\delta_1+1-p} \longrightarrow D_{\delta_1+2(1-p)}$ by setting

$$\varepsilon_5(b_{i,f}) = \begin{cases} b_{i,f} & \text{if } i < j_2,\ i \neq d+1,\ i \neq j_1, \\ b_{j_1,f} & \text{if } i = d+1, \\ b_{d+1,f} & \text{if } i = j_1, j_2, \dots, j_p, \\ b_{i-1,f} & \text{if } j_2 < i < j_3,\ i \neq d+1, \\ \dots & \\ b_{i-p+1,f} & \text{if } j_p < i,\ i \neq d+1, \end{cases}$$

and the epimorphism $\varepsilon_6\colon D_{\delta_1+2(1-p)} \longrightarrow D_{d+1}$ by setting

$$\varepsilon_6(b_{i,f}) = \begin{cases} b_{i,f} & \text{if } i \leq d+1, \\ 1 & \text{if } d+1 < i. \end{cases}$$

Clearly, the degrees of $\varepsilon_6\varepsilon_5\varepsilon_3(x_{1i})$ with respect to the epimorphism $D_{d+1} \longrightarrow D_d$ are less than $n+1$, where $i = 1, \dots, e$.

We denote the homomorphism $\varepsilon_5\varepsilon_4$ by $\widetilde{\varepsilon}_2$ and continue the procedure. For sufficiently large δ_1 at the $(N_{12}-d+1)$th step we get the epimorphism $\widetilde{\varepsilon}_{N_{12}-d+1}\colon D_{\delta_1} \longrightarrow D_{\delta_1+(N_{12}-d+1)(1-p)}$.

Now we consider the epimorphism $\varepsilon'_{N_{12}-d+1}\colon D_{\delta_1+(N_{12}-d+1)(1-p)} \longrightarrow D_{N_{12}}$ defined by the formula

$$\varepsilon'_{N_{12}-d+1}(b_{i,f}) = \begin{cases} b_{i,f} & \text{if } i \leq N_{12}, \\ 1 & \text{if } N_{12} < i. \end{cases}$$

It is clear that the degrees of the elements $\varepsilon'_{N_{12}-d+1}\widetilde{\varepsilon}_{N_{12}-d+1}(x_{1i})$ do not exceed n for all $i = 1, \dots, e$. By induction hypothesis there is an epimorphism $\widetilde{\varepsilon}\colon D_{N_{12}} \longrightarrow D_d$ annihilating the elements $\varepsilon'_{N_{12}-d+1}\widetilde{\varepsilon}_{N_{12}-d+1}(x_{1i})$ for $i = 1, \dots, e$. Then $\varepsilon = \widetilde{\varepsilon}\varepsilon'_{N_{12}-d+1}\widetilde{\varepsilon}_{N_{12}-d+1}$ is the required homomorphism.

The canonical epimorphisms $A_{\delta,c} \longrightarrow A_{\delta',c}$ and $Z_{\delta,c} \longrightarrow Z_{\delta',c}$ induce epimorphisms of the groups $\widehat{A}'_{\delta,c} \longrightarrow \widehat{A}'_{\delta',c}$ and $\widehat{Z}'_{\delta,c} \longrightarrow \widehat{Z}'_{\delta',c}$, which will also be called *canonical* in what follows.

PROPOSITION 5.5.3. *For any natural c, d, e, there is a natural number $N_2 = N_2(c,d,e)$ such that for $\delta \geq N_2$ and for any elements $a'_1, \dots, a'_e$ of the group $\widehat{A}'_{\delta,c+1}$, a canonical homomorphism onto $\widehat{A}'_{d,c+1}$ annihilating all the elements $a'_1, \dots, a'_e$ exists.*

PROOF. We fix an isomorphism α from the group of pth roots of unity onto the group of elements of order p in the group $\mathbb{Q}/\mathbb{Z}$. For every δ and c we define the homomorphism $\beta_\delta\colon Z_{\delta,c} \longrightarrow \widehat{Z}'_{\delta,c}$ by setting $[\beta_\delta(z)](\widehat{z}) = \alpha\widehat{z}(z)$, where $z \in Z_{\delta,c}$, $\widehat{z} \in \widehat{Z}_{\delta,c}$. Clearly, β_δ is an isomorphism. □

One can immediately verify that the following diagram commutes:

$$\begin{array}{ccc} Z_{\delta,c} & \xrightarrow{\beta_\delta} & \widehat{Z}'_{\delta,c} \\ \downarrow{\scriptstyle \varepsilon} & & \downarrow{\scriptstyle \varepsilon'} \\ Z_{\delta',c} & \xrightarrow{\beta_{\delta'}} & \widehat{Z}'_{\delta',c}, \end{array}$$

where ε and ε' are canonical homomorphisms, and ε' is induced by ε. By Lemma 5.5.2, $Z_{\delta,c}$ is a homomorphic image of $D_{\delta,c}$. Consequently, the previous proposition implies that for $\delta \geq N_1(c,\delta',e)$ there is a canonical epimorphism $\widehat{Z}'_{\delta,c} \to \widehat{Z}'_{\delta',c}$ annihilating any e elements (specified beforehand) of the group $\widehat{Z}'_{\delta,c}$.

We consider the commutative diagram

$$\begin{array}{ccccccccc} 1 & \longrightarrow & \widehat{Z}'_{\delta,c+1} & \longrightarrow & \widehat{A}'_{\delta,c+1} & \xrightarrow{\psi_1} & \widehat{Z}'_{\delta,c} & \longrightarrow & 1 \\ & & \downarrow{\scriptstyle \varepsilon'_{c+1}} & & \downarrow{\scriptstyle \varepsilon'} & & \downarrow{\scriptstyle \varepsilon'_c} & & \\ 1 & \longrightarrow & \widehat{Z}'_{\delta',c+1} & \xrightarrow{\psi_2} & \widehat{A}'_{\delta',c+1} & \longrightarrow & \widehat{Z}'_{\delta',c} & \longrightarrow & 1 \end{array}$$

whose rows are exact sequences with natural homomorphisms, and the homomorphisms corresponding to vertical arrows are canonical. For $\delta \geq N_1(c,\delta',e)$, there is a canonical homomorphism that takes $\psi_1(a'_i)$ to 1 for each $i = 1,\dots,e$. Then in $\widehat{Z}'_{\delta',c+1}$ there exist elements $\overline{a}'_i$ such that $\psi_2(\overline{a}'_i) = \varepsilon'(a'_i)$. Therefore, for $\delta' \geq N_1(c+1,d,e)$ we obtain the required result. Thus, $N_2(c,d,e) = N_1(c,N_1(c+1,d,e),e)$.

The canonical epimorphism $\varepsilon\colon A_{\delta,c+1} \to A_{\delta',c+1}$ induces an embedding $\widehat{\varepsilon}$: $\widehat{A}_{\delta',c+1} \to \widehat{A}_{\delta,c+1}$. In turn, the mapping $\widehat{\varepsilon}$ induces the mapping $\psi_{1\varepsilon}\colon H^1(F\cdot B_{\delta',c-1},\widehat{A}_{\delta',c+1}) \to H^1(F\cdot B_{\delta,c-1},\widehat{A}_{\delta,c+1})$ which is the composition of the lifting homomorphism from $H^1(F\cdot B_{\delta',c-1},\widehat{A}_{\delta',c+1})$ to $H^1(F\cdot B_{\delta,c-1},\widehat{A}_{\delta',c+1})$ and the natural mapping from the latter group to $H^1(F\cdot B_{\delta,c-1},\widehat{A}_{\delta,c+1})$.

Let $\overline{\psi}_{1\varepsilon}$ be the mapping dual to $\psi_{1,\varepsilon}$.

PROPOSITION 5.5.4. *For any natural c and d, there is a natural number $N_3 = N_3(c,d)$ such that for $\delta \geq N_3$ and an arbitrary $x \in H^1(F\cdot B_{\delta,c-1},\widehat{A}_{\delta,c+1})'$, there exists a canonical homomorphism ε such that the homomorphism $\overline{\psi}_{x,\varepsilon}$ induced annihilates x.*

PROOF. The fact that the sequence

$$1 \to \widehat{Z}_{\delta,c} \to \widehat{A}_{\delta,c+1} \to \widehat{Z}_{\delta,c+1} \to 1$$

is exact allows us to prove the proposition for $\widehat{Z}_{\delta,c+1}$ and $\widehat{Z}_{\delta,c}$ instead of $\widehat{A}_{\delta,c+1}$. This is similar to what we have done earlier. The group $B_{\delta,c-1}$ acts trivially on $\widehat{Z}_{\delta,c+1}$ and $\widehat{Z}_{\delta,c}$. We consider the commutative diagram

$$\begin{array}{ccccccc} H^1(B_{\delta,c-1},\widehat{Z}_{\delta,c})' & \longrightarrow & H^1(F\cdot B_{\delta,c-1},\widehat{Z}_{\delta,c})' & \longrightarrow & H^1(F,\widehat{Z}_{\delta,c})' & \longrightarrow & 0 \\ \downarrow & & \downarrow & & \downarrow & & \\ H^1(B_{\delta',c-1},\widehat{Z}_{\delta',c})' & \longrightarrow & H^1(F\cdot B_{\delta',c-1},\widehat{Z}_{\delta',c})' & \longrightarrow & H^1(F,\widehat{Z}_{\delta',c})' & \longrightarrow & 0 \end{array}$$

where the rows are exact sequences dual to the Hochschild-Serre sequence and the mappings corresponding to vertical arrows are induced by a canonical homomorphism. □

By Lemma 5.5.2, the sequence of F-modules

$$1 \to \widehat{Z}_{\delta,c+1} \to \widehat{D}_{\delta,c+1} \to \widehat{E}_{\delta,c+1} \to 1$$

is exact. The canonical epimorphism $D_{\delta,c+1} \longrightarrow D_{\delta',c+1}$ induces the commutative diagram

$$\begin{array}{ccccc} H^1(F,\widehat{D}_{\delta,c+1})' & \longrightarrow & H^1(F,\widehat{Z}_{\delta,c+1})' & \xrightarrow{\psi_1} & (\widehat{E}^F_{\delta,c+1})' \\ \downarrow & & \downarrow & & \downarrow \\ H^1(F,\widehat{D}_{\delta',c+1})' & \longrightarrow & H^1(F,\widehat{Z}_{\delta',c+1})' & \longrightarrow & (\widehat{E}^F_{\delta',c+1})' \end{array}$$

The group $D_{\delta,c+1}$ is a free $\mathbb{Z}/p\mathbb{Z}[F]$-module. We observe that the modules $\mathrm{Hom}(\mathbb{Z}[F],\mathbb{Z}/p\mathbb{Z})$ and $(\mathbb{Z}/p\mathbb{Z}[F])'$ are naturally isomorphic. Consequently, by Lemma A.3.3, the module $\widehat{D}_{\delta,c+1}$ is cohomologously trivial.

Now we consider the commutative diagram

$$\begin{array}{ccccc} \widehat{E}'_{\delta,c+1} & \xrightarrow{\psi_2} & (\widehat{E}^F_{\delta,c+1})' & \longrightarrow & 1 \\ \big\downarrow{\scriptstyle\varepsilon} & & \downarrow & & \\ \widehat{E}'_{\delta',c+1} & \longrightarrow & (\widehat{E}^F_{\delta',c+1})' & \longrightarrow & 1 \end{array}$$

with natural homomorphisms. We show that for $\delta \geq N_1(c+1,\delta',1)$, there is a canonical homomorphism that takes any given element from $H^1(F,\widehat{Z}_{\delta,c+1})'$ to 1. Indeed, let $y \in H^1(F,\widehat{Z}_{\delta,c+1})'$. Then $\psi_2^{-1}\psi_1(y)$ is taken to 1 by a canonical homomorphism ε ($\widehat{E}'_{\delta,c+1}$ is a subgroup of the group $\widehat{D}'_{\delta,c+1}$ for which the assertion follows from Proposition 5.5.2). Clearly, the canonical homomorphism ε induces the required mapping $H^1(F,\widehat{Z}_{\delta,c+1})' \to H^1(F,\widehat{Z}_{\delta',c+1})'$.

Now consider the mapping $\psi_{2,\varepsilon}\colon H^1(B_{\delta',c-1},\widehat{Z}_{\delta',c+1})' \longrightarrow H^1(B_{d,c-1},\widehat{Z}_{d,c+1})'$. Since $B_{\delta',c-1}$ acts trivially on the module $\widehat{Z}_{\delta',c+1}$, $\psi_{2,\varepsilon}$ coincides with the natural mapping $\mathrm{Hom}(B_{\delta',1},\mathrm{Hom}(Z_{\delta',c+1},K^*))' \to \mathrm{Hom}(B_{d,1},\mathrm{Hom}(Z_{d,c+1},K^*))'$.

The fact that $\sigma_\delta\colon \mathrm{Hom}(B_{\delta,1},\mathrm{Hom}(Z_{\delta,c+1},K^*)) \longrightarrow \mathrm{Hom}(B_{\delta,1}\otimes Z_{\delta,c+1},K^*)$ defined by the formula $[\sigma_\delta(b')](b\otimes z) = (b'(b))(z)$, where $b' \in \mathrm{Hom}(B_{\delta,1},\widehat{Z}_{\delta,c+1})$, $b \in B_{\delta,1}$, and $z \in Z_{\delta,c+1}$, is an isomorphism can be verified immediately. We may directly check the commutativity of the diagram

$$\begin{array}{ccc} \mathrm{Hom}(B_{\delta',1}\otimes Z_{\delta',c+1},K^*)' & \xrightarrow{\overline{\sigma}_{\delta'}} & \mathrm{Hom}(B_{\delta',1},\mathrm{Hom}(Z_{\delta',c+1},K^*))' \\ \big\downarrow{\scriptstyle\psi_{3,\varepsilon}} & & \big\downarrow{\scriptstyle\psi_{2,\varepsilon}} \\ \mathrm{Hom}(B_{d,1}\otimes Z_{d,c+1},K^*)' & \xrightarrow{\overline{\sigma}_d} & \mathrm{Hom}(B_{d,1},\mathrm{Hom}(Z_{d,c+1},K^*))' \end{array}$$

where $\psi_{3,\varepsilon}$ is a homomorphism induced by the canonical homomorphism ε, and $\overline{\sigma}_{\delta'}$ and $\overline{\sigma}_d$ are the mappings dual to $\sigma_{\delta'}$ and σ_d. We note that the group $B_{\delta',1}\otimes Z_{\delta',c+1}$ is an operator-homomorphic image of $D_{\delta'c+2}$. Consequently, for $\delta' \geq N_1(c+2,d,1)$ there exists a canonical homomorphism ε such that $\psi_{2,\varepsilon}$ annihilates any element of $\mathrm{Hom}(B_{\delta',1},\widehat{Z}_{\delta',c+1})'$, specified beforehand.

Thus, for an element of $H^1(F\cdot B_{\delta,c-1},\widehat{Z}_{\delta,c+1})'$ the value of the constant $N_3(c+1,d)$ is equal to $N_1(c+1,N_1(c+2,d,1),1)$. Then for an element of $H^1(F\cdot B_{\delta,c-1},\widehat{A}_{\delta,c+1})'$ this value is equal to $N_1(c+1,N_1(c+2,N_1(c,N_1(c+1,d,1),1),1),1)$.

Let $h_\delta \in H^2(F\cdot B_{\delta,c-1},A_{\delta,c+1})$ be a cohomology class corresponding to the group extension $F\cdot B_{\delta,c+1}$. For numbers δ and δ' with $\delta' < \delta$ we consider the diagram

$$\begin{array}{ccc} H^2(F\cdot B_{\delta,c-1}, A_{\delta,c+1}) & \xrightarrow{\tau_\varepsilon} & H^2(F\cdot B_{\delta,c-1}, A_{\delta',c+1}) \\ & & \uparrow \lambda_1 \\ & & H^2(F\cdot B_{\delta',c-1}, A_{\delta',c+1}) \end{array}$$

where λ_1 is the lifting homomorphism and τ_ε is the mapping induced by the canonical epimorphism $\varepsilon\colon A_{\delta,c+1} \to A_{\delta',c+1}$. Proposition 1.13.1 and Theorem 3.13.1 immediately imply that $\tau_\varepsilon(h_\delta) = \lambda_1(h_{\delta'})$.

THEOREM 5.5.3. *For natural numbers c and d with $\delta > N_4(c,d)$, in the quasi-Scholzian solution $K_{\delta,c-1}$ of the embedding problem $(K/k, F\cdot B_{\delta,c-1}, B_{\delta,c-1})$ there is a subfield $K_{d,c-1}$ with Galois group $F\cdot B_{d,c-1}$, corresponding to the kernel of the canonical homomorphism, such that the compatibility condition is fulfilled for the embedding problem $(K_{d,c-1}/k, F\cdot B_{d,c+1}, \varphi_{d,c+1}, A_{d,c+1})$.*

PROOF. Proposition 3.14 implies that the compatibility condition for the embedding problem $(K_{d,c-1}/k, F\cdot B_{d,c+1}, \varphi_{d,c+1}, A_{d,c+1})$ is equivalent to the solvability of the associated local problems $(K_{d,c-1}\otimes_k k_\mathfrak{p}/k_\mathfrak{p}, F\cdot B_{d,c+1}, \varphi_{d,c+1}, A_{d,c+1})$ at all points $\mathfrak{p}$ of k. Lemma 3.14 immediately implies that it suffices to consider only the divisors ramified in the extension $K_{d,c-1}/k$.

Let $\mathfrak{p}$ be a critical divisor for $K_{\delta,c-1}/k$, $\mathfrak{P}_{d,c-1}$ a prime divisor of $\mathfrak{p}$ in $K_{d,c-1}$, and let $\mathfrak{P}_{\delta,c-1}$ be a prime divisor of $\mathfrak{P}_{d,c-1}$ in $K_{\delta,c-1}$. We denote by $G_{d,c-1,\mathfrak{p}}$ and $G_{\delta,c-1,\mathfrak{p}}$ the decomposition groups of divisors in the corresponding extensions and by $G_{d,c+1,\mathfrak{p}}$ and $G_{\delta,c+1,\mathfrak{p}}$ the preimages of $G_{d,c-1,\mathfrak{p}}$ and $G_{\delta,c-1,\mathfrak{p}}$ in the groups $F\cdot B_{d,c+1}$ and $F\cdot B_{\delta,c+1}$, respectively. The solvability of the embedding problem $(K_{d,c-1}\otimes_k k_\mathfrak{p}/k_\mathfrak{p}, F\cdot B_{d,c+1}, \varphi_{d,c+1}, A_{d,c+1})$ is equivalent to the solvability of the problem

$$(K_{d,c-1,\mathfrak{P}_{d,c-1}}/k_\mathfrak{p}, G_{d,c+1,\mathfrak{p}}, \varphi_{d,c+1}, A_{d,c+1}).$$

For all δ, let $h_{\delta,\mathfrak{p}} \in H^2(G_{\delta,c-1,\mathfrak{p}}, A_{\delta,c+1})$ be the restriction of the cohomology class $h_\delta \in H^2(F\cdot B_{\delta,c-1}, A_{\delta,c+1})$ corresponding to the extension $F\cdot B_{\delta,c+1}$. Denote by $\lambda_{\delta,\mathfrak{p}}$ the lifting homomorphism $H^2(G_{\delta,c-1,\mathfrak{p}}, A_{\delta,c+1}) \to H^2(\overline{F}_\mathfrak{p}, A_{\delta,c+1})$, where $\overline{F}_\mathfrak{p}$ is the Galois group of the algebraic closure $\overline{k}_\mathfrak{p}$ of the field $k_\mathfrak{p}$. Clearly, we must choose a subfield $K_{d,c-1}$ such that the relation $\lambda_{d,\mathfrak{p}}(h_{d,\mathfrak{p}}) = 0$ is fulfilled.

We consider the commutative diagram

$$\begin{array}{ccccc} & & H^2(G_{\delta,c-1,\mathfrak{p}}, A_{\delta,c+1}) & \xrightarrow{\lambda_{\delta,\mathfrak{p}}} & H^2(\overline{F}_\mathfrak{p}, A_{\delta,c+1}) \\ & & \downarrow \tau_{\mathfrak{p},\varepsilon} & & \downarrow \overline{\tau}_{\mathfrak{p},\varepsilon} \\ H^2(G_{d,c-1,\mathfrak{p}}, A_{d,c+1}) & \xrightarrow{\lambda_2} & H^2(G_{\delta,c-1,\mathfrak{p}}, A_{d,c+1}) & \xrightarrow{\lambda_3} & H^2(\overline{F}_\mathfrak{p}, A_{d,c+1}) \end{array}$$

where λ_2 and λ_3 are lifting homomorphisms, and $\tau_{\mathfrak{p},\varepsilon}$ and $\overline{\tau}_{\mathfrak{p},\varepsilon}$ are the mappings induces by the canonical homomorphism ε. We note that the composition $\lambda_3\lambda_2$ coincides with the homomorphism $\lambda_{d,\mathfrak{p}}$.

Proposition 1.13.1 and Theorem 3.13.1 imply $\tau_{\mathfrak{p},\varepsilon}(h_{\delta,\mathfrak{p}}) = \lambda_2(h_{d,\mathfrak{p}})$. Consequently, if the relation $\overline{\tau}_{\mathfrak{p},\varepsilon}\lambda_{\delta,\mathfrak{p}}(h_{\delta,\mathfrak{p}}) = 0$ is valid for the canonical homomorphism ε, then the embedding problem is solvable at the point $\mathfrak{p}$.

We show that if $\mathfrak{p}$ is a critical divisor for the extension $K_{\delta,c-1}/k$ and $\mathfrak{p}\notin S_0$, then the corresponding local embedding problem is solvable. We note that in this case $\mathfrak{p}$ completely decomposes in the field K. Therefore, $G_{\delta,c+1,\mathfrak{p}}$, is a subgroup of $B_{\delta,c+1}$. Moreover, $K_{\delta,c-1,\mathfrak{P}_{\delta,c-1}}/k_{\mathfrak{p}}$ is a cyclic totally ramified extension.

By Theorem 3.14.1, the solvability of a local embedding problem is equivalent to the solvability of the associated Brauer problems. Let $\chi \in \widehat{A}_{\delta,c+1}$ and let $(K_{\delta,c-1,\mathfrak{P}_{\delta,c-1}}/k_{\mathfrak{p},\chi}, G_\chi, \varphi_\chi, A_{\delta,c+1}/A_{\delta,c+1,\chi})$ be the corresponding Brauer problem, where $A_{\delta,c+1,\chi}$ is the kernel of the character χ, $k_{\mathfrak{p},\chi}$ is the subfield corresponding to the subgroup F_χ that acts trivially on χ, G_χ is the preimage of F_χ in the group $G_{\delta,c+1,\mathfrak{p}}/A_{\delta,c+1,\chi}$, and φ_χ is the corresponding homomorphism. We observe that G_χ is either a cyclic group whose order divides p^{c+1} or the direct product of the group F_χ and the kernel of the Brauer problem. Clearly, in either case the problem is solvable, since $k_{\mathfrak{p}}$ contains all of the p^{c+1}th roots of unity.

Assume that the set S_0 consists of e divisors. We consider the commutative diagram

$$\begin{array}{ccc}
\prod\limits_{\mathfrak{p}\in S_0} H^2(\overline{F}_{\mathfrak{p}}, A_{\delta,c+1}) & \xrightarrow{\prod \overline{\tau}_{\mathfrak{p},\varepsilon}} & \prod\limits_{\mathfrak{p}\in S_0} H^2(\overline{F}_{\mathfrak{p}}, A_{d,c+1}) \\
\big\downarrow{\scriptstyle \prod \omega_{\mathfrak{p},\delta}} & & \big\downarrow{\scriptstyle \prod \omega_{\mathfrak{p},d}} \\
\prod\limits_{\mathfrak{p}\in S_0} \mathrm{Hom}_{\overline{F}_{\mathfrak{p}}}(A_{\delta,c+1}, \overline{k}_{\mathfrak{p}}^*)' & \xrightarrow{\prod \overline{\psi}_{\mathfrak{p},\varepsilon}} & \prod\limits_{\mathfrak{p}\in S_0} \mathrm{Hom}_{\overline{F}_{\mathfrak{p}}}(A_{d,c+1}, \overline{k}_{\mathfrak{p}}^*)',
\end{array}$$

where the mappings corresponding to the vertical arrows are the homomorphisms of Tate's local duality, and $\overline{\psi}_{\mathfrak{p},\varepsilon}$ is induced by the canonical homomorphism ε. We note that if $\mathfrak{p}$ is a finite point, then $\omega_{\mathfrak{p},\delta}$ is an isomorphism, and if $\mathfrak{p}$ is an infinite point, then $\omega_{\mathfrak{p},\delta}$ is injective (see §3 of the Appendix).

For every $\mathfrak{p} \in S_0$ we choose a preimage $x_{\mathfrak{p}}$ from $\widehat{A}'_{\delta,c+1}$ of $\omega_{\mathfrak{p},\delta}\lambda_{\delta,\mathfrak{p}}(h_{\delta,\mathfrak{p}})$ (the preimage is taken under the natural mapping of $\widehat{A}'_{\delta,c+1}$ onto $\mathrm{Hom}_{\overline{F}_{\mathfrak{p}}}(A_{\delta,c+1}, \overline{k}_{\mathfrak{p}}^*)'$). By Proposition 5.5.3, for $\delta > N_2(c,d,e)$ there is a canonical epimorphism $\varepsilon\colon \widehat{A}'_{\delta,c+1} \to \widehat{A}'_{d,c+1}$ taking all of the $x_{\mathfrak{p}}$ to 1. It is clear that ε is the required homomorphism.□

Let $(K_{\delta,c-1}/k, F\cdot B_{\delta,c+1}, \varphi_{\delta,c+1}, A_{\delta,c+1})$ be the embedding problem for which the compatibility condition is fulfilled. We denote by S_δ the set of prime divisors of k, consisting of S_0, of critical divisors for $K_{\delta,c-1}/k$, and of the divisors completely decomposing in $K_{\delta,c-1}$, whose prime divisors generate the p-component of the divisor class group of $K_{\delta,c-1}$.

Let F_{S_δ} be the Galois group of the maximal algebraic extension Ω_δ of k, unramified outside S_δ.

THEOREM 5.5.4 (I. R. Shafarevich). *The semidirect embedding problem* $(K/k, F\cdot B_{d,c}, B_{d,c})$ *has a proper solution.*

PROOF. It suffices to show that for $\delta \geq N_3(c,d)$, we may choose a subfield $K_{d,c-1}$ in $K_{\delta,c-1}/k$, belonging to the kernel of the canonical epimorphism $B_{\delta,c+1} \to B_{d,c+1}$, such that the problem $(K_{d,c-1}/k, F\cdot B_{d,c+1}, \varphi_{d,c+1}, A_{d,c+1})$ is solvable, and we may choose a solution unramified outside S_d. Indeed, applying Theorem 5.5.2 to the latter problem, we obtain a quasi-Scholzian solution $K_{d,c}$ of the problem $(K/k, F\cdot B_{d,c}, B_{d,c})$. In the sequel we shall omit the subscripts $c-1$, c, and $c+1$ for the brevity of notation.

For every δ the exact sequence

$$1 \longrightarrow K_\delta^* \longrightarrow J_{K_\delta,S_\delta} \xrightarrow{\psi_{1,\delta}} C_{K_\delta,S_\delta} \longrightarrow 1$$

induces the exact sequence

$$(1) \qquad 1 \longrightarrow \widehat{A}_\delta \longrightarrow \mathrm{Hom}(A_\delta, J_{K_\delta,S_\delta}) \xrightarrow{\psi_{2,\delta}} \mathrm{Hom}(A_\delta, C_{K_\delta,S_\delta}).$$

We note that $\psi_{2,\delta}$ is surjective. Indeed, let $y_1 \in C_{K_\delta,S_\delta}$ and $y_1^{p_2} = 1$ (the group A_δ is of period p^2). There is an element $y_2 \in J_{K_\delta,S_\delta}$ with $\psi_{1,\delta}(y_2) = y_1$. We put $y_3 = y_2^{p^2}$. Then $y_3 \in K_\delta^*$ and $K_\delta(\sqrt[p^2]{y_3})/K_\delta$ is a splitting algebra (S_δ contains representatives of the p-component of the divisor class group of K_δ). Consequently, $y_3 = y_4^{p^2}$, where $y_4 \in K_\delta^*$, and for $y_2 y_4^{-1}$ we have

$$\psi_{1,\delta}(y_2 y_4^{-1}) = y_1, \qquad (y_2 y_4^{-1})^{p^2} = 1.$$

The sequence (1) induce the exact sequence

$$(2) \qquad H^1(F \cdot B_\delta, \widehat{A}_\delta)' \xrightarrow{\psi_{3,\delta}} \mathrm{Hom}_{F \cdot B_\delta}(A_\delta, C_{K_\delta,S_\delta})' \xrightarrow{\psi_{4,\delta}} \mathrm{Hom}_{F \cdot B_\delta}(A_\delta, J_{K_\delta,S_\delta})'$$

dual to the exact cohomology sequence.

Tate's duality (see §3 of the Appendix) allows us to write the commutative diagram

$$\begin{array}{ccc} H^2(F_{S_\delta}, A_\delta) & \xrightarrow{\tau_{1,\delta}} & \sum_{\mathfrak{p} \in \S_\delta}^{\oplus} H^2(\overline{F}_\mathfrak{p}, A_\delta) \\ \downarrow \tau_{2,\delta} & & \downarrow \tau_{3,\delta} \\ \mathrm{Hom}_{F \cdot B_\delta}(A_\delta, C_{K_\delta,S_\delta})' & \xrightarrow{\psi_{4,\delta}} & \mathrm{Hom}_{F \cdot B_\delta}(A_\delta, J_{K_\delta,S_\delta})'. \end{array}$$

Since the compatibility condition for the embedding problem $(K_\delta/k, F \cdot B_\delta, \varphi_\delta, A_\delta)$ holds, we have $\tau_{1,\delta}\lambda_\delta(h_\delta) = 0$ or, by the diagram, $\psi_{4,\delta}\tau_{2,\delta}\lambda_\delta(h_\delta) = 0$, where λ_δ is the lifting homomorphism $H^2(F \cdot B_\delta, A_\delta) \to H^2(F_{S_\delta}, A_\delta)$.

We denote $\tau_{2,\delta}\lambda_\delta(h_\delta)$ by x_1. Since the sequence (2) is exact, there is $x_2 \in H^1(F \cdot B_\delta, \widehat{A}_\delta)'$ with $\psi_{3,\delta}(x_2) = x_1$. Proposition 5.5.4 implies that for $\delta \geq N_3(c,d)$, there is a canonical homomorphism ε such that the induced mapping

$$\psi_{1,\varepsilon}\colon H^1(F \cdot B_\delta, A_\delta)' \longrightarrow H^1(F \cdot B_d, A_d)'$$

annihilates x_2. □

Now we show that ε is the required homomorphism. Let $S = S_\delta \cup S_d$. First we prove that the embedding problem $(K_d/k, F \cdot B_d, \varphi_d, A_d)$ is solvable, and we may choose a solution unramified outside S.

We consider the commutative diagram

$$\begin{array}{ccccc} & & H^2(F \cdot B_\delta, A_\delta) & \xrightarrow{\lambda_\delta} H^2(F_{S_\delta}, A_\delta) \xrightarrow{\lambda_1} & H^2(F_S, A_\delta) \\ & & \downarrow \psi_{5,\varepsilon} & & \downarrow \psi_{6,\varepsilon} \\ H^2(F \cdot B_d, A_d) & \xrightarrow{\lambda_2} & H^2(F \cdot B_\delta, A_d) & \xrightarrow{\lambda_3} & H^2(F_S, A_d) \end{array}$$

where λ_1, λ_2, and λ_3 are lifting homomorphisms, $\psi_{5,\varepsilon}$ and $\psi_{6,\varepsilon}$ are natural mappings induced by ε. Clearly, our assertion is equivalent to the relation $\lambda_3\psi_{5,\varepsilon}(h_\delta) = 0$ or to $\psi_{6,\varepsilon}\lambda_1\lambda_\delta(h_\delta) = 0$ by the commutativity of the diagram.

Now we consider the diagram

$$\begin{array}{ccccc} H^2(F_{S_\delta}, A_\delta) & \xrightarrow{\lambda_1} & H^2(F_S, A_\delta) & \xrightarrow{\psi_{6,\varepsilon}} & H^2(F_S, A_d) \\ \downarrow{\scriptstyle \tau_{2,\delta}} & & \downarrow & & \downarrow \\ \mathrm{Hom}_{F\cdot B_\delta}(A_\delta, C_{K_\delta,S_\delta})' & \xrightarrow{\psi_1} & \mathrm{Hom}_{F\cdot B_\delta}(A_\delta, C_{K_\delta,S})' & \xrightarrow{\psi_{7,\varepsilon}} & \mathrm{Hom}_{F\cdot B_\delta}(A_d, C_{K_\delta,S})', \end{array}$$

where the mappings corresponding to the vertical arrows are Tate's duality isomorphisms, ψ_1 is induced by the natural epimorphism $C_{K_\delta,S} \to C_{K_\delta,S_\delta}$, and $\psi_{7,\varepsilon}$ is induced by the canonical homomorphism.

Proposition A.3.2 implies the commutativity of the diagram. Therefore, the relation $\psi_{6,\varepsilon}\lambda_1\lambda_\delta(h_\delta) = 0$ is equivalent to $\psi_{7,\varepsilon}\psi_1(x_1) = 1$.

In order to prove the latter relation we consider the commutative diagram

$$\begin{array}{ccccc} H^1(F\cdot B_\delta, \widehat{A}_\delta)' & \xrightarrow{\mathrm{id}} & H^1(F\cdot B_\delta, \widehat{A}_\delta)' & \xrightarrow{\overline{\psi}_{1,\varepsilon}} & H^1(F\cdot B_d, \widehat{A}_d)' \\ \downarrow & & \downarrow & & \downarrow \\ \mathrm{Hom}_{F\cdot B_\delta}(A_\delta, C_{K_\delta,S_\delta})' & \xrightarrow{\psi_1} & \mathrm{Hom}_{F\cdot B_\delta}(A_\delta, C_{K_\delta,S})' & \xrightarrow{\psi_{7,\varepsilon}} & \mathrm{Hom}_{F\cdot B_d}(A_d, C_{K_d,S})', \end{array}$$

where the mappings corresponding to the vertical arrows are homomorphisms dual to the connecting homomorphisms. We note that the groups $\mathrm{Hom}_{F\cdot B_\delta}(A_d, C_{K_\delta,S})$ and $\mathrm{Hom}_{F\cdot B_d}(A_d, C_{K_d,S})$ coincide. Indeed, the kernel W of the canonical epimorphism $B_\delta \to B_d$ acts trivially on A_d. Moreover, we have $C^W_{K_\delta,S} = C_{K_d,S}$ (see Lemma A.3.11). It is clear that $\psi_{7,\varepsilon}\psi_1$ annihilates x_1, i.e., the embedding problem $(K_d/k, F\cdot B_d, \varphi_d, A_d)$ has a solution unramified outside S.

To complete the proof of the theorem, it remains to show that the lifting homomorphism $\lambda_4\colon H^2(F_{S_d}, A_d) \to H^2(F_S, A_d)$ is injective. Indeed, in this case $\lambda_d(h_d) = 0$, where λ_d is the lifting homomorphism $H^2(F\cdot B_d, A_d) \to H^2(F_{S_d}, A_d)$ (we have already shown that $\lambda_4\lambda_d(h_d) = 0$).

We set $\overline{S}_d = S\setminus S_d$ and $\overline{V}_{\overline{S}_d} = \{\{x_\mathfrak{P}\} \in J_{K_d} : x_\mathfrak{P} \in V_\mathfrak{P} \text{ for } \mathfrak{P} \in \overline{S}_d \text{ and } x_\mathfrak{P} = 1 \text{ for } \mathfrak{P} \notin \overline{S}_d\}$. We note that the set $\overline{S}_d$ consists of the divisors of k that completely decompose in the field K_d. Since the intersection of $\overline{V}_{\overline{S}_d}$ and K_d^* in the group J_{K_d} is trivial, the sequence

$$1 \to \overline{V}_{\overline{S}_d} \to C_{K_d,S} \to C_{K_d,S_d} \to 1$$

is exact.

This sequence induces the exact sequence

$$1 \to \mathrm{Hom}(A_d, \overline{V}_{\overline{S}_d}) \to \mathrm{Hom}(A_d, C_{K_d,S}) \xrightarrow{\psi_2} \mathrm{Hom}(A_d, C_{K_d,S_d}). \tag{3}$$

The fact that ψ_2 is surjective easily follows from the commutative diagram with natural homomorphisms:

$$\begin{array}{ccccc} \mathrm{Hom}(A_d, J_{K_d,S}) & \longrightarrow & \mathrm{Hom}(A_d, J_{K_d,S_d}) & \longrightarrow & 1 \\ \downarrow & & \downarrow & & \\ \mathrm{Hom}(A_d, C_{K_d,S}) & \xrightarrow{\psi_2} & \mathrm{Hom}(A_d, C_{K_d,S_d}) & & \\ & & \downarrow & & \\ & & 1 & & \end{array}$$

The sequence (3) induces the exact cohomology sequence

$$\mathrm{Hom}_{F\cdot B_d}(A_d, C_{K_d,S}) \xrightarrow{\psi_3} \mathrm{Hom}_{F\cdot B_d}(A_d, C_{K_d,S_d}) \longrightarrow H^1(F\cdot B_d, \mathrm{Hom}(A_d, \overline{V}_{\overline{S}_d})).$$

Lemma A.3.4 implies that the group $H^1(F\cdot B_d, \mathrm{Hom}(A_d, \overline{V}_{\overline{S}_d}))$ is trivial, i.e., ψ_3 is surjective. It remains to observe that Proposition A.3.2 implies that the injectivity of λ_4 is equivalent to the surjectivity of ψ_3.

THEOREM 5.5.5. *If the embedding problem $(K/k, G, \varphi, B)$ with the nilpotent kernel for algebraic number fields has a solution, then it has a proper solution.*

PROOF. Let L be a solution of the problem, and let $L = \sum_i^{\oplus} L_0 e_i$ be a decomposition into a direct sum of fields. If H is the Galois group of L_0/k, then G is representable as a product (generally speaking, not semidirect) of the subgroups H and B. We may speak of the action of H on B. □

Let $\overline{G}$ be the semidirect extension of B by H with a given action of H on B. The group G is a homomorphic image of $\overline{G}$ under the mapping $\overline{\varphi}$ given by the formula $\overline{\varphi}((h,b)) = hb$, where $h \in H$ and $b \in B$.

By Theorem 5.5.4, the semidirect embedding problem $(L_0/k, \overline{G}, B)$ has a proper solution $\overline{L}$. It remains to note that the subfield $\overline{L}^{\mathrm{Ker}\,\overline{\varphi}}$ is a solution of the embedding problem $(K/k, G, \varphi, B)$.

§6. The inverse problem for solvable groups in Galois theory

We deduce some consequences of the main theorem of the previous section.

1°. THEOREM 5.6.1 (I. R. Shafarevich). *For any algebraic number field k and any finite solvable group B, there is a solution of the inverse problem in Galois theory.*

We prove the theorem by induction on the order of B. By Theorem A.5.2, there are proper subgroups C_1 and C_2 of B with $B = C_1C_2$, one of them, e.g., C_1, being a nilpotent normal subgroup.

We note that the conjugation of the elements in C_1 by the elements in C_2 determines an action of C_2 on C_1. We consider a semidirect extension G of C_1 by C_2 and define the epimorphism $\eta\colon G \longrightarrow B$ by setting $\eta((c_2, c_1)) = c_2c_1$, where $c_1 \in C_1$ and $c_2 \in C_2$.

By the induction hypothesis there is an extension K/k whose Galois group is isomorphic to C_2. By Theorem 5.5.4, there is a proper solution L of the semidirect embedding problem $(K/k, G, \varphi, C_1)$, where φ_1 is the natural mapping of G onto C_2. Clearly, the subfield of invariant elements of L with respect to the subgroup $\mathrm{Ker}\,\eta$ is a Galois extension over k with group B.

Let B be a finite solvable group with generators $b_1, \dots, b_m$, $N = (n_1, \dots, n_c)$ a collection of natural numbers. We consider the following decreasing series of subgroups of B:

$$B_1 = B, \quad B_2 = [B_1, B_1]B_1^{n_1}, \quad \dots, \quad B_{i+1} = [B_i, B_i]B_i^{n_i}, \quad \dots,$$

where square brackets stand for the commutator subgroup of the corresponding group. If $B_c \neq 1$ and $B_{c+1} = 1$, then we say that B is a *group of class c with respect to N*.

For a prime number p we denote by $q_{j,p}$ the maximal power of p that divides the number n_j, and by $N_{j,p}$ the collection of numbers $(n_1, \dots, n_{j-1}, q_{j,p}, \dots, q_{c,p})$. For an arbitrary group, we may consider a series of subgroups corresponding to $N_{j,p}$. We shall denote by $B_{j,p,i}$ the terms of this series for the group B.

Consider another decreasing series of subgroups of an arbitrary group B. This series is defined for N and natural j not exceeding c as follows: $B_{j,i} = \bigcap_p B_{j,p,i}$. We say that the group B is *of class c with respect to N and j* if $B_{j,c} \neq 1$ and $B_{j,c+1} = 1$.

Suppose B is an F-operator group. We consider a free F-operator group U with the generators $u_{i,f}$, where $i = 1, \dots, m$ and $f \in F$. The action of F on the elements $u_{i,f}$ is defined by the formula $(u_{i,f})^{f'} = u_{i,ff'}$. In the sequel we shall denote the quotient groups U/U_{c+1}, $U/U_{j,p,c+1}$, and $U/U_{j,c+1}$ by D, $D_{(j),p}$, and $D_{(j)}$, respectively. Clearly, these groups are finite.

PROPOSITION 5.6.1. *Every F-operator group B of class c with respect to N with m generators is an operator-homomorphic image of D.*

PROOF. We consider the epimorphism $\psi_1 \colon U \longrightarrow B$ by setting $\psi_1(u_{i,f}) = b_i^f$. It suffices to show that the kernel R_{c+1} of the mapping ψ_1 contains the subgroup U_{c+1}. We prove this by induction on c. For $c = 1$ the assertion is trivial. Assume that it is proved for $c = i$ and R_{i+1}, R_{i+2} are the kernels of the mappings of the group U onto the groups B/B_{i+1} and B/B_{i+2}, respectively. By assumption, U_{i+1} is contained in R_{i+1}. Since B_{i+1}/B_{i+2} is an Abelian group of period dividing n_{i+1}, we have $R_{i+1}^{n_{i+1}}(R_{i+1}, R_{i+1}) \subset R_{i+2}$. Consequently, $U_{i+2} = U_{i+1}^{n_{i+1}}(U_{i+1}, U_{i+1}) \subset R_{i+2}$.

We observe that the subgroups U_j, $U_{j,i}$, and $U_{j,p,i}$ coincide as far as the jth term. Therefore, the epimorphism $D_{(j),p} \longrightarrow U/U_j$ is well defined. □

LEMMA 5.6.1. *The group $D_{(j)}$ is isomorphic to the direct product $\overline{D}_{(j)}$ of the groups $D_{(j),p}$ with amalgamated quotient group U/U_j.*

PROOF. We observe that it suffices to consider the prime numbers p dividing at least one of the numbers $n_j, \dots, n_c$, since otherwise $D_{(j),p}$ is isomorphic to U/U_j.

Consider the mapping $\psi_2 \colon D_{(j)} \longrightarrow \prod_p D_{(j),p}$ defined as follows:

$$\psi_2\Big(u \bmod \Big(\bigcap_p U_{j,p,c+1}\Big)\Big) = \prod_p u \bmod U_{j,p,c+1}.$$

It is immediately verified that ψ_2 is a monomorphism and its image is contained in the subgroup $\overline{D}_{(j)}$. Note that for distinct p the orders of $U_{j,p,j}/U_{j,p,c+1}$ are relatively prime. Moreover, for every p the mapping ψ_2 induces the epimorphism $U_{j,j}/U_{j,c+1} \longrightarrow U_{j,p,j}/U_{j,p,c+1}$. Consequently, ψ_2 maps $U_{j,j}/U_{j,c+1}$ onto the group $\prod_p U_{j,p,j}/U_{j,p,c+1}$. □

We note that from the proof of the lemma it follows that the kernel E_j of the natural epimorphism $D_{(j+1)} \longrightarrow D_{(j)}$ is a nilpotent group (it is contained in the kernel of the natural mapping of $D_{(j+1)}$ onto U/U_{j+1}, which is the direct product of the groups $U_{j+1}/U_{j+1,p,c+1}$).

The following assertion can be immediately verified.

LEMMA 5.6.2. *If B is a group of class c with respect to N ($N_{j,p}$, N, and j), then the class of its subgroups and quotient groups with respect to N ($N_{j,p}$, N, and j, respectively) does not exceed c.*

LEMMA 5.6.3. *The group $D_{(j),p}$ is maximal among the groups of class c with respect to $N_{j,p}$. More precisely, let*

$$\begin{array}{ccc} U & \xrightarrow{\psi_1} & X \\ & \psi_3 \searrow & \downarrow \psi_2 \\ & & D_{(j),p} \end{array}$$

be a commutative triangle, where ψ_1, ψ_2, and ψ_3 are surjective. Then ψ_2 is an isomorphism.

We prove this lemma by induction on c. For $c = 1$ the assertion is trivial. Assume that it is valid for $c = i$ and the groups $X/X_{j,p,i+1}$ and $U/U_{j,p,i+1}$ are isomorphic. It remains to observe that the commutativity of the triangle

$$\begin{array}{ccc} U_{j,p,i+1} & \xrightarrow{\psi_4} & X_{j,p,i+1}/X_{j,p,i+2} \\ & \psi_6 \searrow & \downarrow \psi_5 \\ & & U_{j,p,i+1}/U_{j,p,i+2} \end{array}$$

implies that ψ_5 is an isomorphism, which proves the assertion for $c = i+1$.

LEMMA 5.6.4. *Assume that the group X of class c with respect to N and $j+1$ is an extension of a group C by $D_{(j)}$. Then X is a semidirect extension.*

We prove this lemma by induction on the order of C. Let C_0 be a minimal normal subgroup of X, contained in C. By induction, X/C_0 is a semidirect extension, and it contains a subgroup $\widetilde{D}_j$ isomorphic to $D_{(j)}$ and disjoint with C/C_0. Let X_0 be the full preimage of $\widetilde{D}_j$ in X. We have the group extension

$$1 \to C_0 \to X_0 \to \widetilde{D}_{(j)} \to 1,$$

where X_0 is a group of class c with respect to N and $j+1$, and C_0 is an elementary Abelian p-group (C_0 is a minimal normal subgroup of X). By induction, X_0 is a semidirect extension, whence the extension X is also semidirect.

Thus, it is sufficient to consider the case where C and C_0 coincide. Let H be a kernel of the natural mapping of $D_{(j)}$ onto $D_{(j),p}$. Clearly, the order of H is relatively prime with p. The group C is an H-module. Since C is a minimal normal subgroup in X, only two cases may occur, namely, either C is a trivial H-module and $C^H = C$, or $C^H = 1$.

First we consider the case $C^H = 1$. Let $h \in H^2(D_{(j)}, C)$ be the cohomology class corresponding to the extension X. Clearly, the groups $H^i(H, C)$ are trivial (the orders of H and C are relatively prime). By Proposition 5.2 of Chapter IV from [**1**], we have the exact sequence

$$0 \to H^2(D_{(j),p}, C^H) \xrightarrow{\lambda} H^2(D_{(j)}, C) \xrightarrow{\rho} H^2(H, C) = 0,$$

where λ is the lifting homomorphism and p is the restriction homomorphism. Clearly, the relation $C^H = 1$ implies that the group $H^2(D_{(j)}, C)$ is trivial, i.e., X is a semidirect extension.

Now we consider the case $C^H = C$. Then the full preimage $\widetilde{H}$ of the subgroup H in X is the direct product of the subgroups H and C. Clearly, H is normal in

X. We consider the exact sequence $1 \longrightarrow C \longrightarrow X/H \xrightarrow{\psi} D_{(j),p} \longrightarrow 1$. Lemma 3.13.2 implies that the cohomology class corresponding to the extension X is the lifting of the class corresponding to the extension X/H. Clearly, it suffices to show that X/H is a semidirect extension of C by $D_{(j),p}$.

It is immediately verified that X/H is a group of class c with respect to $N_{j,p}$. Let $d_{i,f}$ be the generators of $D_{(j),p}$ that are the images of the generators $u_{i,f}$ of the free group U. We denote by $\overline{d}_{i,f}$ preimages of $d_{i,f}$ in X/H with respect to ψ and by $\overline{D}_{j,p}$ the subgroup generated by the elements $\overline{d}_{i,f}$. Then $\psi(\overline{D}_{j,p}) = D_{(j),p}$ and, by Lemma 5.6.3, the groups $\overline{D}_{j,p}$ and $D_{(j),p}$ are isomorphic, which means that X/H is a semidirect extension of C by $D_{(j),p}$.

2°. In conclusion, we consider a special case of the embedding problem $(K/k, G, \varphi, B)$ with solvable kernel for algebraic number fields, where the orders of the kernel and the Galois group F of the field to be embedded are relatively prime.

In this case, by Schur's theorem, G is a semidirect extension of the subgroup B by F, and we may assume that B is an F-operator group. Taking into account that B is solvable, the proof of this fact is relatively simple, and we give it here. We proceed by induction on the order of B. If B is Abelian, then the assertion follows from the fact that $H^2(F, B)$ is trivial. If B is non-Abelian, then its commutator subgroup B_1 is normal in G. Then G/B_1 is a semidirect extension of B/B_1 by F, and there is an embedding β of F in G/B_1. Let G_1 be the full preimage of the group $\beta(F)$ in G with respect to the natural mapping of G onto G/B_1. We have the exact sequence of groups $1 \longrightarrow B_1 \longrightarrow G_1 \longrightarrow F \longrightarrow 1$. By induction, it splits, i.e., there is an inclusion of F in G_1. Clearly, it may be extended to an inclusion of F in G, and this subgroup and the subgroup B are disjoint.

Let B be a group of solvability class c with m generators. We put n_i to be equal to the period of the group B_i/B_{i+1}, where B_i is the $(i-1)$th commutator subgroup of B. Let N be the collection of numbers $(n_1, \dots, n_c)$. Then B is a group of class c with respect to N. By Proposition 5.6.1, the group B is an operator-homomorphic image of the corresponding group D.

Let $\overline{G}$ be a semidirect extension by F of a solvable group $\overline{B}$ of class c with respect to N and j (as before, we assume that the orders of the groups $\overline{B}$ and F are relatively prime). In turn, let $\overline{B}$ be an extension of a group H by $D_{(j)}$. The following sequence of groups is exact:

$$1 \longrightarrow H \longrightarrow \overline{G} \longrightarrow F \cdot D_{(j)} \longrightarrow 1. \tag{1}$$

PROPOSITION 5.6.2. *The sequence* (1) *splits.*

PROOF. We prove the proposition by induction on the order of H. Let H_1 be a minimal normal subgroup of $\overline{G}$, contained in H. Clearly, H_1 is an elementary Abelian p-group. Then, by induction, the extension

$$1 \longrightarrow H/H_1 \longrightarrow \overline{G}/H_1 \longrightarrow F \cdot D_{(j)} \longrightarrow 1$$

splits, and there is an inclusion β of the group $F \cdot D_{(j)}$ in $\overline{G}/H_1$. Let $\overline{G}_1$ be the full preimage in $\overline{G}$ of the subgroup $\beta(F \cdot D_{(j)})$. We have the group extension

$$1 \longrightarrow H_1 \longrightarrow \overline{G}_1 \xrightarrow{\psi} F \cdot D_{(j)} \longrightarrow 1.$$

Denote by $\overline{B}_1$ the full preimage of the subgroup $D_{(j)}$ with respect to ψ. Clearly, $\overline{B}_1$ is a group of class c with respect to N and $j+1$. By Lemma 5.6.4, $\overline{B}_1$ is a semidirect extension of the subgroup H_1 by $D_{(j)}$.

Let $h \in H^2(F \cdot D_{(j)}, H_1)$ be the cohomology class corresponding to the extension $\overline{G}_1$. Then the restriction $\rho(h)$ of the class h to the subgroup $D_{(j)}$ (the image of the restriction homomorphism $\rho\colon H^2(F \cdot D_{(j)}, H_1) \longrightarrow H^2(D_{(j)}, H_1)$) corresponds to the extension $\overline{B}_1$. Therefore, $\rho(h) = 0$. It remains to note that ρ is injective (the orders of F and H_1 are relatively prime), i.e., $h = 0$, which completes the proof. □

The group-theoretic results obtained are useful in studying the embedding problem.

THEOREM 5.6.2. *The embedding problem $(K/k, F \cdot B, B)$, in which the orders of the groups F and B are relatively prime, has a proper solution provided that B is solvable.*

PROOF. It suffices to show that there is a proper solution of the problem $(K/k, F \cdot D, D)$, for which the original problem is an associated one.

For $j \leq c$ we consider the embedding problems $(K/k, F \cdot D_{(j)}, D_{(j)})$ in succession. We observe that $D_{(1)}$ is a nilpotent group. Therefore, by Theorem 5.5.4, there is a proper solution L_1 of the problem $(K/k, F \cdot D_{(1)}, D_{(1)})$. Now let L_j be a proper solution of the problem $(K/k, F \cdot D_{(j)}, D_{(j)})$. We consider the embedding problem $(L_j/k, F \cdot D_{(j+1)}, \varphi_j, E_{j+1})$, where φ_j is the natural mapping of the group $F \cdot D_{j+1}$ onto $F \cdot D_{(j)}$. As was mentioned above, E_{j+1} is a nilpotent group. By Proposition 5.6.2, the group extension

$$1 \longrightarrow E_{j+1} \longrightarrow F \cdot D_{(j+1)} \xrightarrow{\varphi_j} F \cdot D_{(j)} \longrightarrow 1$$

splits. To complete the proof of the theorem, we again use Theorem 5.5.4. □

Appendix

§1. Associative algebras

THEOREM A.1.1. *Let A/k be an associative algebra with unit containing a subalgebra B/k isomorphic to the algebra of matrices of order m over k, and let the unit elements of A and B coincide. Then A is the algebra of matrices of order m over the subalgebra of elements that commute with the subalgebra B.*

PROOF. Let $\varepsilon_{ij} \in B$, $i, j = 1, 2, \dots, m$, be the matrix units of the algebra B. Then $\sum_i \varepsilon_{ii} = 1$. Consider the linear mappings of A into itself: $S(a) = \varepsilon_{11} a \varepsilon_{11}$, $T(a) = \sum_i \varepsilon_{i1} a \varepsilon_{1i}$ for $a \in A$. These mappings are projections: $S(S(a)) = S(a)$, $T(T(a)) = T(a)$. Moreover, $S(T(a)) = S(a)$, $T(S(a)) = T(a)$. It is clear that $S(A)$ is a subalgebra of A with the unit element ε_{11}. We prove that $T(A)$ is a subalgebra isomorphic to $S(A)$. Indeed, $T(a)T(b) = \sum_{i,j} \varepsilon_{i1} a \varepsilon_{1i} \varepsilon_{j1} b \varepsilon_{1j} = \sum_i \varepsilon_{i1} a \varepsilon_{11} b \varepsilon_{1i} = T(S(a)S(b))$. Therefore, $S(a) \to T(S(a)) = T(a)$ is a homomorphism and $T(a) \to S(T(a)) = S(a)$ is inverse to it. The unit element of $T(a)$ is equal to $1 = T(1)$.

Now we show that $T(A)$ is the centralizer of B in A. Indeed, if $x \in A$ commutes with all matrix units ε_{ij}, then $T(x) = \sum_i \varepsilon_{i1} x \varepsilon_{1i} = \sum_i x \varepsilon_{i1} \varepsilon_{1i} = \sum_i x \varepsilon_{ii} = x$, i.e., $x \in T(A)$. Conversely, if $x = T(y)$, then $x \varepsilon_{kl} - \varepsilon_{kl} x = \sum_i \varepsilon_{i1} y \varepsilon_{1i} \varepsilon_{kl} - \varepsilon_{kl} \sum_i \varepsilon_{i1} y \varepsilon_{1i} = \varepsilon_{k1} y \varepsilon_{1l} - \varepsilon_{k1} y \varepsilon_{1l} = 0$.

Now we set $x_{kl} = \sum_i \varepsilon_{ik} x \varepsilon_{li}$ for $x \in A$ $(k, l = 1, 2, \dots, m)$. Then $T(x_{kl}) = x_{kl}$, so that $x_{kl} \in T(A)$ and

$$\sum_{k,l} x_{kl} \varepsilon_{kl} = \sum_{k,l,i} \varepsilon_{ik} x \varepsilon_{li} \varepsilon_{kl} = \sum_{k,l} \varepsilon_{kk} x \varepsilon_{ll} = x.$$

Thus, A is the matrix algebra over $T(A)$. □

THEOREM A.1.2. *If an associative algebra A/k with unit decomposes into a direct sum of right ideals isomorphic as A-modules, then A is isomorphic to the matrix algebra over the subalgebra eAe, where e is the component of unit in any of the direct summands.*

PROOF. Let $A = B_1 \oplus B_2 \oplus \cdots \oplus B_m$. Then we have a decomposition $1 = e_1 + e_2 + \cdots + e_m$, where $e_i = B_i$. The elements e_i are orthogonal idempotents, since $e_i = 1 \cdot e_i = e_1 e_i + \cdots + e_i e_i + \cdots + e_m e_i$ but $e_k e_k \in B_k$, so $e_i^2 = e_i$, $e_k e_i = 0$ for $k \neq i$.

Let ψ_k, $k = 1, \dots, m$, be isomorphisms mapping B_1 onto B_k (ψ_1 is identical) as right A-modules. Then $\psi_k(x_1 y) = \psi_k(x_1) y$ for $x_1 \in B_1$, $y \in A$; $\psi_k^{-1}(x_k y) = \psi_k^{-1}(x_k) y$ for $x_k \in B_k$, $y \in A$. We set $\varepsilon_{k1} = \psi_k(e_1)$, $\varepsilon_{1k} = \psi_k^{-1}(e_k)$. Then $\varepsilon_{k1} \varepsilon_{1k} = \psi_k(e_1) \varepsilon_{1k} = \psi_k(\varepsilon_{1k}) = e_k$, $\varepsilon_{1k} \varepsilon_{k1} = \psi_k^{-1}(e_k \varepsilon_{k1}) = \psi_k^{-1}(\varepsilon_{k1}) = e_1$, since e_k is a left unit in B_k. Moreover, we set $\varepsilon_{ij} = \varepsilon_{i1} \varepsilon_{1j}$. These definitions agree because $\varepsilon_{i1}\varepsilon_{11} = \psi_i(e_1)e_1 = \varepsilon_{i1}$, $\varepsilon_{11}\varepsilon_{1j} = \varepsilon_{1j}$. Now we have $\varepsilon_{ij}\varepsilon_{kl} = \varepsilon_{i1}\varepsilon_{1j}\varepsilon_{k1}\varepsilon_{1l} = \varepsilon_{i1}\varepsilon_{1j}e_je_k\varepsilon_{k1}\varepsilon_{1l} = 0$ for $k \neq j$, $\varepsilon_{ij}\varepsilon_{jl} = \varepsilon_{i1}\varepsilon_{1j}\varepsilon_{j1}\varepsilon_{1l} = \varepsilon_{i1}e_1\varepsilon_{1l} = \varepsilon_{i1}\varepsilon_{1l} = \varepsilon_{il}$, i.e., the elements ε_{ij} are multiplied as matrix units. Moreover, $\varepsilon_{11} + \varepsilon_{22} + \cdots + \varepsilon_{mm} = e_1 + e_2 + \cdots + e_m = 1$. By Theorem A.1.1, A is the algebra of matrices of order m over $T(A)$, but $T(A)$ is isomorphic to $S(A) = e_1Ae_1$. □

THEOREM A.1.3. *If an associative algebra A with unit decomposes into a direct sum of m isomorphic right ideals in two ways, then there exists an inner automorphism taking one decomposition to the other.*

PROOF. Let $A = e_1A \oplus e_2A \oplus \cdots \oplus e_mA = \varepsilon_1A \oplus \cdots \oplus \varepsilon_mA$ be two decompositions of A, let e_i, ε_i be idempotents of the corresponding components. By the Krull–Schmidt theorem, the direct summands of the decompositions are isomorphic. Let ψ_i be an isomorphism of e_iA onto ε_iA. We put $c_i = \psi_i^{-1}(\varepsilon_i) \in e_iA$, $\gamma_i = \psi_i(e_i) \in \varepsilon_iA$. Then $c_i\gamma_i = \psi_i^{-1}(\varepsilon_i)\gamma_i = \psi_i^{-1}(\varepsilon_i\gamma_i) = \psi_i^{-1}(\gamma_i) = e_i$, $\gamma_ic_i = \psi_i(e_i)c_i = \psi_i(e_ic_i) = \psi_i(c_i) = \varepsilon_i$, since e_i and ε_i are left units of the corresponding components. Moreover, $c_i\varepsilon_i = c_i$, $\gamma_ie_i = \gamma_i$, so $c_i\varepsilon_iA = c_iA \subset e_iA$; $\gamma_ie_iA = \gamma_iA \subset \varepsilon_iA$. For $i \neq j$ we have $c_i\varepsilon_jA = \psi_i^{-1}(\varepsilon_i)\varepsilon_jA = \psi_i^{-1}(\varepsilon_i\varepsilon_j)A = 0$. Similarly, $\gamma_i\varepsilon_jA = 0$. Therefore, $c_i\gamma_j = \gamma_ic_j = 0$ for $i \neq j$.

We set $c = \sum_{i=1}^m c_i$, $\gamma = \sum_{i=1}^m \gamma_i$. Then $c\gamma = \sum_i c_i\gamma_i + \sum_{i\neq j} c_i\gamma_j = \sum_i e_i = 1$, and in the same way $\gamma c = 1$. In addition,

$$\gamma e_i c = \sum_{j,k} \gamma_j e_i c_k = \sum_k \gamma_i e_i c_k = \sum_k \gamma_i c_k = \gamma_i c_i = \varepsilon_i,$$

and the theorem is proved. □

THEOREM A.1.4. *Let an algebra A with unit be an algebra of matrices of order k over a subalgebra B and also an algebra of matrices of order mk over a subalgebra C. Then B is isomorphic to the algebra of matrices of order m over C.*

PROOF. We have $A = \sum_{i=1}^{mk\,\oplus} e_iA$, where e_iA are isomorphic right ideals, and $C \approx e_1Ae_1$. Moreover, $A = \sum_{j=1}^{k\,\oplus} E_jA$, and $B \approx E_1AE_1$. By the Krull–Schmidt theorem, the ideal E_1A is isomorphic to the sum of m ideals $\sum_{i=1}^{m\,\oplus} e_iA$. Therefore, there exists an inner automorphism of A taking E_1 to $\sum_{i=1}^m e_i$. Therefore, we may assume that $E_1 = \sum_{i=1}^m e_i$ and $E_1A = \sum_{i=1}^{m\,\oplus} e_iA$. But then $B \approx E_1AE_1 \approx \sum_{i=1}^{m\,\oplus} e_iAE_1$, i.e., B decomposes into the direct sum of m isomorphic right ideals e_iAE_1. By Theorem A.1.2, we find that B is isomorphic to the algebra of matrices of order m over the algebra $e_1AE_1e_1 = e_1Ae_1$, which is isomorphic to C. □

THEOREM A.1.5. *Let an algebra A with unit contain rings of matrices of relatively prime orders k and m, and let the unit of A be the unit of each of the rings. Then A is the algebra of matrices of order mk over a subalgebra.*

Let e_{ij} $(i, j = 1, \ldots, m)$ and $\varepsilon_{\alpha\beta}$ $(\alpha, \beta = 1, \ldots, k)$ be the matrix units. Consider the algebra A as a right A-module. We have $A = e_{11}A \oplus \cdots \oplus e_{mm}A$, $A = \varepsilon_{11}A \oplus \cdots \oplus \varepsilon_{kk}A$. Let $B_1, \ldots, B_n$ be all nonisomorphic indecomposable right A-modules. We decompose $e_{11}A$ into indecomposable modules, $e_{11}A \approx t_1B_1 \oplus \cdots \oplus t_nB_n$, where $t_i \geq 0$ is the multiplicity of the occurrence of B_i in $e_{11}A$. Then $A \approx mt_1B_1 \oplus \cdots \oplus mt_nB_n$. Similarly, we have $\varepsilon_{11}A \approx s_1B_1 \oplus \cdots \oplus s_nB_n$ $(s_i \geq 0)$

and $A = ks_1B_1 \oplus ks_2B_2 \oplus \cdots \oplus ks_nB_n$. By the Krull–Schmidt theorem, we have $mt_i = ks_i$, whence $t_i = kq_i$, $q_i \in \mathbb{Z}$, since m and k are relatively prime. Therefore, $e_{11}A \approx k(q_1B_1 \oplus \cdots \oplus q_nB_n)$ and $A \approx km(q_1B_1 \oplus \cdots \oplus q_nB_n)$, i.e., A decomposes into a direct sum of km ideals isomorphic to $q_1B_1 \oplus \cdots \oplus q_nB_n$. It remains to apply Theorem A.1.2.

§2. Simple algebras

1°. It is well known (see [**6**]) that a maximal subfield of a simple algebra is a splitting field of it; in other words, if A is a simple algebra with center k, and K is a maximal subfield in A, then $A \otimes_k K$ is the full matrix algebra over K. It is less known that this fact admits the following conversion.

THEOREM A.2.1 (see [**5**]). *Let A be a simple algebra over the center k of dimension n^2 and let K be a splitting field of A of degree n over k. Then A contains a subfield $\widetilde{K}$ isomorphic to K.*

PROOF. Consider the algebra k_n of matrices of order n over the field k. As is known, any extension of the field k of degree n can be embedded in such a matrix algebra. We embed the field K in the algebra k_n. This embedding induces an embedding i of the algebra $K \otimes_k A$ in the algebra $k_n \otimes_k A$, which is the algebra of matrices of order n over A. At the same time, $K \otimes_k A \approx K_n \approx k_n \otimes_k K$. This isomorphism, together with i, induces an embedding $i^*: k_n \otimes_k K \longrightarrow A_n$. Let B be the image of k_n under the embedding i^*, and let $\widetilde{A}$ be the centralizer of B in the algebra A_n. Since A_n and B are simple algebras with common unit, and $B \approx k_n$, $A_n \approx k_n \otimes_k A$, then $\widetilde{A}$ is isomorphic to A. Since K is the center of the algebra $k_n \otimes_k K$, the image of K under the embedding i^* is contained in the centralizer $\widetilde{A}$. Therefore, a field isomorphic to K is contained in the algebra $\widetilde{A}$ isomorphic to A. The theorem is proved. □

2°. Let k be a field containing a primitive nth root ε of unity (the characteristic of k does not divide n) and $a, b \in k^*$. We denote by $k(a, b)$ the associative central algebra of dimension n^2 over k with basis $\xi^i\eta^j$ $(i, j = 0, 1, \ldots, n-1)$, where the elements ξ and η satisfy the relations $\xi^n = a$, $\eta^n = b$, $\eta^{-1}\xi\eta = \xi\varepsilon$. These relations give the multiplication table of the basis elements. For $p = 2$, $a = b = -1$ we obtain the ordinary quaternion algebra. For this reason, the algebra $K(a, b)$ is also called the *generalized quaternion algebra.*

We fix the extension $K = k(\xi)$. This is a cyclic radical extension of the field k. Let $F = \mathrm{Gal}(K/k) \approx \mathbb{Z}/n\mathbb{Z}$ and let f be a generator of the group F. The mapping $(f^i, f^j) \longrightarrow \delta_{ij}$ $(i, j = 0, 1, \ldots, n-1)$, where $\delta_{ij} = 1$ for $i + j < n$ and $\delta_{ij} = b$ for $i + j \geq n$, is a multiplicative two-dimensional cocycle in $Z^2(F, K^*)$. The crossed product of the field K and the group F with factor set $\{\delta_{ij}\}$ consists of sums of the form $\sum_{i=0}^{n-1} v_{f^i}x_i$, where $x_i \in K$ and $xv_f = v_fx^f$, $v_f^n = b$. Therefore, this crossed product is isomorphic to the generalized quaternion algebra $k(a, b)$ defined above.

THEOREM A.2.2 (Hasse). *The algebra $k(a, b)$ splits completely* (*i.e., is isomorphic to the algebra of matrices of order n over k*) *if and only if b is the norm of an element from $k(\sqrt[p]{a})$ over k.*

PROOF. First let $k(\xi) = K$ be a field, i.e., $t^n - a$ is an irreducible polynomial over k. As is known (see [**6**]), the condition that the crossed product splits completely is equivalent to the homologous triviality of the factor set. In other words, there exist elements $z_0, z_1, \dots, z_{n-1} \in K^*$ for which $z_{i+j} = z_i^{f^j} z_j$ if $i+j < n$; $z_{i+j-n} = z_i^{f^j} z_j b^{-1}$ if $i+j \geq n$. Hence $z_0 = 1$, $z_2 = z_1^{1+f}, \dots, z_{n-1} = z_1^{1+f+\dots+f^{n-2}}$; $1 = z_1^{1+f+\dots+f^{n-1}} b^{-1}$, i.e., $Nz_1 = b$. Conversely, if $Nz_1 = b$, then replacing in the crossed product the representative v_f by $v_f z_1^{-1}$, we obtain the unit factor set.

Now let $k(\xi)$ be a Galois algebra over k with the field-kernel K_1 and idempotents $e_1, \dots, e_d$, where d/n. Then the algebra $A = k(a,b)$ decomposes into a direct sum of isomorphic right ideals $e_i A$. This isomorphism is given by multiplication of $e_i x$ ($x \in A$) by η^j ($j = 0, 1, \dots, d-1$) from the left, since $e_1, \dots, e_d$ are permuted by inner automorphisms induced by η^j. Therefore, A is isomorphic to the algebra of matrices of order d over its subalgebra eAe (Theorem A.1.2). This subalgebra is again a generalized quaternion algebra that is defined by the constants $a^{1/d} \in k^*$ (since $\xi^{n/d} = a^{1/d}$) and $b = (\eta^d)^{n/d}$. By Theorem A.1.4, the algebra A splits completely if and only if the algebra eAe splits completely, i.e., if b is the norm of an element from $K_1 \subset k(\xi)$.

The image of the algebra $k(a,b)$ in the Brauer group $\mathrm{Br}\, k$ will be denoted by $k[a,b]$ (or merely $[a,b]$). By the product $[a_1,b_1][a_2,b_2]$, we mean the image in $\mathrm{Br}\, k$ of the tensor product $k(a_1,b_1) \otimes_k k(a_2,b_2)$. □

THEOREM A.2.3. *The symbols $[a,b]$ (for fixed k and n) are subject to the following relations*:

a) $[a,b_1][a,b_2] = [a,b_1b_2]$; $[a_1,b][a_2,b] = [a_1a_2,b]$;
b) $[a,b] = [b,a^{-1}]$;
c) $[a,b][b,a] = 1$;
d) $[a,1-a] = 1$ *for* $a \neq 1$.

PROOF. a) Consider the tensor product $A = k(a,b_1) \otimes k(a,b_2)$. Let ξ_1, η_1 be generators of the first algebra, ξ_2, η_2 generators of the second. Consider the algebra generated over k by the elements η_1 and $\xi_1^{-1}\xi_2$. This subalgebra B is the generalized quaternion algebra with constants b_1 and 1, i.e., this is a matrix algebra. Therefore, A is a matrix algebra over the centralizer of the subalgebra B. This centralizer is generated by the elements ξ_2 and $\eta_1\eta_2$, and, thus, is isomorphic to the algebra $k(a,b_1b_2)$. The second part is similar.

b) Let $A = k(a,b)$ and let ξ, $\eta \in A$, $\xi^n = a$, $\eta^n = b$, $\eta^{-1}\xi\eta = \xi\varepsilon$. After the change of variables $\xi_1 = \eta$, $\eta_1 = \xi^{-1}$, we have $\xi_1^n = b$, $\eta_1^n = a^{-1}$, $\eta_1^{-1}\xi_1\eta_1 = \xi_1\varepsilon$, i.e., $A = k(b,a^{-1})$.

c) By the above properties, we have $[a,b][b,a] = [a,b][a,b^{-1}] = [a,1]$, but $k(a,1)$ is a matrix algebra.

d) This follows from the relation $1 - a = N(1-\xi)$, where $\xi^n = a$. □

Thus, we see that the symbols $[a,b]$ have the familiar properties of the Hilbert symbol.

We note that in the case of local and global number fields any central simple algebra has a cyclic splitting field (see [**1**]). Therefore, if the field k contains a necessary root of unity, such an algebra is isomorphic to a matrix algebra over a generalized quaternion algebra. Indeed, let K/k be a cyclic splitting field of the algebra A/k. One may assume that A is a division algebra (otherwise, we have a matrix algebra over a division algebra). Then $K = k(\xi)$, where $\xi^m = a \in k^*$. Let f be a generator of the group $\mathrm{Gal}(K/k) \approx \mathbb{Z}/m\mathbb{Z}$. Find an element $\eta \in A$ such that $\eta^{-1}\xi\eta = \xi\varepsilon$. (This element exists by the Noether–Skulem theorem (see [**6**, p. 421]).)

It is clear that ξ and η generate the entire algebra A, and η^m is a central element of A. Therefore, $\eta^m = b$ and $A = k(a,b)$.

3°. Now we consider the case $n = 2$ in more detail. In this case $-1 \in k$, so that any central simple algebra of dimension 4 over k (we assume that $\operatorname{char} k \neq 2$) is a generalized quaternion algebra $k(a,b)$. The condition of a splitting of this algebra means that $b = N\theta$, $\theta \in k(\sqrt{a})$, i.e., the equation $x^2 - ay^2 = b$ has a solution in the field k. Otherwise, $k(a,b)$ is a division algebra (i.e., a division ring).

We denote by 1, i, j, k the basis elements of $A = k(a,b)$, so that $i^2 = a$, $j^2 = b$, $k = ij = -ji$, whence $k^2 = -ab$. The quaternion $x - yi - zj - tk$ is conjugate to $x+yi+zj+tk \in A$, so that $(x+yi+zj+tk)(x-yi-zj-tk) = x^2-ay^2-bz^2+abt^2$ is the norm of a quaternion. It is clear that for $\alpha, \beta \in A$, we have $N\alpha\beta = N\alpha N\beta$. The elements of the form $yi+zj+tk$ are said to be *pure quaternions*. If $\beta = yi+zj+tk$ is a pure quaternion, then $\beta^2 = ay^2 + bz^2 - abt^2 \in k$. For $\alpha = x + \beta$, where $x \in k$ and β is a pure quaternion, we have $\alpha^2 = x^2 + \beta^2 + 2x\beta$; therefore, if $x \neq 0$, $\beta \neq 0$, then $(x+\beta)^2 \notin k$.

Now we try to answer the following question. When are $k_1(a_1,b_1)$ and $k(a_2,b_2)$ isomorphic algebras? For this, it is necessary and sufficient that there exist elements α and β in the algebra $k(a_2,b_2)$ such that $\alpha^2 = a_1$, $\beta^2 = b_1$, $\beta\alpha = -\alpha\beta$. We have seen that α and β must be pure quaternions in $k(a_2,b_2)$. Therefore, $\alpha = y_1i+z_1j+t_1k$, $\beta = y_2i+z_2j+t_2k$, where 1, i, j, k is the standard basis in $k(a_2,b_2)$. The condition $\alpha\beta + \beta\alpha = 0$ means that $a_2y_1y_2 + b_2z_1z_2 - a_2b_2t_1t_2 = 0$. Thus, the isomorphicity of $k(a_1,b_1)$ and $k(a_2,b_2)$ means that in the field k the following system of equations is consistent:

$$\begin{aligned} a_2y_1^2 + b_2z_1^2 - a_2b_2t_1^2 &= a_1, \\ a_2y_2^2 + b_2z_2^2 - a_2b_2t_2^2 &= b_1, \\ a_2y_1y_2 + b_2z_1z_2 - a_2b_2t_1t_2 &= 0 \end{aligned}$$

(clearly, one may write an equivalent dual system with a_2, b_2 on the right-hand side).

Now we consider the tensor product $A_1 \otimes_k A_2$, where A_1 and A_2 are generalized quaternion algebras over k. This product may be either the full matrix algebra over k (of order 4), or a division ring, or an algebra of matrices of order 2 over a division ring. The case of the full matrix algebra is realized if and only if A_1 and A_2 are isomorphic. The following theorem by Albert enables one to separate the second and the third cases.

THEOREM A.2.4. *The algebra $A = A_1 \otimes A_2$, where A_1, A_2 are generalized quaternion algebras over k, contains zero divisors if and only if there exists a quadratic extension of k that is embeddable in both algebras A_1 and A_2.*

PROOF. If K is an extension mentioned above, $K = k(\sqrt{c})$, z_1 and z_2 are the images of $\sqrt{c}$ in A_1 and A_2, then $(z_1\otimes 1 - 1\otimes z_2)(z_1\otimes 1 + 1\otimes z_2) = 0$, and sufficiency is proved.

Now let $A_1 \otimes A_2$ contain zero divisors. If either A_1 or A_2 is a matrix algebra, then our assertion follows from the fact that any quadratic extension is embeddable in the algebra of matrices of order 2. Therefore, we may assume that A_1 and A_2 are division algebras.

We denote by i, j, k the standard basis of the algebra A_2, where $i^2 = a$, $j^2 = b$, $k^2 = -ab$. The elements of the algebra $A = A_1 \otimes A_2$ are linear combinations of the elements 1, i, j, k over the division ring A_1.

Let $\alpha = \alpha_1 + \alpha_2 i + \alpha_3 j + \alpha_4 k$ be a zero divisor in A ($\alpha_i \in A_1$). At least one element of α_2, α_3, α_4 is different from zero. Assume that $\alpha_4 \neq 0$. Multiplying by α_4^{-1} one can find that a zero divisor α exists with $\alpha_4 = 1$. Then the first component of the element $\beta = \alpha(ab + \alpha_1 k)$ is zero and β is also a zero divisor (a three-termed zero divisor).

Assume that A contains a two-termed zero divisor. It can be reduced, as before, to the form $\alpha_1 + k$ (or $\alpha_1 + i$, $\alpha_1 + j$), where $\alpha_1 \in A_1$. But $(\alpha_1 - k)(\alpha_1 + k) = \alpha_1^2 + ab \in A_1$ is a zero divisor, whence $\alpha_1^2 + ab = 0$. Therefore, $k(\sqrt{-ab})$ can be embedded both in A_1 and in A_2 (since $k^2 = -ab$). Now we assume that A is free of nontrivial two-termed zero divisors.

Let $\alpha = \alpha_1 i + \alpha_2 j + k$ be a zero divisor in A ($\alpha_i \in A$). Then $\alpha^2 = (a\alpha_1^2 + b\alpha_2^2 - ab) + (\alpha_1\alpha_2 - \alpha_2\alpha_1)k$ is a two-termed zero divisor, so $\alpha^2 = 0$. Therefore, $\alpha_1\alpha_2 = \alpha_2\alpha_1$ and $a\alpha_1^2 + b\alpha_2^2 = ab$. The elements α_1 and α_2 cannot simultaneously belong to k; otherwise, $\alpha \in A_2$. Let $\alpha_1 \notin k$. Consider the field $K = k(\alpha_1)$; K is a quadratic extension of k which can be embedded in A_1. Since $\alpha_2\alpha_1 = \alpha_1\alpha_2$, then $\alpha_2 \in K$, so $\alpha = \alpha_1 i + \alpha_2 j + k$ is contained in the algebra $A_2 \otimes K = K(a, b)$. Therefore, K is a splitting field for the algebra A_2, and, thus, is embeddable in the algebra A_2. The theorem is proved. □

§3. The Tate duality

In this section we present some facts of Galois cohomology theory of number and local fields needed for the study of the embedding problem.

1°. Let F be a finite group. For F-modules A and B we consider the operations $\otimes$ and Hom over $\mathbb{Z}$, assuming that the results are F-modules with the action of F by the rules $(a \otimes b)^f = a^f \otimes b^f$ and $\varphi^f(a) = (\varphi(a^{f^{-1}}))^f$ for $\varphi \in \operatorname{Hom}(A, B)$.

For cohomology we use $\cup$-product that maps $H^p(F, A) \otimes H^q(F, B)$ to $H^{p+q}(F, A \otimes B)$. We regard the cohomology groups in the sense of Tate with positive and negative numbers (see [**1**], Chapter IV, §6, The Tate groups). Properties of $\cup$-product are described in the same book (Chapter IV, §7). Moreover, if there is an F-homomorphism $\psi\colon A \otimes B \to C$, then it induces a homomorphism $\psi_{p+q}\colon H^{p+q}(A \otimes B) \to H^{p+q}(C)$. The application of ψ_{p+q} following $\cup$-product yields a homomorphism

$$\psi_{p+q}\colon H^p(F, A) \otimes H^q(F, B) \longrightarrow H^{p+q}(F, C).$$

This homomorphism is called the $\cup$-product relative to ψ.

Define a natural mapping $\psi\colon A \otimes \operatorname{Hom}(A, B) \to B$ by the formula $\psi(a \otimes \beta) = \beta(a)$ for $a \in A$, $\beta \in \operatorname{Hom}(A, B)$. It is easy to see that ψ is an F-homomorphism. It generates a homomorphism $\psi_{p,q}$ under applying $\cup$-product relative to ψ:

$$\psi_{p,q}\colon H^p(F, A) \otimes H^q(F, \operatorname{Hom}(A, B)) \longrightarrow H^{p+q}(F, B).$$

Let $T = Q/\mathbb{Z}$ be regarded as an F-module with trivial operators. The F-module $A' = \operatorname{Hom}(A, T)$ is called *dual* to an F-module A. If A is a group of period m, then it is natural to identify $\operatorname{Hom}(A, \mathbb{Z}_m)$ with A', where $\mathbb{Z}_m$ is the cyclic group of order m.

In the last formula we replace B with T and set $q = -p-1$. We obtain

$$\psi_{p,-p-1}\colon H^p(F,A)\otimes H^{-p-1}(F,A')\to H^{-1}(F,T).$$

We recall that $H^{-1}(F,T) = \operatorname{Ker}(T\to mT)\approx \mathbb{Z}_m$, where m is the order of the group F. Thus, there is a pairing of the groups $H^p(F,A)$ and $H^{-p-1}(F,A')$ with values in $\mathbb{Z}_m$. This pairing turns out to be nondegenerate. This means that each of the groups $H^p(F,A)$ and $H^{-p-1}(F,A')$ is dual to the other. For the proof of this important result, see [**26**, Chapter XII, Theorem 6.4] or Appendix, §6.

Now let an F-module A be a free Abelian group of finite rank, i.e., a module of the integer representation of the group F (written in the multiplicative form).

The exact sequence

$$0\to\mathbb{Z}\to\mathbb{Q}\to T\to 0$$

implies the exact sequence

$$1\to\operatorname{Hom}(A,\mathbb{Z})\to\operatorname{Hom}(A,\mathbb{Q})\to\operatorname{Hom}(A,T)\to 1.$$

The modules $\mathbb{Q}$ and $\operatorname{Hom}(A,\mathbb{Q})$ admit a unique division by an integer. Therefore, $H^i(F,\mathbb{Q}) = 0$ and $H^i(F,\operatorname{Hom}(A,\mathbb{Q})) = 0$, and the connecting homomorphisms for the cohomology groups $H^i(F,\operatorname{Hom}(A,T))$ and $H^{i+1}(F,\operatorname{Hom}(A,\mathbb{Z}))$, as well as for the cohomology groups $H^i(F,T)$ and $H^{i+1}(F,\mathbb{Z})$, are isomorphisms. Therefore, there exists a nondegenerate pairing of the groups $H^p(F,A)$ and $H^{-p}(F,A^*)$, where $A^* = \operatorname{Hom}(A,\mathbb{Z})$.

Now let P be a multiplicatively written module of the regular integer representation of the group F, i.e., a free group with generators e_f, $f\in F$, and with the action of F by the rule $e_f^g = e_{fg}$, $g\in F$. We note that P is a free module for any subgroup F_0 of the group F. There is a natural homomorphism $P\to\mathbb{Z}$ that maps $\prod e_f^{a_f}$ to $\sum a_f$. We denote by I the kernel of this homomorphism. It consists of $\prod e_f^{a_f}$ with $\sum a_f = 0$. The kernel I, as a free Abelian group, is generated by the elements $e_f e_1^{-1}$ for $f\neq 1$. The exact sequence $1\to I\to P\to\mathbb{Z}\to 1$ generates the exact sequence

$$\to H^i(F,P)\to H^i(F,\mathbb{Z})\xrightarrow{\delta} H^{i+1}(F,I)\to H^{i+1}(F,P)\to .$$

The connecting homomorphism δ is an isomorphism, since $H^i(F,P) = 0$ for all i. In particular, $H^1(F,I)\approx H^0(F,\mathbb{Z})$, so that $H^1(F,I)$ is the cyclic group of order $m = (F:1)$. For a cocycle from a generator of the group $H^1(F,I)$, it is natural to take the result of applying $f-1$ to any element from P with the sum of exponents equal to 1. It is simpler to take e_1, and then it is natural to take $c_f = e_f e_1^{-1}$ as a generator cocycle for $H^1(F,I)$.

Now we proceed with description of a construction that is important in the sequel. Let C be a given F-module and $h(f_1,f_2)\in K$ a cocycle. We construct a new module C_1 that is the extension of C by I. Namely, consider the direct product of C by I and introduce the action of F on this product by the rule

$$\left(c,\prod e_f^{a_f}\right)^g = \left(c^g\prod_f h(f,g)^{a_f},\prod e_{fg}^{a_f}\right) \tag{1}$$

for $g \in F$. It is easy to see that the action of g does not destroy the structure of an Abelian group. We show that

$$\Big(c, \prod e_f^{a_f}\Big)^{g_1 g_2} = \Big(\Big(c, \prod e_f^{a_f}\Big)^{g_1}\Big)^{g_2}.$$

Indeed,

$$\begin{aligned}\Big(c, \prod e_f^{a_f}\Big)^{g_1 g_2} &= \Big(c^{g_1 g_2} \prod h(f, g_1 g_2)^{a_f}, \prod e_{f g_1 g_2}^{a_f}\Big) \\ &= \Big(c^{g_1 g_2} \prod (h(f, g_1 g_2) h(g_1, g_2))^{a_f}, \prod e_{f g_1 g_2}^{a_f}\Big).\end{aligned}$$

We can add the factor $h(g_1, g_2)^{a_f}$, since $\sum a_f = 0$. On the other hand,

$$\begin{aligned}\Big(\Big(c, \prod e_f^{a_f}\Big)^{g_1}\Big)^{g_2} &= \Big(c^{g_1} \prod h(f, g_1)^{a_f}, \prod e_{f g_1}^{a_f}\Big)^{g_2} \\ &= \Big(c^{g_1 g_2} \prod (h(f, g_1)^{g_2} h(f g_1, g_2))^{a_f}, \prod e_{f g_1 g_2}^{a_f}\Big).\end{aligned}$$

Equality (1) is proved because $h(f_1, f_2)$ is a cocycle.

It is obvious that the mapping $(c, \prod e_f^{a_f}) \to \prod e_f^{a_f}$ is an F-homomorphism of C_1 onto I with kernel $(C, 1)$ naturally isomorphic to C.

Let C be a class formation (the term is introduced in [**71**]), i.e., an F-module that is subject to the following conditions:

1. $H^1(F_0, C) = 0$ for any subgroup F_0 of F.

2. $H^2(F_0, C)$ is a cyclic group of order $(F_0 : 1)$ with canonical generator h_{F_0}, and the restriction homomorphism $H^2(F, C) \to H^2(F_0, C)$ maps h_F to h_{F_0}.

We construct C_1 by means of the cocycle $h(g_1, g_2)$ from the generator of the group $H^2(F, C)$ and compute the cohomology groups $H^i(F_0, C_1)$ for any subgroup F_0 of F. The exact sequence $1 \to C \to C_1 \to I \to 1$ implies the exact sequence

$$\begin{aligned}0 = H^1(F_0, C) &\to H^1(F_0, C_1) \to H^1(F_0, I) \\ &\xrightarrow{\delta} H^2(F_0, C) \to H^2(F_0, C_1) \to H^2(F_0, I).\end{aligned}$$

It follows from the exact sequence $1 \to I \to P \to \mathbb{Z} \to 1$ that $H^i(F_0, \mathbb{Z})$ and $H^{i+1}(F_0, I)$ are isomorphic because P is a free F_0-module, so that $H^2(F_0, I) = 0$ and $H^1(F_0, I)$ is the cyclic group of order $(F_0 : 1)$. One can take $z_f = e_f e_1^{-1}$, $f \in F_0$, as a cocycle in the generator of the group $H^1(F_0, I)$.

We recall that the connecting homomorphism δ in the exact sequence

$$\to H^i(F_0, C) \to H^i(F_0, C_1) \to H^i(F_0, I) \xrightarrow{\delta} H^{i+1}(F_0, C) \to$$

is defined as follows. One must take a cocycle in a cohomology class of $H^i(F_0, I)$, take its inverse image in C_1, and then apply the coboundary map. For $i = 1$ we have

$$\begin{aligned}(\delta(z_f))(g_1, g_2) &= (1, e_{g_1} e_1^{-1})^{g_2} (1, e_{g_2} e_1^{-1})(1, e_{g_1 g_2}^{-1} e_1) \\ &= (h(g_1, g_2), e_{g_1 g_2} e_{g_2}^{-1})(1, e_{g_2} e_1^{-1})(1, e_{g_1 g_2}^{-1} e_1) = (h(g_1, g_2), 1).\end{aligned}$$

Thus, δ is an isomorphism of $H^1(F_0, I)$ and $H^2(F_0, C)$. Therefore, $H^1(F_0, C_1) = 0$ and $H^2(F_0, C_1) = 0$. Hence it follows (Theorem 9.4, Chapter IV in [**1**]) that C_1 is a cohomologously trivial module.

LEMMA A.3.1. *Let C_1 be a cohomologously trivial module, let A be an integer lattice that is an F-module. Then* $\operatorname{Hom}(A, C_1)$ *is also a cohomologously trivial module.*

PROOF. Let $1 \to C_2 \to C_3 \to C_1 \to 1$ be the presentation of C_1 as a quotient module of a free F-module C_3. It is clear that the modules C_3 and C_2 are cohomologously trivial. Moreover, C_2 and C_3 are free Abelian groups. Since A is a lattice, the sequence of F-modules

$$1 \to \operatorname{Hom}(A, C_2) \to \operatorname{Hom}(A, C_3) \to \operatorname{Hom}(A, C_1) \to 1$$

is exact. First we show that for any free module C_3, the module $\operatorname{Hom}(A, C_3)$ is also free. Indeed, let $\{x_i\}$ be a base of the module C_3, and let a_j be generators of the module A. It is easily verified that the elements $\chi_{ij} \in \operatorname{Hom}(A, C_3)$ defined by relations

$$\chi_{ij}(a_m) = \begin{cases} x_i & \text{if } m = j, \\ 1 & \text{if } m \neq j, \end{cases}$$

form a base of the module $\operatorname{Hom}(A, C_3)$.

We note that, by Theorem 9.3 of Chapter IV in [**1**], the module C_2 is projective. Then there exists a module $\overline{C}_2$ such that $\overline{C}_2 \times C_2$ is a free module. Therefore, the module $\operatorname{Hom}(A, C_2) \times \operatorname{Hom}(A, \overline{C}_2)$ is also free, and so $\operatorname{Hom}(A, C_2)$ is projective. Thus, we have shown that the modules $\operatorname{Hom}(A, C_2)$ and $\operatorname{Hom}(A, C_3)$ are cohomologously trivial, and so is $\operatorname{Hom}(A, C_1)$. □

THEOREM A.3.1 (Nakayama–Tate). *Let C be a class formation and let A be an integer lattice, then for all i, $\cup$-product that takes $H^i(F, A) \otimes H^{2-i}(F, \operatorname{Hom}(A, C))$ to $H^2(F, C)$ is nondegenerate.*

PROOF. Consider the commutative diagram

$$\begin{array}{ccc}
H^i(F,A) \otimes H^{-i}(F, \operatorname{Hom}(A, \mathbb{Z})) & \xrightarrow{\psi_2} & H^0(F, \mathbb{Z}) \\
\downarrow \text{id} \qquad\qquad \downarrow \psi_8 & & \downarrow \psi_9 \\
H^i(F,A) \otimes H^{1-i}(F, \operatorname{Hom}(A, I)) & \xrightarrow{\psi_{10}} & H^1(F, I) \\
\downarrow \text{id} \qquad\qquad \downarrow \psi_{11} & & \downarrow \psi_7 \\
H^i(F,A) \otimes H^{2-i}(F, \operatorname{Hom}(A, C)) & \xrightarrow{\psi_{12}} & H^2(F, C),
\end{array}$$

where ψ_{10} and ψ_{12} are $\cup$-products, and ψ_8, ψ_9, ψ_{11} are connecting homomorphisms which are isomorphisms in our case. Then the nondegeneracy of ψ_{12} follows from the nondegeneracy of ψ_2. □

2°. Let G be a finite group, H a normal subgroup of G, $F = G/H$, and let A be an F-module. For any subgroup $G_0 \subset G$ and for any positive integer i, the restriction homomorphism $\rho \colon H^i(G, A) \to H^i(G_0, A)$ is well defined.

LEMMA A.3.2. *The restriction homomorphism commutes with lifting; more precisely, the following diagram is commutative:*

$$\begin{array}{ccc}
H^i(G, A) & \xrightarrow[\rho]{} & H^i(G_0, A) \\
\uparrow \lambda & & \uparrow \lambda_0 \\
H^i(F, A) & \xrightarrow[\rho_0]{} & H^i(G_0/H \cap G_0, A),
\end{array}$$

where λ_0 *and* λ *are lifting homomorphisms,* ρ *and* ρ_0 *are restriction homomorphisms.*

PROOF. The diagram of cocycles

$$\begin{array}{ccc} Z^i(G,A) & \xrightarrow[\overline{\rho}]{} & Z^i(G_0,A) \\ \uparrow{\scriptstyle\overline{\lambda}} & & \uparrow{\scriptstyle\overline{\lambda}_0} \\ Z^i(F,A) & \xrightarrow[\overline{\rho}_0]{} & Z^i(G_0/H\cap G_0,A) \end{array}$$

corresponds to the initial diagram. It is clear that it is sufficient to verify the commutativity of the latter diagram. Let $\sigma_1,\dots,\sigma_i\in G_0$, $\overline{f}\in Z^i(F,A)$. Then

$$(\overline{\rho}\,\overline{\lambda}\,\overline{f})(\sigma_1,\dots,\sigma_i)=\overline{f}(\sigma_1 \bmod H\cap G_0,\dots,\sigma_i \bmod H\cap G_0).$$

On the other hand,

$$(\overline{\lambda}_0\overline{\rho}_0\overline{f})(\sigma_1,\dots,\sigma_i)=\overline{\lambda}_0\overline{f}(\sigma_1,\dots,\sigma_i)=\overline{f}(\sigma_1 \bmod H\cap G_0,\dots,\sigma_i \bmod H\cap G_0).$$

Let F_0 be a subgroup of the group F and let $\mathbb{Z}[F/F_0]$ be a free Abelian group with generators e_g that are in one-to-one correspondence with the cosets of F with respect to F_0. We convert $\mathbb{Z}[F/F_0]$ into an F-module by setting $e^f_{F_0\overline{g}}=e_{F_0\leq\overline{gf}}$, where $\overline{g}$ is a representative of the class g. □

LEMMA A.3.3. *For all* $i\geq 0$, *a natural isomorphism*

$$H^i(F,\operatorname{Hom}(\mathbb{Z}[F/F_0],A))\to H^i(F_0,A)$$

exists.

PROOF. Let P_i be the ith term of the standard F-resolvent of the group $\mathbb{Z}$; then one may regard P_i as the corresponding term of the F_0-resolvent. It follows readily that the mapping

$$\varphi\colon \operatorname{Hom}_F(P_i,\operatorname{Hom}(\mathbb{Z}[F/F_0],A))\to\operatorname{Hom}_{F_0}(P_i,A)$$

given by the rule $(\varphi\chi)(p)=\chi(p)(e_1)$, with $\chi\in\operatorname{Hom}_F(P_i,\operatorname{Hom}(\mathbb{Z}[F/F_0],A))$, $p\in P_i$, is an isomorphism, which proves the assertion.

Let K/k be a normal extension of algebraic number fields with Galois group F, A an F-module, $\mathfrak{p}$ a prime divisor of the field k, $\mathfrak{P}$ a prime divisor in the field K, $F_{\mathfrak{P}}$ its decomposition group. We denote by $\overline{V}_{\mathfrak{p}}$ the projection of the subgroup of units of the idèle group of K to the components corresponding to the prime divisors of the divisor $\mathfrak{p}$ in the field K, i.e., $\overline{V}_{\mathfrak{p}}=\prod_{\mathfrak{P}/\mathfrak{p}}V_{\mathfrak{P}}$, where $V_{\mathfrak{P}}$ is the group of $\mathfrak{P}$-adic units. □

LEMMA A.3.4. *For all* i *there exists an isomorphism*

$$H^i(F,\operatorname{Hom}(A,\overline{V}_{\mathfrak{p}}))\to H^i(F_{\mathfrak{P}},\operatorname{Hom}(A,V_{\mathfrak{P}})).$$

PROOF. Let, as before, P_i be the ith term of the standard F-resolvent for $\mathbb{Z}$. It is immediate that the mappings

$$\varphi\colon \operatorname{Hom}_F(P_i, \operatorname{Hom}(A, \overline{V}_{\mathfrak{p}})) \longrightarrow \operatorname{Hom}_{F_{\mathfrak{P}}}(P_i, \operatorname{Hom}(A, V_{\mathfrak{P}})),$$
$$\psi\colon P_i \otimes_F \operatorname{Hom}(A, \overline{V}_{\mathfrak{p}}) \longrightarrow P_i \otimes_{F_{\mathfrak{P}}} \operatorname{Hom}(A, V_{\mathfrak{P}})$$

given by the rules $(\varphi\chi)(p)(a) = \operatorname{pr}_{\mathfrak{P}}(\chi(p)(a)), \psi(p \otimes \mu) = p \otimes \widetilde{\mu}$ are isomorphisms, where $p \in P$, $a \in A$, $\chi \in \operatorname{Hom}_F(P_i, \operatorname{Hom}(A, \overline{V}_{\mathfrak{p}}))$, $\mu \in \operatorname{Hom}(A, \overline{V}_{\mathfrak{p}})$, and $\widetilde{\mu} \in \operatorname{Hom}(A, V_{\mathfrak{P}})$, with $\widetilde{\mu}(a) = \operatorname{pr}_{\mathfrak{P}} \mu(a)$.

For any G-module B we denote by N_G the endomorphism of the module B that maps the element $b \in B$ to the product $\prod_{g \in G} b^g$, and we denote by $I(G)B$ the submodule generated by the elements b^{g-1}, where $g \in G$.

For $i = 0, -1$ and for a G-module B, we define the deflation mapping

$$\operatorname{def}_i\colon H^i(G, \operatorname{Hom}(A, B)) \longrightarrow H^i(F, \operatorname{Hom}(A, B^H))$$

by setting

$$\operatorname{def}_{-1}(\overline{f}_{-1} \bmod I(G) \operatorname{Hom}(A, B)) = (N_H \overline{f}_{-1}) \bmod I(F) \operatorname{Hom}(A, B^H)$$

and

$$\operatorname{def}_0(\overline{f}_0 \bmod N_G \operatorname{Hom}(A, B)) = \overline{f}_0 \bmod N_F \operatorname{Hom}(A, B^H),$$

where $\overline{f}_{-1} \in \operatorname{Ker} N_G \operatorname{Hom}(A, B)$ and $\overline{f}_0 \in \operatorname{Hom}_G(A, B)$. □

LEMMA A.3.5. *The deflation mapping is dual to the lifting homomorphism or, more precisely, for $i = 2, 3$ the following diagram is commutative:*

$$\begin{array}{ccc} H^i(G, A) & \xrightarrow{\theta_1} & H^{2-i}(G, \operatorname{Hom}(A, B))' \\ \uparrow{\scriptstyle\lambda} & & \uparrow{\scriptstyle\overline{\operatorname{def}}} \\ H^i(F, A) & \xrightarrow{\theta_2} & H^{2-i}(F, \operatorname{Hom}(A, B^H))', \end{array}$$

where $\overline{\operatorname{def}}$ is the mapping dual to the deflation mapping, and θ_1, θ_2 are the homomorphisms induced by $\cup$-product.

PROOF. Let $i = 2$. Consider the diagram

$$\begin{array}{ccc} H^2(G, A) \otimes H^0(G, \operatorname{Hom}(A, B)) & \xrightarrow{\theta_3} & H^2(G, B) \\ \uparrow{\scriptstyle\lambda} \qquad \uparrow{\scriptstyle\operatorname{def}_0} & & \uparrow{\scriptstyle\lambda_1} \\ H^2(F, A) \otimes H^0(F, \operatorname{Hom}(A, B^H)) & \xrightarrow{\theta_4} & H^2(F, B^H), \end{array}$$

where θ_3 and θ_4 are $\cup$-products, and λ_1 is the lifting homomorphism. One may directly verify that the relation $\theta_3(\lambda(f) \otimes g) = \lambda_1\theta_4(f \otimes \operatorname{def}_0(g))$ holds, where $f \in H^2(F, A)$, $g \in H^0(G, \operatorname{Hom}(A, B))$, and, thus, the initial diagram is commutative for $i = 2$.

Further, in the proof we use the shift of dimension. Let $I_1 = \operatorname{Hom}(\mathbb{Z}[F], A)$ be an F-module with natural action of F. Consider the inclusion $\theta_5\colon A \longrightarrow I_1$ defined by the relations $\theta_5(a)(e_\sigma) = a$ for all $\sigma \in F$.

The exact sequence $1 \longrightarrow A \xrightarrow{\theta_5} I_1 \longrightarrow I_2 \longrightarrow 1$ splits over $\mathbb{Z}$. Indeed, we define the mapping $\overline{\theta}_5 \colon I_1 \longrightarrow A$ by setting $\theta_5(\mu) = \mu(e_1)$ for $\mu \in I_1$. It is easy to check that $\overline{\theta}_5$ is a homomorphism and $\overline{\theta}_5\theta_5 = \text{id}$. Therefore, for any G-module D the following sequence is exact:

$$1 \longrightarrow \operatorname{Hom}(I_2, D) \longrightarrow \operatorname{Hom}(I_1, D) \longrightarrow \operatorname{Hom}(A, D) \longrightarrow 1.$$

Consider the diagram with natural homomorphisms

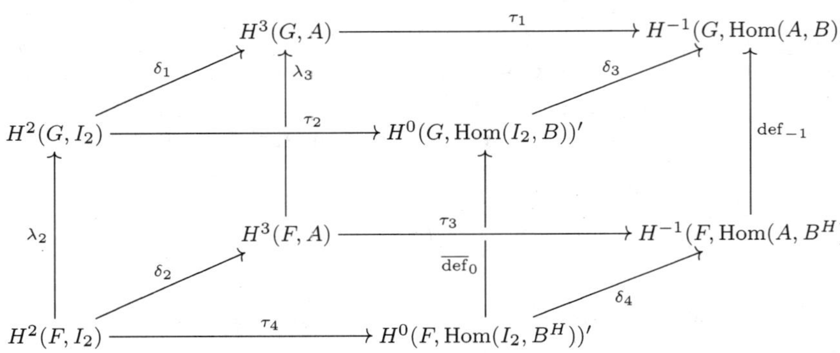

We have already seen that $\tau_2\lambda_2 = \overline{\operatorname{def}}_0\, \tau_4$. It is clear that I_1 and $\operatorname{Hom}(I_1, B^H)$ are cohomologously trivial modules, so δ_2 and $\overline{\delta}_4$ are isomorphisms. We have $\overline{\delta}_3\tau_2 = \tau_1\delta_1$ and $\tau_3\delta_2 = \overline{\delta}_4\tau_4$. Note that lifting and deflation commute with the connecting homomorphisms. Then we have

$$\begin{aligned} \tau_1\lambda_3 &= \tau_1\delta_1\lambda_2\delta_2^{-1} = \overline{\delta}_3\tau_2\lambda_2\delta_2^{-1} \\ &= \overline{\delta}_3\, \overline{\operatorname{def}}_0\, \tau_4\delta_2^{-1} = \overline{\operatorname{def}}_{-1}\, \overline{\delta}_4\tau_4\delta_2^{-1} = \overline{\operatorname{def}}_{-1}\, \tau_3\delta_2\delta_2^{-1} = \overline{\operatorname{def}}_{-1}\, \tau_3. \end{aligned}$$

3°. We recall the definition of the cohomology groups of profinite groups. Define the notion of a projective limit of groups. Let Γ be a directed set of indices, and we assume that a group A_α corresponds to each $\alpha \in \Gamma$. Assume that for each pair α, $\beta \in \Gamma$ with $\alpha \leq \beta$ a homomorphism $\varphi_\alpha^\beta \colon A_\beta \longrightarrow A_\alpha$ is given so that if $\alpha = \beta$ then $\varphi_\alpha^\beta = \text{id}$, and if $\alpha \leq \beta \leq \gamma$ then $\varphi_\alpha^\beta\varphi_\beta^\gamma = \varphi_\alpha^\gamma$. Then the projective limit $A = \varprojlim A_\alpha$ of the groups A_α relative to the homomorphisms $\{\varphi_\alpha^\beta\}$ is a subset of the direct product $\prod_{\alpha\in\Gamma} A_\alpha$ consisting of elements $\{a_\alpha\}_{\alpha\in\Gamma}$ such that for $\alpha \leq \beta$, $\varphi_\alpha^\beta(a_\beta) = a_\alpha$. It is easy to check that A is a subgroup of $\prod_{\alpha\in\Gamma} A_\alpha$.

In the definition of a profinite group A, all A_α are finite groups. When we consider the projective limit of compact groups, we shall always assume that φ_α^β are continuous maps. It is easy to see that the projective limit of compact groups is a compact group in natural topology.

Let $F = \varprojlim F_\alpha$ be a profinite group and H_α the kernel of the projection of F to the component with index α. We define the cohomology groups for discrete F-modules A, i.e., for modules that can be represented in the form $\bigcup_{\alpha\in\Gamma} A^{H_\alpha}$. In the sequel we shall consider only discrete modules. For $\alpha \leq \beta$, $i \geq 1$, the lifting homomorphism $\lambda_\alpha^\beta \colon H^i(F_\alpha, A^{H_\alpha}) \longrightarrow H^i(F_\beta, A^{H_\beta})$ is well defined. It is clear that the groups $\{H^i(F_\alpha, A^{H_\alpha})\}_{\alpha\in\Gamma}$ form an inductive system relative to the homomorphisms λ_α^β (see [**1**, Chapter V, §2]). The inductive limit $\varinjlim H^i(F_\alpha, A^{H_\alpha})$ is called the *ith cohomology group* $H^i(F, A)$.

Let H be a subgroup of finite index in F. We set $\widetilde{H}_\alpha = H \cap H_\alpha$ and $\widetilde{F}_\alpha = F/\widetilde{H}_\alpha$. It is immediate that the natural mapping $\varinjlim H^i(F_\alpha, A^{H_\alpha}) \to \varinjlim H^i(\widetilde{F}_\alpha, A^{\widetilde{H}_\alpha})$ is an isomorphism.

It follows directly from the properties of the inductive limit that an exact sequence $1 \to A_1 \to A_2 \to A_3 \to 1$ of F-modules induces the exact cohomology sequence

$$H^i(F, A_1) \to H^i(F, A_2) \to H^i(F, A_3) \to H^{i+1}(F, A_1) \to \dots$$

for $i \geq 1$.

Now let A be an F-module finitely generated as an Abelian group. Then there exists a subgroup of finite index $H \subset F$ such that the elements of H act trivially on the elements of the module A. It is clear that the groups $\{\widetilde{F}_\alpha\}$ form a projective system and $\varprojlim \widetilde{F}_\alpha = F$.

Let C be an F-module such that C^{H_α} for $\alpha \in F$ is a class formation. As an example of such a module, one may take the multiplicative group $\overline{k}^*$ of the algebraic closure of a local field.

For $\alpha \leq \beta$ and $i = 0, -1$, the deflation homomorphism is well defined:

$$\mathrm{def}_i\colon H^i(\widetilde{F}_\beta, \mathrm{Hom}(A, C^{\widetilde{H}_\beta})) \to H^i(\widetilde{F}_\alpha, \mathrm{Hom}(A, C^{\widetilde{H}_\alpha})).$$

It is clear that the groups $\{H^i(\widetilde{F}_\alpha, \mathrm{Hom}(A, C^{\widetilde{H}_\alpha}))\}_{\alpha\in\Gamma}$ form a projective system. We call the projective limit $\varprojlim H^i(\widetilde{F}_\alpha, \mathrm{Hom}(A, C^{\widetilde{H}_\alpha}))$ the *ith cohomology group of the module* $\mathrm{Hom}(A, C)$ for $i = 0, -1$.

We prove several assertions needed in the sequel. For arbitrary $\alpha \in F$ let exact sequences of groups $1 \to A_\alpha \xrightarrow{\psi_{1\alpha}} B_\alpha \xrightarrow{\psi_{2\alpha}} D_\alpha \to 1$ exist. Assume that for $\alpha \leq \beta$, the following diagram is commutative:

$$\begin{array}{ccccccccc}
1 & \longrightarrow & A_\beta & \xrightarrow{\psi_{1\beta}} & B_\beta & \xrightarrow{\psi_{2\beta}} & D_\beta & \longrightarrow & 1 \\
 & & \downarrow{\scriptstyle\varphi_\alpha^\beta} & & \downarrow{\scriptstyle\psi_\alpha^\beta} & & \downarrow{\scriptstyle\theta_\alpha^\beta} & & \\
1 & \longrightarrow & A_\alpha & \xrightarrow{\psi_{1\alpha}} & B_\alpha & \xrightarrow{\psi_{2\alpha}} & D_\alpha & \longrightarrow & 1.
\end{array}$$

It is clear that we may consider the sequence

$$\varprojlim A_\alpha \xrightarrow{\widetilde{\psi}_1} \varprojlim B_\alpha \xrightarrow{\widetilde{\psi}_2} \varprojlim D_\alpha.$$

It is immediate that this sequence is exact and $\widetilde{\psi}_1$ is injective. □

LEMMA A.3.6. *If the groups A_α and B_α are compact for all $\alpha \in \Gamma$ and if all the maps considered are continuous, then $\widetilde{\psi}_2$ is surjective.*

PROOF. Let $d \in \varprojlim D_\alpha$, let d_α be the projection of d onto the component with index α, and let b_α be the inverse image of d_α in B_α. We note that the cosets $E_\alpha = b_\alpha \psi_{1\alpha}(A_\alpha)$ are compact sets. Therefore, $E = \prod_{\alpha\in\Gamma} E_\alpha$ is a compact subspace of the space $\prod_{\alpha\in\Gamma} B_\alpha$.

Let Γ_β be a subset of elements in Γ consisting of α's with $\alpha \leq \beta$. We set $\overline{E}_\beta = \prod_{\alpha\in\Gamma_\beta} \psi^\beta_\alpha(E_\alpha) \times \prod_{\alpha\in\Gamma\setminus\Gamma_\beta} E_\alpha$. It is clear that the sets $\{\overline{E}_\beta\}$ form a centered system of compact sets (i.e., any finite intersection is nonempty); therefore, $\bigcap_{\beta\in\Gamma} \overline{E}_\beta$ is nonempty. Thus, in the group $\varprojlim B_\alpha$ there exists an element b such that $\widetilde{\psi}_2(b) = d$. The lemma is proved. □

For the module $\operatorname{Hom}_{\widetilde{H}_\alpha}(A, C)$ we set

$$\overline{N}_{\widetilde{H}_\alpha} \operatorname{Hom}(A, C) = \bigcap_{\alpha\leq\beta} N_{\widetilde{H}_\alpha/\widetilde{H}_\beta} \operatorname{Hom}_{\widetilde{H}_\beta}(A, C).$$

LEMMA A.3.7. *If the modules* $\operatorname{Hom}_{\widetilde{H}_\alpha}(A, C)$ *are compact for all* α, *then the following relations hold*:

$$H^{-1}(F, \operatorname{Hom}(A, C)) = \varprojlim H^{-1}(\widetilde{F}_\alpha, \overline{N}_{\widetilde{H}_\alpha} \operatorname{Hom}_{\widetilde{H}_\alpha}(A, C)),$$
$$H^0(F, \operatorname{Hom}(A, C)) = \operatorname{Hom}_F(A, C)/\overline{N}_F \operatorname{Hom}(A, C).$$

PROOF. It is clear that for all $\alpha \in \Gamma$ the groups $\operatorname{Ker} N_{\widetilde{F}_\alpha} \operatorname{Hom}(A, C^{\widetilde{H}_\alpha})$ and $I(\widetilde{F}_\alpha) \operatorname{Hom}(A, C^{\widetilde{H}_\alpha})$ are compact. Passing to the projective limit in the exact sequence

$$1 \to I(\widetilde{F}_\alpha) \operatorname{Hom}(A, C^{\widetilde{H}_\alpha}) \to \operatorname{Ker} N_{\widetilde{F}_\alpha} \operatorname{Hom}(A, C^{\widetilde{H}_\alpha})$$
$$\to H^{-1}(\widetilde{F}_\alpha, \operatorname{Hom}(A, C^{\widetilde{H}_\alpha})) \to 0$$

we have

$$1 \to \varprojlim I(\widetilde{F}_\alpha) N_{\widetilde{H}_\alpha} \operatorname{Hom}(A, C^{\widetilde{H}_\alpha}) \to \varprojlim \operatorname{Ker} N_{\widetilde{F}_\alpha} (\overline{N}_{H_\alpha} \operatorname{Hom}(A, C^{\widetilde{H}_\alpha}))$$
$$\to \varprojlim H^{-1}(\widetilde{F}_\alpha, \operatorname{Hom}(A, C^{\widetilde{H}_\alpha})) \to 0.$$

The first assertion of the lemma follows from the latter exact sequence. The second assertion is proved in a similar way. □

Let $\{A_\alpha\}$ be an inductive system of groups relative to the maps $\{\varphi^\beta_\alpha\}$. It is clear that the groups $\{A'_\alpha\}$ form a projective system relative to the dual maps $\{\overline{\varphi}^\beta_\alpha\}$.

LEMMA A.3.8. *If* $\varinjlim A_\alpha$ *is the inductive limit of the groups* A_α *with discrete topology, then there exists a natural isomorphism* $\varinjlim A_\alpha \to (\varprojlim A'_\alpha)'$.

PROOF. Let $\tau_1\colon \sum_{\alpha\in\Gamma}^{\oplus} A_\alpha \to (\prod_{\alpha\in\Gamma} A'_\alpha)'$ be a natural topological isomorphism, namely,

$$\tau_1(a_{\alpha_1} + \cdots + a_{\alpha_n})\Big(\prod_{\alpha\in\Gamma} b_\alpha\Big) = \prod_{i=1}^{n} b_{\alpha_i}(a_{\alpha_i}),$$

where $a_{\alpha_i} \in A_{\alpha_i}$, $b_{\alpha_i} \in A'_{\alpha_i}$.

We note that $\varinjlim A_\alpha = (\sum_{\alpha\in\Gamma}^{\oplus} A_\alpha)/M$, where M is a subgroup generated by the elements $\varphi^\beta_\alpha(a_\alpha) - a_\alpha$. Then $\tau_1(a_\alpha - \varphi^\beta_\alpha(a_\alpha))(\prod b_\alpha) = b_\alpha(a_\alpha)(\overline{\varphi}^\beta_\alpha(b_\beta))(a_\alpha^{-1})$.

Therefore, if $\prod b_\alpha \in \varprojlim A'_\alpha$, then $\tau_1(a_\alpha - \varphi^\beta_\alpha(a_\alpha))(\prod b_\alpha) = 1$, i.e., τ_1 induces $\tau_2\colon \varinjlim A_\alpha \to (\varprojlim A'_\alpha)'$. It is clear that τ_2 is a homomorphism. It follows from the commutative diagram

$$\begin{array}{ccc} \varinjlim A_\alpha & \longrightarrow & (\varprojlim A'_\alpha)' \\ \uparrow & & \uparrow \\ \sum\limits_\alpha^{\oplus} A_\alpha & \longrightarrow & \left(\prod\limits_\alpha A'_\alpha\right)' \end{array}$$

that τ_2 is surjective.

Now consider the map $\overline{\tau}_1\colon \prod A'_\alpha \to (\sum A_\alpha{}^{\oplus})'$ dual to τ_1. Let $\chi \in (\varinjlim A_\alpha)'$, φ_α a natural homomorphism of A_α to $\varinjlim A_\alpha$, and let $\overline{\varphi}_\alpha\colon (\varinjlim A_\alpha)' \to A'_\alpha$ be a homomorphism dual to φ_α. We note that $\overline{\varphi}^\beta_\alpha \overline{\varphi}_\beta(\chi) = \overline{\varphi}_\alpha(\chi)$ for $\alpha \le \beta$, i.e., $\prod \overline{\varphi}_\alpha(\chi) \in \varprojlim A'_\alpha$, so $\overline{\tau}_1$ maps the subgroup $\varprojlim A'_\alpha$ onto $(\varinjlim A_\alpha)'$, and τ_2 is an isomorphism. $\square$

Now we prove an analog of Lemma A.3.3 for profinite groups.

LEMMA A.3.9. *Let F be a profinite group, H a subgroup of finite index in F, A an F-module. Then for $i \ge 1$ there exists an isomorphism*

$$H^i(F, \mathrm{Hom}(\mathbb{Z}[F/H], A)) \to H^i(H, A).$$

PROOF. Let $F = \varprojlim F/H_\alpha$, then

$$H^i(F, \mathrm{Hom}(\mathbb{Z}[F/H], A)) = \varinjlim H^i(F/H_\alpha, \mathrm{Hom}_{H_\alpha}(\mathbb{Z}[F/H], A)).$$

As we have observed, one may assume that the system of subgroups $\{H_\alpha\}$ is such that $H_\alpha \subset H$ for all α. Then $\mathrm{Hom}_{H_\alpha}(\mathbb{Z}[F/H], A) = \mathrm{Hom}(\mathbb{Z}[F/H], A^{H_\alpha})$. The lemma now follows from the fact that the groups

$$H^i(F/H_\alpha, \mathrm{Hom}(\mathbb{Z}[F/H], A^{H_\alpha})) \quad \text{and} \quad H^i(H/H_\alpha, A^{H_\alpha})$$

are isomorphic. $\square$

4°. Let k be a local field (a finite extension of the p-adic number field), $\overline{k}$ the algebraic closure of k and $\overline{F}$ the Galois group of the extension $\overline{k}/k$. For a finite normal subextension k_α/k with Galois group F_α, the multiplicative group k^*_α of the field is a class formation, so for an F_α-module A the homomorphism $\psi_\alpha\colon H^i(F_\alpha A) \to H^{2-i}(F_\alpha, \mathrm{Hom}(A, k^*_\alpha))'$ is well defined.

Let A be an F-module finitely generated as an Abelian group. Then there exists a subgroup H of finite index in $\overline{F}$, whose elements act trivially on the module A. As we have noted in 3°, the cohomology group $H^i(\overline{F}, A)$ can be defined by means of the inductive system $\{H^i(\widetilde{F}_\alpha, A)\}$, where $\widetilde{F}_\alpha$ runs over the Galois groups of extensions $\widetilde{k}_\alpha/k$ containing the field of invariants $\overline{k}^H$.

Lemma A.3.5 implies that for $i = 2, 3$ and $\alpha \le \beta$, the following diagram is commutative:

$$\begin{array}{ccc} H^i(\widetilde{F}_\beta, A) & \longrightarrow & H^{2-i}(\widetilde{F}_\beta, \mathrm{Hom}(A, k^*_\beta))' \\ \uparrow{\scriptstyle\lambda} & & \uparrow{\scriptstyle\overline{\mathrm{def}}} \\ H^i(\widetilde{F}_\alpha, A) & \longrightarrow & H^{2-i}(\widetilde{F}_\alpha, \mathrm{Hom}(A, k^*_\beta))'. \end{array}$$

By Lemma A.3.8, the passage to the limit yields a homomorphism ν: $H^i(\overline{F}, A) \to H^{2-i}(\overline{F}, \operatorname{Hom}(A, \overline{k}^*))'$. We note that from the Nakayama–Tate theorem it follows that if A is a lattice, then ν is an isomorphism.

LEMMA A.3.10. *If A is a lattice, then $H^3(\overline{F}, A) = 0$.*

PROOF. It is sufficient to prove the dual assertion, namely,

$$\varprojlim H^{-1}(\widetilde{F}_\alpha, \operatorname{Hom}(A, k_\alpha^*)) = 0.$$

For all α, we have the exact sequence $1 \to V_\alpha \to k_\alpha^* \xrightarrow{\omega_\alpha} \mathbb{Z} \to 0$, where ω_α maps every element $x \in k_\alpha^*$ to its value of the exponent, and V_α is the group of units.

The exact sequence

$$1 \to \operatorname{Hom}(A, V_\alpha) \to \operatorname{Hom}(A, k_\alpha^*) \to \operatorname{Hom}(A, \mathbb{Z}) \to 1$$

induces the exact cohomology sequence

$$H^{-1}(\widetilde{F}_\alpha, \operatorname{Hom}(A, V_\alpha)) \to H^{-1}(\widetilde{F}_\alpha, \operatorname{Hom}(A, k_\alpha^*)) \to H^{-1}(\widetilde{F}_\alpha, \operatorname{Hom}(A, \mathbb{Z})).$$

For $\alpha \leq \beta$ we consider the diagram

$$\begin{array}{ccccc} H^{-1}(\widetilde{F}_\beta, \operatorname{Hom}(A, V_\beta)) & \longrightarrow & H^{-1}(\widetilde{F}_\beta, \operatorname{Hom}(A, k_\beta^*)) & \longrightarrow & H^{-1}(\widetilde{F}_\beta, \operatorname{Hom}(A, \mathbb{Z})) \\ \Big\downarrow \delta_\alpha^\beta & & \Big\downarrow \varepsilon_\alpha^\beta & & \Big\downarrow \theta_\alpha^\beta \\ H^{-1}(\widetilde{F}_\alpha, \operatorname{Hom}(A, V_\alpha)) & \longrightarrow & H^{-1}(\widetilde{F}_\alpha, \operatorname{Hom}(A, k_\alpha^*)) & \longrightarrow & H^{-1}(\widetilde{F}_\alpha, \operatorname{Hom}(A, \mathbb{Z})). \end{array}$$

Here the maps δ_α^β and θ_α^β are induced by the map $N_{H_{\alpha,\beta}}$, where $H_{\alpha,\beta}$ is the kernel of the natural homomorphism of the group $\widetilde{F}_\beta$ onto $\widetilde{F}_\alpha$, the map θ_α^β is induced by multiplication of the module $\operatorname{Hom}(A, \mathbb{Z})$ by the inertia degree of the extension $\widetilde{k}_\beta/\widetilde{k}_\alpha$. The commutativity of the diagram is immediate. Passing to the limit, we have the exact sequence

$$\begin{aligned} \varprojlim H^{-1}(\widetilde{F}_\alpha, \operatorname{Hom}(A, V_\alpha)) \to \varprojlim H^{-1}(\widetilde{F}_\alpha, \operatorname{Hom}(A, k_\alpha^*)) \\ \to \varprojlim H^{-1}(\widetilde{F}_\alpha, \operatorname{Hom}(A, \mathbb{Z})). \end{aligned}$$

We observe that $\operatorname{Hom}(A, V_\alpha)$ is a compact group, so, by Lemma A.3.7,

$$\varprojlim H^{-1}(\widetilde{F}_\alpha, \operatorname{Hom}(A, V_\alpha)) = \varprojlim H^{-1}(\widetilde{F}_\alpha, \overline{N}_{\widetilde{H}_\alpha} \operatorname{Hom}(A, V_\alpha)).$$

Since A is a lattice, $\overline{N}_{\widetilde{H}_\alpha} \operatorname{Hom}(A, V_\alpha) = \operatorname{Hom}(A, \overline{N}_{\widetilde{H}_\alpha} V_\alpha)$. It follows from local class field theory (see [**1**, Chapter VI]) that $\overline{N}_{\widetilde{H}_\alpha} V_\alpha = 1$, so

$$\varprojlim H^{-1}(\widetilde{F}_\alpha, \operatorname{Hom}(A, V_\alpha)) = 0.$$

Now we compute $\varprojlim H^{-1}(\widetilde{F}_\alpha, \operatorname{Hom}(A, \mathbb{Z}))$. Let $x \in \varprojlim H^{-1}(\widetilde{F}_\alpha, \operatorname{Hom}(A, \mathbb{Z}))$, $x_\alpha = \operatorname{pr}_\alpha x$ be the projection of x to the component with index α, and $\overline{x}_\alpha$ the representative of x_α in the group $\operatorname{Ker} N_{\widetilde{F}_\alpha} \operatorname{Hom}(A, \mathbb{Z})$. Let the order of the group $H^{-1}(\widetilde{F}_\alpha, \operatorname{Hom}(A, \mathbb{Z}))$ be equal to n and let $\widetilde{k}_\beta/\widetilde{k}_\alpha$ be an unramified extension of degree n. If $x_\beta = \operatorname{pr}_\beta x$ and $\overline{x}_\beta$ is the inverse image of x_β in the group $\operatorname{Ker} N_{\widetilde{F}_\beta} \operatorname{Hom}(A, \mathbb{Z})$, then $\overline{x}_\alpha = \overline{x}_\beta^n \bmod I(\widetilde{F}_\alpha) \operatorname{Hom}(A, \mathbb{Z})$. We note that from the last equality it follows that $N_{\widetilde{F}_\alpha} \overline{x}_\beta = 1$, i.e., $\overline{x}_\alpha \in I(\widetilde{F}_\alpha) \operatorname{Hom}(A, \mathbb{Z})$ and $x_\alpha = 1$. □

THEOREM A.3.2 (Tate's local duality). *For a finite F-module A, the map $\psi\colon H^2(\overline{F},A) \longrightarrow \operatorname{Hom}_{\overline{F}}(A,\overline{k}^*)'$ is an isomorphism.*

PROOF. First we show that ν defined above is an isomorphism. Let $1 \longrightarrow A_1 \longrightarrow A_2 \longrightarrow A \longrightarrow 1$ be a representation of A as the quotient module of a lattice. To the exact sequence

$$H^2(\overline{F},A_1) \longrightarrow H^2(\overline{F},A_2) \longrightarrow H^2(\overline{F},A) \longrightarrow H^3(\overline{F},A_1) = 0$$

there corresponds the sequence

$$H^0(\overline{F},\operatorname{Hom}(A_1,\overline{k}^*))' \longrightarrow H^0(\overline{F},\operatorname{Hom}(A_2,\overline{k}^*))' \longrightarrow H^0(\overline{F},\operatorname{Hom}(A,\overline{k}^*))'.$$

Consider the commutative diagram with natural homomorphisms

$$\begin{array}{ccccc}
 & & H^0(\overline{F},\operatorname{Hom}(A_2,\mathbb{Z})) & \xrightarrow{\tau_1} & H^0(\overline{F},\operatorname{Hom}(A_1,\mathbb{Z})) \\
 & & \uparrow & & \uparrow \\
H^0(\overline{F},\operatorname{Hom}(A,\overline{k}^*)) & \xrightarrow{\tau_2} & H^0(\overline{F},\operatorname{Hom}(A_2,\overline{k}^*)) & \longrightarrow & H^0(\overline{F},\operatorname{Hom}(A_1,\overline{k}^*)) \\
\uparrow{\scriptstyle \mathrm{id}} & & \uparrow{\scriptstyle \tau_3} & & \uparrow{\scriptstyle \tau_4} \\
\varprojlim H^0(\widetilde{F}_\alpha,\operatorname{Hom}(A,V_\alpha)) & \longrightarrow & \varprojlim H^0(\widetilde{F}_\alpha,\operatorname{Hom}(A_2,V_\alpha)) & \longrightarrow & \varprojlim H^0(\widetilde{F}_\alpha,\operatorname{Hom}(A_1,V_\alpha)).
\end{array}$$

We note that the middle vertical sequence can be obtained by passing to the projective limit in the exact sequence

$$H^0(\widetilde{F}_\alpha,\operatorname{Hom}(A_2,V_\alpha)) \longrightarrow H^0(\widetilde{F}_\alpha,\operatorname{Hom}(A_2,\widetilde{k}_\alpha^*)) \longrightarrow H^0(\widetilde{F}_\alpha,\operatorname{Hom}(A_2,\mathbb{Z})).$$

We prove that τ_3 is injective by using the relation $\varprojlim H^{-1}(\widetilde{F}_\alpha,\operatorname{Hom}(A_2,\mathbb{Z})) = 0$. Similarly we prove that τ_4 is injective. Observe that τ_1 is an isomorphism (this follows from the isomorphism of the groups $\operatorname{Hom}(A_2,\mathbb{Z})$ and $\operatorname{Hom}(A_1,\mathbb{Z})$). The bottom sequence of the diagram coincides with the exact sequence

$$\operatorname{Hom}_{\overline{F}}(A,V) \longrightarrow \operatorname{Hom}_{\overline{F}}(A_2,V) \longrightarrow \operatorname{Hom}_{\overline{F}}(A_1,V).$$

Indeed, $\operatorname{Hom}(A,V_\alpha)$ is a compact module, so

$$\varprojlim H^0(\widetilde{F}_\alpha,\operatorname{Hom}(A,V_\alpha)) = \operatorname{Hom}_{\overline{F}}(A,V)/\overline{N}_{\overline{F}}\operatorname{Hom}_{\overline{F}}(A,V),$$

where V is a subgroup of units of $\overline{k}$. It remains to note that

$$\overline{N}_{\overline{F}}\operatorname{Hom}_{\overline{F}}(A,V) \subset N_{\widetilde{F}_\alpha}\operatorname{Hom}(A,\overline{N}_{\widetilde{H}_\alpha}V_\alpha) = 1$$

(it follows from local class field theory that $\overline{N}_{\widetilde{H}_\alpha}V_\alpha = 1$). We can show similarly that for $i = 1,2$,

$$\varprojlim H^0(\widetilde{F}_\alpha,\operatorname{Hom}(A_i,V_\alpha)) = \operatorname{Hom}_{\overline{F}}(A_i,V).$$

Diagram chase yields the exactness of the middle horizontal sequence of the diagram and proves that τ_2 is injective. Hence, ν is an isomorphism. It remains to note that

$$H^0(\overline{F},\operatorname{Hom}(A,\overline{k}^*)) = \varprojlim H^0(\widetilde{F}_\alpha,\operatorname{Hom}(A,V_\alpha)) = \operatorname{Hom}_{\overline{F}}(A,V) = \operatorname{Hom}_{\overline{F}}(A,\overline{k}^*)$$

(we use the finiteness of the module A). □

Now consider the duality map in the case where k is the field of real numbers. Let $\mathbb{C}$ be the field of complex numbers, F the Galois group of the extension $\mathbb{C}/k$, and A an F-module. Observe that $\mathbb{C}^*$ is a class formation, and, thus, the homomorphism

$$\varphi_i\colon H^i(F,A) \longrightarrow H^{2-i}(F,\operatorname{Hom}(A,\mathbb{C}^*))'$$

is well defined.

THEOREM A.3.3. *If A is a finitely generated F-module, then φ_i is an isomorphism for arbitrary i.*

PROOF. It follows from the Nakayama–Tate theorem that φ_i is an isomorphism if A is a lattice. Let $1 \to A_1 \to A_2 \to A \to 1$ be a representation of A as the quotient module of a lattice.

Consider the diagram

$$\begin{array}{ccccc} H^i(F, A_1) & \longrightarrow & H^i(F, A_2) & \longrightarrow & H^i(F, A) \\ \downarrow{\scriptstyle \varphi_{i,1}} & & \downarrow{\scriptstyle \varphi_{i,2}} & & \downarrow{\scriptstyle \varphi_i} \\ H^{2-i}(F, \mathrm{Hom}(A_1, \mathbb{C}^*))' & \longrightarrow & H^{2-i}(F, \mathrm{Hom}(A_2, \mathbb{C}^*))' & \longrightarrow & H^{2-i}(F, \mathrm{Hom}(A, \mathbb{C}^*))' \end{array}$$

$$\begin{array}{ccccc} \longrightarrow & H^{i+1}(F, A_1) & \longrightarrow & H^{i+1}(F, A_2) \\ & \downarrow{\scriptstyle \varphi_{i+1,1}} & & \downarrow{\scriptstyle \varphi_{i+1,2}} \\ \longrightarrow & H^{1-i}(F, \mathrm{Hom}(A_1, \mathbb{C}^*))' & \longrightarrow & H^{1-i}(F, \mathrm{Hom}(A_2, \mathbb{C}^*))', \end{array}$$

where the bottom row is the exact sequence dual to the cohomology sequence for the triple:

$$1 \to \mathrm{Hom}(A, \mathbb{C}^*) \to \mathrm{Hom}(A_2, \mathbb{C}^*) \to \mathrm{Hom}(A_1, \mathbb{C}^*) \to 1.$$

It is clear that the diagram is commutative and the theorem follows from the fact that $\varphi_{i,1}$ and $\varphi_{i,2}$ are isomorphisms for all i (A_1 and A_2 are lattices). □

5°. Let k be an algebraic number field, S the set of points of k, Ω the maximal algebraic extension of k unramified outside S, i.e., the compositum of all finite normal extensions k_α/k in which only the divisors of the set S can ramify. Denote by F_S the Galois group of Ω/k, by $V_{S,\alpha}$ the subgroup of units of the idèle group J_α of k_α whose projection to the components corresponding to the divisors from S is trivial. Note that a natural embedding $V_{S,\alpha} \to C_\alpha$ exists, where C_α is the idèle class group of k_α. We will denote by $C_{S,\alpha}$ the quotient group $C_\alpha/V_{S,\alpha}$.

Observe that C_α is a class formation. It follows from [**1**, Chapter 6, Proposition 1.1 and Chapter 7, Proposition 7.2] that $V_{S,\alpha}$ is a cohomologously trivial module. The exact sequence

$$H^i(F_\alpha, V_{S,\alpha}) \to H^i(F_\alpha, C_\alpha) \to H^i(F_\alpha, C_{S,\alpha}) \to H^{i+1}(F_\alpha, V_{S,\alpha})$$

implies that the cohomology groups of the modules $C_{S,\alpha}$ and C_α coincide, i.e., $C_{S,\alpha}$ is a class formation.

LEMMA A.3.11. *For $\beta \geq \alpha$, $C_{S,\beta}^{H_{\alpha,\beta}} = C_{S,\alpha}$, where $H_{\alpha,\beta}$ is the kernel of the natural homomorphism of the group F_β onto F_α.*

PROOF. The exact sequence of $H_{\alpha,\beta}$-modules

$$1 \to V_{S,\beta} \to C_\beta \to C_{S,\beta} \to 1$$

induces the exact cohomology sequence

$$1 \to V_{S,\beta}^{H_{\alpha,\beta}} \to C_\beta^{H_{\alpha,\beta}} \to C_{S,\beta}^{H_{\alpha,\beta}} \to H^1(H_{\alpha,\beta}, V_{S,\beta}) = 0$$

or $1 \longrightarrow V_{S,\alpha} \longrightarrow C_\alpha \longrightarrow C_{S,\beta}^{H_{\alpha,\beta}} \longrightarrow 1$, whence the lemma follows. □

Let A be an F_S-module, finitely generated as an Abelian group. As we have noted before, there exists a subgroup H of finite index in F_S, whose elements act trivially on A. As before, we define the cohomology groups for the group F_S by using the Galois groups $\widetilde{F}_\alpha$ of the extensions $\widetilde{k}_\alpha/k$ containing the field of invariants Ω^H.

For arbitrary $\alpha \in \Gamma$, the functorial homomorphism

$$H^i(\widetilde{F}_\alpha, A) \longrightarrow H^{2-i}(\widetilde{F}_\alpha, \mathrm{Hom}(A, C_{S,\alpha}))'$$

is well defined. For $i = 2, 3$, one may pass to the limit and obtain the homomorphism

$$\psi_S \colon H^i(F_S, A) \longrightarrow H^{2-i}(F_S, \mathrm{Hom}(A, C_S))',$$

where C_S is the inductive limit $\varinjlim C_{S,\alpha}$ relative to the inclusions $C_{S,\alpha} \to C_{S,\beta}$ for $\alpha \le \beta$. We note that the Nakayama–Tate theorem implies that for lattices, ψ_S is an isomorphism.

Let $\overline{J}_\alpha$ be a subgroup of the idèle group of the field $\widetilde{k}_\alpha$ consisting of elements $x = \{x_\mathfrak{p}\}$ such that $\prod_\mathfrak{p} |x_\mathfrak{p}|_\mathfrak{p} = 1$, where $|\ |_\mathfrak{p}$ is a valuation at the point $\mathfrak{p}$. Then the group $J_\alpha/\overline{J}_\alpha$ is isomorphic to the multiplicative group $\mathbb{R}^*$ of positive real numbers. It is clear that $\widetilde{k}_\alpha^* V_{S,\alpha} \subset \overline{J}_\alpha$, so that $\overline{J}_\alpha/\widetilde{k}_\alpha^* V_{S,\alpha}$ is a subgroup of the group $C_{S,\alpha}$ which we will denote by $\overline{C}_{S,\alpha}$ in the sequel. Note that Theorem 16.1 in [**1**, Chapter II] implies that $\overline{C}_{S,\alpha}$ is a compact group.

LEMMA A.3.12. *If A is a finitely generated $\widetilde{F}_\alpha$-module, then the natural mapping*

$$H^i(\widetilde{F}_\alpha, \mathrm{Hom}(A, \overline{C}_{S,\alpha})) \longrightarrow H^i(\widetilde{F}_\alpha, \mathrm{Hom}(A, C_{S,\alpha}))$$

is an isomorphism for all i.

PROOF. Let A_1 be the torsion subgroup of the group A. Then $A_2 = A/A_1$ is a lattice. Consider the commutative diagram with natural homomorphisms

$$\begin{array}{ccccccc}
1 & \longrightarrow & \mathrm{Hom}(A_1, \overline{C}_{S,\alpha}) & \longrightarrow & \mathrm{Hom}(A_1, C_{S,\alpha}) & \longrightarrow & 1 \\
 & & \uparrow & & \uparrow & & \uparrow \\
1 & \longrightarrow & \mathrm{Hom}(A, \overline{C}_{S,\alpha}) & \longrightarrow & \mathrm{Hom}(A, C_{S,\alpha}) & \xrightarrow{\tau_1} & \mathrm{Hom}(A, \mathbb{R}^*) \\
 & & \uparrow & & \uparrow & & \uparrow{\scriptstyle \tau_3} \\
1 & \longrightarrow & \mathrm{Hom}(A_2, \overline{C}_{S,\alpha}) & \longrightarrow & \mathrm{Hom}(A_2, C_{S,\alpha}) & \xrightarrow{\tau_2} & \mathrm{Hom}(A_2, \mathbb{R}^*).
\end{array}$$

It is clear that τ_1 is surjective.

Note that in the group $\mathrm{Hom}(A, \mathbb{R}^*)$ the equation $x^n = a$ has a unique solution for any natural n; therefore, $\mathrm{Hom}(A, \mathbb{R}^*)$ is a cohomologically trivial module, and the lemma follows from the exact sequence

$$\begin{aligned} H^{i-1}(\widetilde{F}_\alpha, \mathrm{Hom}(A, \mathbb{R}^*)) &\to H^i(\widetilde{F}_\alpha, \mathrm{Hom}(A, \overline{C}_{S,\alpha})) \\ &\to H^i(\widetilde{F}_\alpha, \mathrm{Hom}(A, C_{S,\alpha})) \to H^i(\widetilde{F}_\alpha, \mathrm{Hom}(A, \mathbb{R}^*)). \end{aligned}$$

□

LEMMA A.3.13. *The group* $\overline{N}_{\widetilde{H}_\alpha} C_{S,\alpha}$ *is a division group.*

PROOF. Let $\{S_i\}$ be a sequence of sets of prime divisors of $\widetilde{k}_\alpha$ with the properties:

1. S_i is a finite set;
2. $S_i \subset S_{i+1}$;
3. $\bigcup_i S_i = S$.

For a prime p and natural i, j, we denote by k_{ijp} the class field to the subgroup $J_\alpha^{p^j} V_{S_i,\alpha} \widetilde{k}_\alpha^*$. It is clear that $k_{ijp} \subset \Omega$, and the union of the fields k_{ijp} coincides with the maximal Abelian extension of the field $\widetilde{k}_\alpha$ contained in Ω. Then

$$\overline{N}_{\widetilde{H}_\alpha} C_{S,\alpha} = \bigcap_p \bigcap_j \bigcap_i J_\alpha^{p^j} V_{S_i,\alpha} \widetilde{k}_\alpha^* / V_{S,\alpha} \widetilde{k}_\alpha^*.$$

Compute $\bigcap_i J^{p^j} V_{S_i,\alpha} \widetilde{k}_\alpha^* / V_{S,\alpha} \widetilde{k}_\alpha^*$ for p and j fixed. Note that

$$\bigcap_i J_\alpha^{p^j} V_{S_i,\alpha} \widetilde{k}_\alpha^* / V_{S,\alpha} \widetilde{k}_\alpha^* = \varprojlim J_\alpha^{p^j} V_{S_i,\alpha} \widetilde{k}_\alpha^* / V_{S,\alpha} \widetilde{k}_\alpha^*,$$

where the projective limit is taken with respect to the embeddings, or more precisely, for $i_1 \le i_2$, the natural mapping

$$J_\alpha^{p^j} V_{S_{i_2},\alpha} \widetilde{k}_\alpha^* / V_{S,\alpha} \widetilde{k}_\alpha^* \longrightarrow J_\alpha^{p^j} V_{S_{i_1},\alpha} \widetilde{k}_\alpha^* / V_{S,\alpha} \widetilde{k}_\alpha^*$$

is injective.

Consider the exact sequence

$$1 \longrightarrow C_{S,k_\alpha}^{p^j} \xrightarrow{\tau} \varprojlim J_\alpha^{p^j} V_{S_i,\alpha} \widetilde{k}_\alpha^* / V_{S,\alpha} \widetilde{k}_\alpha^* \longrightarrow \varprojlim \Big(\prod_{\mathfrak{p} \in S \setminus S_i} V_{\mathfrak{p},\alpha} / \widetilde{k}_\alpha^* J_\alpha^{p^j} \cap \prod_{\mathfrak{p} \in S \setminus S_i} V_{\mathfrak{p},\alpha} \Big).$$

Since $\varprojlim \prod_{\mathfrak{p} \in S \setminus S_i} V_{\mathfrak{p},\alpha} = 1$, Lemma A.3.6 implies that τ is an isomorphism. Therefore,

$$\overline{N}_{\widetilde{H}_\alpha} C_{S,\alpha} = \bigcap_p \bigcap_j C_{S,\alpha}^{p^j}$$

and, thus, it is a division group. □

Note that similar arguments show that $\overline{N}_{\widetilde{H}_\alpha} \overline{C}_{S,\alpha}$ also is a division group.

THEOREM A.3.4 (Tate's global duality). *If* A *is an* F_S*-module finitely generated as an Abelian group, then the mapping*

$$\psi_S \colon H^2(F_S, A) \longrightarrow H^0(F_S, \mathrm{Hom}(A, C_S))'$$

is an isomorphism.

PROOF. Let $1 \longrightarrow A_1 \longrightarrow A_2 \longrightarrow A \longrightarrow 1$ be a representation of A as the quotient module of a lattice. Consider the diagram with natural homomorphisms

$$\begin{array}{ccccc}
H^2(F_S, A_1) & \longrightarrow & H^2(F_S, A_2) & \longrightarrow & H^2(F_S, A) \\
\downarrow{\scriptstyle \psi_S} & & \downarrow{\scriptstyle \psi_S} & & \downarrow{\scriptstyle \psi_S} \\
H^0(F_S, \mathrm{Hom}(A_1, C_S))' & \longrightarrow & H^0(F_S, \mathrm{Hom}(A_2, C_S))' & \longrightarrow & H^0(F_S, \mathrm{Hom}(A, C_S))'
\end{array}$$

$$\begin{array}{ccccc}
\xrightarrow{\ \delta\ } & H^3(F_S, A_1) & \longrightarrow & H^3(F_S, A_2) \\
& \downarrow{\scriptstyle \psi_S} & & \downarrow{\scriptstyle \psi_S} \\
\xrightarrow{\ \overline{\delta}_1\ } & H^{-1}(F_S, \mathrm{Hom}(A_1, C_S))' & \longrightarrow & H^{-1}(F_S, \mathrm{Hom}(A_2, C_S))'.
\end{array}$$

Note that, by Lemma A.3.12, one can replace the module C_S in the diagram with the module $\overline{C}_S = \varinjlim \overline{C}_{S,\alpha}$.

We describe the map $\overline{\delta}_1$ in more detail. Let $x \in H^{-1}(F_S, \mathrm{Hom}(A_1, \overline{C}_S))$. Since $\mathrm{Hom}(A_1, \overline{C}_{S,\alpha})$ is a compact group for all $\alpha \in \Gamma$, we have

$$H^{-1}(F_S, \mathrm{Hom}(A_1, \overline{C}_S)) = \varprojlim H^{-1}(\widetilde{F}_\alpha, \mathrm{Hom}(A_1, N_{\widetilde{H}_\alpha} \overline{C}_{S,\alpha})).$$

The exact sequence

$$1 \longrightarrow \mathrm{Hom}(A, N_{\widetilde{H}_\alpha} \overline{C}_{S,\alpha}) \longrightarrow \mathrm{Hom}(A_2, N_{\widetilde{H}_\alpha} \overline{C}_{S,\alpha}) \longrightarrow \mathrm{Hom}(A_1, N_{\widetilde{H}_\alpha} \overline{C}_{S,\alpha}) \longrightarrow 1$$

induces the exact cohomology sequence. Then for all α the homomorphism

$$\delta_\alpha \colon H^{-1}(\widetilde{F}_\alpha, \mathrm{Hom}(A_1, N_{\widetilde{H}_\alpha} \overline{C}_{S,\alpha})) \longrightarrow H^0(\widetilde{F}_\alpha, \mathrm{Hom}(A, N_{\widetilde{H}_\alpha} \overline{C}_{S,\alpha}))$$

is well defined. Moreover, there exists a natural homomorphism

$$\varepsilon_\alpha \colon H^0(\widetilde{F}_\alpha, \mathrm{Hom}(A, N_{\widetilde{H}_\alpha} \overline{C}_{S,\alpha})) \longrightarrow H^0(\widetilde{F}_\alpha, \mathrm{Hom}(A, \overline{C}_{S,\alpha})).$$

Let $\delta_{1\alpha}$ be the composition of the homomorphisms δ_α and ε_α. It is immediate that for $\alpha \le \beta$ the relation $\mathrm{def}_0 \, \delta_{1\beta} = \delta_{1\alpha} \, \mathrm{def}_{-1}$ holds. Therefore, the passage to the projective limit yields a homomorphism

$$\delta_1 \colon \varprojlim H^{-1}(\widetilde{F}_\alpha, \mathrm{Hom}(A_1, \overline{C}_{S,\alpha})) \longrightarrow \varprojlim H^0(\widetilde{F}_\alpha, \mathrm{Hom}(A, \overline{C}_{S,\alpha})).$$

Then $\overline{\delta}_1$ is the homomorphism dual to δ_1.

The commutativity of the diagram is verified directly. We verify the exactness of the bottom row by considering the dual sequence.

Let $x_1 \in \varprojlim H^{-1}(\widetilde{F}_\alpha, \mathrm{Hom}(A_1, \overline{C}_{S,\alpha}))$ and $\delta_1(x_1) = 0$. Note that $x_{1\alpha} = \mathrm{pr}_\alpha \, x_1 \in H^{-1}(\widetilde{F}_\alpha, N_{\widetilde{H}_\alpha} \mathrm{Hom}(A_1, \overline{C}_{S,\alpha}))$, and from the relation $\delta_{1\alpha}(x_{1\alpha}) = 0$ for all $\alpha \in \Gamma$ it follows that there exist elements $x_{2\alpha} \in H^{-1}(\widetilde{F}_\alpha, \mathrm{Hom}(A_2, N_{\widetilde{H}_\alpha} \overline{C}_{S,\alpha}))$ such that $\omega_\alpha(x_{2\alpha}) = x_{1\alpha}$, where

$$\omega_\alpha \colon H^{-1}(\widetilde{F}_\alpha, \mathrm{Hom}(A_2, N_{\widetilde{H}_\alpha} \overline{C}_{S,\alpha})) \longrightarrow H^{-1}(\widetilde{F}_\alpha, \mathrm{Hom}(A_1, N_{\widetilde{H}_\alpha} \overline{C}_{S,\alpha}))$$

is a natural homomorphism. One can show that the elements $x_{2\alpha}$ can be chosen in such a way that

$$\prod_{\alpha} x_{2\alpha} \in H^{-1}(F_S, \operatorname{Hom}(A_2, \overline{C}_S))$$

(the proof is similar to that of Lemma A.3.6).

Consider the exact sequence

$$1 \longrightarrow \operatorname{Hom}_{F_S}(A, \overline{C}_S) \xrightarrow{\overline{\eta}} \operatorname{Hom}_{F_S}(A_2, \overline{C}_S) \xrightarrow{\overline{\theta}} \operatorname{Hom}_{F_S}(A_1, \overline{C}_S)$$

and let

$$\eta\colon H^0(F_S, \operatorname{Hom}(A, \overline{C}_S)) \longrightarrow H^0(F_S, \operatorname{Hom}(A_2, \overline{C}_S))$$

and

$$\theta\colon H^0(F_S, \operatorname{Hom}(A_2, \overline{C}_S)) \longrightarrow H^0(F_S, \operatorname{Hom}(A_1, \overline{C}_S))$$

be the homomorphisms induced by $\overline{\eta}$ and $\overline{\theta}$, respectively.

Let $y_1 \in H^0(F_S, \operatorname{Hom}(A, \overline{C}_S))$, where $\overline{y}_1$ is a representative of y_1 in the group $\operatorname{Hom}_{F_S}(A, \overline{C}_S)$ and $\eta(y_1) = 0$. Then $\overline{\eta}(\overline{y}_1) \in N_{\widetilde{F}_\alpha} \operatorname{Hom}(A_2, N_{\widetilde{H}_\alpha} \overline{C}_{S,\alpha})$, i.e., $\overline{\eta}(\overline{y}_1) \in \operatorname{Hom}_{\widetilde{F}_\alpha}(A_2, \overline{N}_{\widetilde{H}_\alpha} \overline{C}_{S,\alpha})$, so $\overline{y}_1 \in \operatorname{Hom}_{\widetilde{F}_\alpha}(A, \overline{N}_{\widetilde{H}_\alpha} \overline{C}_{S,\alpha})$ and there exists an element $y_2 \in H^{-1}(F_S, \operatorname{Hom}(A_1, \overline{C}_S))$ such that $\delta_1(y_2) = y_1$.

Now we assume that $z_1 \in H^0(F_S, \operatorname{Hom}(A_2, \overline{C}_S))$, $\overline{z}_1$ is a representative of z_1 in $\operatorname{Hom}_{F_S}(A_2, \overline{C}_S)$ and $\theta(z_1) = 0$. Then $\overline{\theta}(\overline{z}_1) \in N_{\widetilde{F}_\alpha} \operatorname{Hom}(A_1, \overline{N}_{\widetilde{H}_\alpha} \overline{C}_{S,\alpha})$ for all α, and in the group $N_{\widetilde{F}_\alpha} \operatorname{Hom}(A_2, \overline{N}_{\widetilde{H}_\alpha} \overline{C}_{S,\alpha})$ there exists an element z_2 such that $\overline{\theta}(z_2) = \overline{\theta}(\overline{z}_1)$. Note that the elements $\overline{z}_1$ and $\overline{z}_1 z_2^{-1}$ lie in the same class, and $\overline{\theta}(\overline{z}_1 z_2^{-1}) = 1$. Therefore, there exists an element $z_3 \in \operatorname{Hom}_{F_S}(A, \overline{C}_S)$ such that $\overline{\eta}(z_3) = \overline{z}_1 z_2^{-1}$. It remains to observe that the isomorphism of the groups $H^2(F_S, A)$ and $H^0(F_S, \operatorname{Hom}(A, \overline{C}_S))'$ follows directly from the exactness of the bottom row and the commutativity of the diagram. The theorem is proved. □

6°. In this section we consider a relationship between the global and local duality. Let k be a field of algebraic numbers, S a set of points of k, Ω the maximal algebraic extension of k unramified outside S, $k_{\mathfrak{p}}$ the completion of k at the point $\mathfrak{p}$, $K_{\mathfrak{p}}$ an algebraic extension of $k_{\mathfrak{p}}$ containing Ω, F_S and $F_{\mathfrak{p}}$ the Galois groups of the extensions Ω/k and $K_{\mathfrak{p}}/k_{\mathfrak{p}}$, respectively.

If A is an F_S-module, then its structure induces the structure of an $F_{\mathfrak{p}}$-module on A. Indeed, the embedding $\Omega \longrightarrow K_{\mathfrak{p}}$ induces the map $\delta\colon F_{\mathfrak{p}} \longrightarrow F_S$. By setting $a^f = a^{\delta(f)}$ for $f \in F_{\mathfrak{p}}$, we convert A into an $F_{\mathfrak{p}}$-module.

Consider the map δ in more detail. Let $\{k\alpha/k\}_{\alpha\in\Gamma_1}$ be an inductive system of finite normal extensions with Galois groups F_α that determines the field Ω and let $\{k_{\mathfrak{p},\beta}/k_{\mathfrak{p}}\}_{\beta\in\Gamma_2}$ be a similar system of local extensions with groups F_β that determines the field $K_{\mathfrak{p}}$.

We fix the point $\mathfrak{P}$ of Ω over $\mathfrak{p}$ that is fully determined by the choice of points $\mathfrak{p}_\alpha$ of the fields k_α over $\mathfrak{p}$, and if $\alpha_1 \le \alpha_2$, i.e., $k_{\alpha_1} \subseteq k_{\alpha_2}$, then the point $\mathfrak{p}_{\alpha_2}$ lies over $\mathfrak{p}_{\alpha_1}$.

For any $\alpha \in \Gamma_1$ an embedding $k_\alpha \longrightarrow k_{\alpha,\mathfrak{p}_\alpha}$ exists. The Galois group $F_{\mathfrak{p}_\alpha}$ of the extension $k_{\alpha,\mathfrak{p}_\alpha}/k_{\mathfrak{p}}$ (the decomposition group of the divisor $\mathfrak{p}_\alpha$) can be embedded in F_α. Let $\beta = \Phi(\alpha) \in \Gamma_2$ be the index corresponding to the field $k_{\alpha,\mathfrak{p}_\alpha}$, i.e., the mapping $\Phi\colon F_1 \longrightarrow F_2$ is well defined, and for every $\alpha \in \Gamma_1$ the mapping $F_{\Phi(\alpha)} \longrightarrow F_\alpha$ is given. We observe that for $\alpha_1 \leq \alpha_2$, the diagram

$$\begin{array}{ccc} F_{\Phi(\alpha_2)} & \longrightarrow & F_{\alpha_2} \\ \downarrow & & \downarrow \\ F_{\Phi(\alpha_1)} & \longrightarrow & F_{\alpha_1} \end{array}$$

is commutative. Therefore, the homomorphism

$$\delta\colon F_{\mathfrak{p}} = \varprojlim F_\beta \longrightarrow \varprojlim F_\alpha = F_S$$

is well defined.

For $i \geq 1$ and an arbitrary F_S-module A the homomorphism δ induces a mapping

$$\varepsilon_{\mathfrak{p}}: H^i(F_S, A) \longrightarrow H^i(F_{\mathfrak{p}}, A).$$

Indeed, the embedding $F_{\Phi(\alpha)} \longrightarrow F_\alpha$ induces the homomorphism

$$\varepsilon_\alpha\colon H^i(F_\alpha, A^{H_\alpha}) \longrightarrow H^i(F_{\Phi(\alpha)}, A^{H_{\Phi(\alpha)}})$$

that is the composition of the restriction homomorphism

$$H^i(F_\alpha, A^{H_\alpha}) \longrightarrow H^i(F_{\Phi(\alpha)}, A^{H_\alpha})$$

and the natural homomorphism

$$H^i(F_{\Phi(\alpha)}, A^{H_\alpha}) \longrightarrow H^i(F_{\Phi(\alpha)}, A^{H_{\Phi(\alpha)}}),$$

where H_α and $H_{\Phi(\alpha)}$ are the kernels of the natural mappings $F_S \longrightarrow F_\alpha$ and $F_{\mathfrak{p}} \longrightarrow F_{\Phi(\alpha)}$, respectively.

By Lemma A.3.2, restriction commutes with lifting, so the passage to the inductive limit yields a homomorphism

$$\varepsilon_{\mathfrak{p}}\colon H^i(F_S, A) = \varinjlim H^i(F_\alpha, A^{H_\alpha}) \longrightarrow \varinjlim H^i(F_\beta, A^{H_\beta}) = H^i(F_{\mathfrak{p}}, A),$$

where H_β is the kernel of the mapping $F_{\mathfrak{p}} \longrightarrow F_\beta$.

Now let A be a finite F_S-module, $H_{\overline{\alpha}}$ the subgroup of the group F_S consisting of all elements which act trivially on A, $H_{\overline{\beta}}$ the same subgroup of the group $F_{\mathfrak{p}}$, let $k_{\overline{\alpha}} = \Omega^{H_{\overline{\alpha}}}$ and $k_{\overline{\beta}} = K_{\mathfrak{p}}^{H_{\overline{\beta}}}$ be the fields of invariants for $H_{\overline{\alpha}}$ and $H_{\overline{\beta}}$. We set

$$\begin{gathered} \mathrm{Gal}(k_{\overline{\alpha}}/k) = G \quad \text{and} \quad \mathrm{Gal}(k_{\overline{\beta}}/k_{\mathfrak{p}}) = G_{\mathfrak{p}}; \\ \widetilde{k}_\alpha = k_{\overline{\alpha}} k_\alpha, \qquad \widetilde{k}_\beta = k_{\overline{\beta}} k_\beta; \\ \widetilde{F}_\alpha = \mathrm{Gal}(\widetilde{k}_\alpha/k) \quad \text{and} \quad \widetilde{F}_\beta = \mathrm{Gal}(\widetilde{k}_\beta/k_{\mathfrak{p}}). \end{gathered}$$

For $i = 0, -1$, one can define a mapping

$$\theta_S\colon \prod_{\mathfrak{p} \in S} H^i(F_{\mathfrak{p}}, \mathrm{Hom}(A, K_{\mathfrak{p}}^*)) \longrightarrow H^i(F_S, \mathrm{Hom}(A, C_S))$$

in the following way. The embedding $\widetilde{F}_{\Phi(\alpha)} \to \widetilde{F}_\alpha$ induces the homomorphism

$$\theta_{S,\alpha}\colon \prod_{\mathfrak{p}\in S} H^i(\widetilde{F}_{\Phi(\alpha)}, \mathrm{Hom}(A, \widetilde{k}^*_{\Phi(\alpha)})) \xrightarrow{\theta_{1\alpha}} \prod_{\mathfrak{p}\in S} H^i(\widetilde{F}_\alpha, \mathrm{Hom}(A, J_{\widetilde{k}_\alpha,\mathfrak{p}}))$$
$$\to H^i(\widetilde{F}_\alpha, \mathrm{Hom}(A, J_{\widetilde{k}_\alpha,S})) \to H^i(\widetilde{F}_\alpha, \mathrm{Hom}(A, C_{S,\alpha}))$$

which is a composition mapping for the sequence written above, where $\theta_{1\alpha}$ is an isomorphism (see Lemma A.3.4).

One can verify directly that for $\alpha_1 \le \alpha_2$, the following diagram commutes:

$$\begin{array}{ccc}
\prod_{\mathfrak{p}\in S} H^i(\widetilde{F}_{\Phi(\alpha_2)}, \mathrm{Hom}(A, \widetilde{k}^*_{\Phi(\alpha_2)})) & \xrightarrow{\theta_{S,\alpha_2}} & H^i(\widetilde{F}_{\alpha_2}, \mathrm{Hom}(A, C_{S,\alpha_2})) \\
\downarrow{\scriptstyle \mathrm{def}} & & \downarrow{\scriptstyle \mathrm{def}} \\
\prod_{\mathfrak{p}\in S} H^i(\widetilde{F}_{\Phi(\alpha_1)}, \mathrm{Hom}(A, \widetilde{k}^*_{\Phi(\alpha_1)})) & \xrightarrow{\theta_{S,\alpha_1}} & H^i(\widetilde{F}_{\alpha_1}, \mathrm{Hom}(A, C_{S,\alpha_1})).
\end{array}$$

Therefore, the passage to the projective limit yields the homomorphism θ_S.

Let $K_\mathfrak{p}$ be the algebraic closure $\overline{k}_\mathfrak{p}$ of $k_\mathfrak{p}$ and $\overline{F}_\mathfrak{p} = \mathrm{Gal}(\overline{k}_\mathfrak{p}/k_\mathfrak{p})$. As was shown above, if A is a finite F_S-module, hence an $\overline{F}_\mathfrak{p}$-module, then there exist the duality isomorphisms

$$\varphi\colon H^2(F_S, A) \to H^0(F_S, \mathrm{Hom}(A, C_S))',$$
$$\varphi_\mathfrak{p}\colon H^2(\overline{F}_\mathfrak{p}, A) \to H^0(\overline{F}_\mathfrak{p}, \mathrm{Hom}(A, \overline{k}^*_\mathfrak{p}))'.$$

Denote by ε_S the natural homomorphism

$$H^2(F_S, A) \to \sum_{\mathfrak{p}\in S}^{\oplus} H^2(\overline{F}_\mathfrak{p}, A).$$

PROPOSITION A.3.1. *The diagram*

$$\begin{array}{ccc}
H^2(F_S, A) & \xrightarrow{\varepsilon_S} & \sum_{\mathfrak{p}\in S}^{\oplus} H^2(\overline{F}_\mathfrak{p}, A) \\
\downarrow{\scriptstyle \varphi} & & \downarrow \\
H^0(F_S, \mathrm{Hom}(A, C_S))' & \xrightarrow{\overline{\theta}_S} & \left(\prod_{\mathfrak{p}\in S} H^0(\overline{F}_\mathfrak{p}, \mathrm{Hom}(A, \overline{k}^*_\mathfrak{p}))\right)'
\end{array}$$

is commutative, where $\overline{\theta}_S$ is the homomorphism dual to θ_S.

PROOF. It is clear that the following diagram commutes:

$$\begin{array}{ccc}
H^2(\widetilde{F}_\alpha, A) & \longrightarrow & \sum_{\mathfrak{p}\in S}^{\oplus} H^2(\widetilde{F}_{\Phi(\alpha)}, A) \\
\downarrow & & \downarrow \\
H^0(\widetilde{F}_\alpha, \mathrm{Hom}(A, C_{S,\alpha}))' & \longrightarrow & \prod_{\mathfrak{p}\in S} H^0(\widetilde{F}_{\Phi(\alpha)}, \mathrm{Hom}(A, \widetilde{k}^*_{\Phi(\alpha)}))'.
\end{array}$$

By Lemma A.3.5, lifting is dual to deflation, therefore the passage to the limit completes the proof of the proposition.

We transform the mapping $\overline{\theta}_S$. Note that $\mathrm{Hom}(A, C_{S,\alpha})$ is a compact group, so the relation

$$H^0(F_S, \mathrm{Hom}(A, C_S)) = \mathrm{Hom}_{F_S}(A, C_S)/\overline{N}_{F_S} \mathrm{Hom}_{F_S}(A, C_S)$$

holds. The group $H^0(\overline{F}_{\mathfrak{p}}, \mathrm{Hom}(A, \overline{k}^*_{\mathfrak{p}}))$ coincides with the group $\mathrm{Hom}_{\overline{F}_{\mathfrak{p}}}(A, \overline{k}^*_{\mathfrak{p}})$ or $\mathrm{Hom}_{G_{\mathfrak{p}}}(A, k^*_{\overline{\beta}})$ which, in turn, is isomorphic to the group $\mathrm{Hom}_G(A, J_{k_{\overline{\alpha}},\mathfrak{p}})$. Thus, one can identify the group $\prod_{\mathfrak{p}\in S} H^0(\overline{F}_{\mathfrak{p}}, \mathrm{Hom}(A, \overline{A}^*_{\mathfrak{p}}))$ with $\mathrm{Hom}_G(A, J_{k_{\overline{\alpha}},S})$ and the homomorphism $\overline{\theta}_S$ with the mapping dual to the natural homomorphism

$$\mathrm{Hom}_G(A, J_{k_{\overline{\alpha}},S}) \longrightarrow \mathrm{Hom}_G(A, C_{S,\overline{\alpha}})/\overline{N}_{F_S} \mathrm{Hom}_{F_S}(A, C_S).$$

Let $k_{\alpha,S}$ be the subgroup of the group k^*_α consisting of elements x such that in the decomposition of the principal divisor (x) into the prime factors only the divisors of the set S occur; let $E_{\alpha,S}$ be the projection of the idèle group of the field k_α to the S-components. For arbitrary $\alpha \in \Gamma_1$ an embedding $k_{\alpha,S} \longrightarrow E_{\alpha,S}$ exists; we denote its cokernel by $D_{\alpha,S}$. It is clear that $D_{\alpha,S}$ is a subgroup of the group $C_{S,\alpha}$.

For $\alpha_1 \leq \alpha_2$ there exist embeddings

$$k_{\alpha_1,S} \longrightarrow k_{\alpha_2,S}, \qquad E_{\alpha_1,S} \longrightarrow E_{\alpha_2,S}, \qquad D_{\alpha_1,S} \longrightarrow D_{\alpha_2,S};$$

therefore the passage to the inductive limit with respect to the embeddings in the exact sequence

$$1 \longrightarrow k_{\alpha,S} \longrightarrow E_{\alpha,S} \longrightarrow D_{\alpha,S} \longrightarrow 1$$

yields the exact sequence of F_S-modules

$$1 \longrightarrow \Omega_S \longrightarrow E_S \longrightarrow D_S \longrightarrow 1.$$

We observe that if n is a natural number and S contains all the divisors of n and infinite divisors, then Ω_S is a group with division by n. Indeed, let $x \in \Omega_S$. Then $x \in k_{\alpha,S}$ for some $\alpha \in \Gamma_1$. It is clear that $k_\alpha(\sqrt[n]{x}) \subset \Omega$, i.e., x is the nth power in Ω_S.

Let $\widehat{A}$ be the character group of $\mathrm{Hom}(A, \Omega^*)$. The following homomorphism

$$\eta_S\colon H^1(F_S, \widehat{A}) \longrightarrow \prod_{\mathfrak{p}\in S} H^1(\overline{F}_{\mathfrak{p}}, \widehat{A})$$

is well defined. □

THEOREM A.3.5 (J. Tate). *Let A be a finite F_S-module and let S contain all infinite divisors and all prime divisors of order equal to the order of the group A. Then the kernel of the mapping*

$$\varepsilon_S\colon H^2(F_S, A) \longrightarrow \sum_{\mathfrak{p}\in S}^{\oplus} H^2(\overline{F}_{\mathfrak{p}}, A)$$

is dual to the kernel of η_S.

PROOF. The exact sequence of F_S-modules

$$1 \longrightarrow \widehat{A} = \operatorname{Hom}(A, \Omega_S) \longrightarrow \operatorname{Hom}(A, E_S) \longrightarrow \operatorname{Hom}(A, D_S) \longrightarrow 1$$

induces the cohomology sequence

$$\operatorname{Hom}_{F_S}(A, E_S) \longrightarrow \operatorname{Hom}_{F_S}(A, D_S) \longrightarrow H^1(F_S, \widehat{A}) \xrightarrow{\nu} H^1(F_S, \operatorname{Hom}(A, E_S)).$$

We note that for arbitrary $\alpha \in \Gamma_1$ the kernels of the homomorphisms

$$H^1(\widetilde{F}_{\alpha_1}, \widehat{A}) \to \prod_{\mathfrak{p} \in S} H^1(\widetilde{F}_{\Phi(\alpha_1)}, \widehat{A})$$

and

$$H^1(\widetilde{F}_{\alpha_1}, \widehat{A}) \to H^1(\widetilde{F}_{\alpha_1}, \operatorname{Hom}(A, E_{\alpha_1, S}))$$

coincide; therefore, the homomorphisms η_S and ν have the same property.

We show that there exists a natural homomorphism ω: $\operatorname{Ker} \varepsilon_S \to (\operatorname{Ker}_{\eta_S})'$. Consider the commutative diagram

(1)

$$\begin{array}{ccccccc}
0 & \longrightarrow & \operatorname{Ker} \varepsilon_S & \longrightarrow & H^2(F_S, A) & \longrightarrow & \sum\limits_{\mathfrak{p} \in S}^{\oplus} H^2(\overline{F}_{\mathfrak{p}}, A) \\
 & & & & \downarrow{\scriptstyle \varphi} & & \downarrow{\scriptstyle \varphi_S} \\
 & & & & H^0(F_S, \operatorname{Hom}(A, C_S))' & \longrightarrow & \left(\prod\limits_{\mathfrak{p} \in S} H^0(\overline{F}_{\mathfrak{p}}, \operatorname{Hom}(A, \overline{k}_{\mathfrak{p}}^*))\right)' \\
 & & & & \downarrow{\scriptstyle \tau_1} & & \downarrow{\scriptstyle \tau_2} \\
0 & \longrightarrow & (\operatorname{Ker} \eta_S)' & \longrightarrow & \operatorname{Hom}_{F_S}(A, C_S)' & \longrightarrow & \operatorname{Hom}_{F_S}(A, E_S)'.
\end{array}$$

The mappings φ, φ_S, τ_1, τ_2 are injective, so the injective homomorphism

$$\omega\colon \operatorname{Ker} \varepsilon_S \longrightarrow (\operatorname{Ker} \eta_S)'$$

is well defined.

First consider the case where F_S acts trivially on the modules A and $\widehat{A}$. It is clear that it suffices to set $A = \mathbb{Z}/n\mathbb{Z}$.

The commutative diagram

$$\begin{array}{ccccccccc}
0 & \longrightarrow & \mathbb{Z}/n\mathbb{Z} & \longrightarrow & \Omega_S & \longrightarrow & \Omega_S & \longrightarrow & 1 \\
 & & \downarrow{\scriptstyle \mathrm{id}} & & \downarrow & & \downarrow & & \\
0 & \longrightarrow & \mathbb{Z}/n\mathbb{Z} & \longrightarrow & \overline{k}_{\mathfrak{p}}^* & \longrightarrow & \overline{k}_{\mathfrak{p}}^* & \longrightarrow & 1
\end{array}$$

induces the commutative diagram

$$\begin{array}{ccccccc}
H^1(F_S, \Omega_S) & \longrightarrow & H^1(F_S, \Omega_S) & \longrightarrow & H^2(F_S, \mathbb{Z}/n\mathbb{Z}) & \longrightarrow & H^2(F_S, \Omega_S) \\
 & & \downarrow & & \downarrow{\scriptstyle \varepsilon_S} & & \downarrow{\scriptstyle \mu} \\
 & & 0 & \longrightarrow & \sum\limits_{\mathfrak{p} \in S}^{\oplus} H^2(\overline{F}_{\mathfrak{p}}, \mathbb{Z}/n\mathbb{Z}) & \longrightarrow & \sum\limits_{\mathfrak{p} \in S}^{\oplus} H^2(\overline{F}_{\mathfrak{p}}, \overline{k}^*).
\end{array}$$

We observe that μ is the composition of the homomorphism

$$\mu_1 \colon H^2(F_S, \Omega_S) \to H^2(F_S, E_S) \approx \sum_{\mathfrak{p}\in S}^{\oplus} H^2(F_\mathfrak{p}, \Omega_{\overline{\mathfrak{p}}}^*)$$

and the collection of the lifting homomorphisms

$$\lambda_\mathfrak{p} \colon H^2(F_\mathfrak{p}, \Omega_{\overline{\mathfrak{p}}}^*) \to H^2(\overline{F}_\mathfrak{p}, \overline{k}_\mathfrak{p}^*).$$

Since $\lambda_\mathfrak{p}$ is an injective mapping, the injectivity of μ_1 implies the injectivity of μ. We show that $D_S = C_S$. Indeed, $D_{\alpha,S} \subset C_{S,\alpha}$ for all $\alpha \in \Gamma_1$, so $D_S \subset C_S$. Let $x \in C_S$ and let $x_1 \in J_S$ be the inverse image of x. Then $x_1 \in J_{k_\alpha,S}$ for some $\alpha \in \Gamma_1$. If k_{α_2} is the maximal Abelian unramified extension of the field k_α, then in the field k_{α_2} there exists an element x_2 such that $x_2^{-1}x \in E_S$ (in the field k_{α_2} any divisor of k_α is principal), so $D_S = C_S$ and $H^1(F_S, D_S) = 0$. Thus we have

$$\operatorname{Ker} \varepsilon_S \approx H^1(F_S, \Omega_S)/nH^1(F_S, \Omega_S).$$

Now compute the group $H^1(F_S, \Omega_S)$ in the exact sequence

$$\Omega_S^{F_S} \to E_S^{F_S} \to D_S^{F_S} \to H^1(F_S, \Omega_S) \to H^1(F_S, E_S).$$

We note that

$$H^1(F_S, E_S) = \varinjlim H^1(F_\alpha, E_{\alpha,S}) = \varinjlim \prod_{\mathfrak{p}\in S} H^1(F_\alpha, J_{k_\alpha,\mathfrak{p}}).$$

Therefore, the sequence is of the form

$$1 \to k_S \to E_{k,S} \to C_{k,S} \to H^1(F_S, \Omega_S) \to 0,$$

i.e., $H^1(F_S, \Omega_S) = \mathrm{Cl}/\mathrm{Cl}_S$, where Cl is the divisor class group of the field k and Cl_S is the subgroup generated by the divisors from the set S.

Consider the group $\operatorname{Ker} \eta_S$. Note that from class field theory it follows that $\operatorname{Ker} \eta_S$ coincides with the kernel of the mapping dual to the natural homomorphism $E_{k,S}/E_{k,S}^n \to J_{k,S}/J_{k,S}^n k^*$. It is clear that $\operatorname{Ker} \eta_S \approx (\mathrm{Cl}/\mathrm{Cl}^n \mathrm{Cl}_S)'$, i.e., $\operatorname{Ker} \varepsilon_S \approx (\operatorname{Ker} \eta_S)'$.

Thus, the following sequence is exact:

$$\prod_{\mathfrak{p}\in S} H^0(\overline{F}_\mathfrak{p}, \widehat{\mathbb{Z}/n\mathbb{Z}}) \to H^0(F_S, \operatorname{Hom}(\mathbb{Z}/n\mathbb{Z}, C_S)) \to \operatorname{Ker} \eta_S \to 0.$$

It is clear that the sequence in which we replace the group F_S with H_α and $\overline{F}_\mathfrak{p}$ with $H_{\Phi(\alpha)}$ is exact. For $\alpha_1 \le \alpha_2$, we have the commutative diagram

$$\begin{array}{ccccccc}
\prod_{\mathfrak{p}\in S} H^0(H_{\Phi(\alpha_2)}, \widehat{\mathbb{Z}/n\mathbb{Z}}) & \longrightarrow & H^0(H_{\alpha_2}, \operatorname{Hom}(\mathbb{Z}/n\mathbb{Z}, C_S)) & \longrightarrow & \operatorname{Ker} \eta_{S,\alpha_2} & \longrightarrow & 0 \\
\downarrow & & \downarrow & & \downarrow{\scriptstyle \mu_{\alpha_1}^{\alpha_2}} & & \\
\prod_{\mathfrak{p}\in S} H^0(H_{\Phi(\alpha_1)}, \widehat{\mathbb{Z}/n\mathbb{Z}}) & \longrightarrow & H^0(H_{\alpha_1}, \operatorname{Hom}(\mathbb{Z}/n\mathbb{Z}, C_S)) & \longrightarrow & \operatorname{Ker} \eta_{S,\alpha_1} & \longrightarrow & 0,
\end{array}$$

where the vertical mappings are induced by the norm mapping $N_{H_{\alpha_1}/H_{\alpha_2}}$. Passing to the projective limit, we have

$$\varprojlim H^0(H_\alpha, \mathrm{Hom}(\mathbb{Z}/n\mathbb{Z}, C_S)) = 0,$$

therefore $\varprojlim \mathrm{Ker}\, \eta_{S,\alpha} = 0$.

Now for $\alpha \in \Gamma_1$ consider the exact sequences

$$\prod_{\mathfrak{p}\in S} \mathrm{Hom}_{H_{\Phi(\alpha)}}(\mathbb{Z}/n\mathbb{Z}, k_{\mathfrak{p}}^*) \longrightarrow \mathrm{Hom}_{H_\alpha}(\mathbb{Z}/n\mathbb{Z}, C_S) \longrightarrow \mathrm{Ker}\, \eta_{S,\alpha} \longrightarrow 0.$$

Passing to the projective limit with respect to the norm mappings and the homomorphisms $\mu_{\alpha_1}^{\alpha_2}$, we obtain

$$\varprojlim \prod_{\mathfrak{p}\in S} \widehat{\mathbb{Z}/n\mathbb{Z}} \longrightarrow \varprojlim \mathrm{Hom}_{H_\alpha}(\mathbb{Z}/n\mathbb{Z}, C_S) \longrightarrow \varprojlim \mathrm{Ker}\, \eta_{S,\alpha} \longrightarrow 0.$$

Since $\varprojlim \prod \widehat{\mathbb{Z}/n\mathbb{Z}} = 0$, we have $\varprojlim \mathrm{Hom}(\mathbb{Z}/n\mathbb{Z}, C_{S,\alpha}) = 0$, i.e., for each α we have $\overline{N}_{H_\alpha} \mathrm{Hom}(\mathbb{Z}/n\mathbb{Z}, C_{S,\alpha}) = 0$.

Now consider the general case. Clearly we have the relation

$$\overline{N}_{F_S} \mathrm{Hom}(A, C_S) = N_G \overline{N}_H \mathrm{Hom}_H(A, C_S) = 1.$$

Therefore, in (1) τ_1 is an isomorphism. Hence it follows that ω is also an isomorphism. The theorem is proved. □

COROLLARY. *If A is a finite module, then the diagram*

$$\begin{array}{ccc} H^2(F_S, A) & \longrightarrow & \sum\limits_{\mathfrak{p}\in S}^{\oplus} H^2(\overline{F}_{\mathfrak{p}}, A) \\ \downarrow & & \downarrow \\ \mathrm{Hom}_G(A, C_{S,\overline{\alpha}})' & \longrightarrow & \mathrm{Hom}_G(A, J_{k_{\overline{\alpha}},S})' \end{array}$$

is commutative.

In conclusion, we consider the relationship between the duality mappings under an increase in the number of points of the set S. Let S and T be sets of prime divisors of the field k with $S \subset T$, let Ω and Δ be maximal algebraic extensions of the field k unramified outside S and T, respectively, $F_S = \mathrm{Gal}(\Omega/k)$, $F_T = \mathrm{Gal}(\Delta/k)$.

Let A be a finite F_S-module (hence an F_T-module) and H the subgroup of F_S consisting of elements acting trivially on A, $K = \Omega^H$, $G = \mathrm{Gal}(K/k)$.

Note that $\Omega \subset \Delta$; therefore, the homomorphism $F_T \longrightarrow F_S$ that induces the lifting homomorphism $\lambda\colon H^2(F_S, A) \longrightarrow H^2(F_T, A)$ is well defined. The natural mapping $C_{K,S} \longrightarrow C_{K,T}$ induces the mapping $\gamma\colon \mathrm{Hom}_G(A, C_{K,T}) \longrightarrow \mathrm{Hom}_G(A, C_{K,S})$.

PROPOSITION A.3.2. *If A is a finite module, S contains all infinite divisors and all prime divisors of order equal to the order of A, then the following diagram commutes*:

$$\begin{array}{ccc} H^2(F_S, A) & \xrightarrow{\lambda} & H^2(F_T, A) \\ \downarrow{\scriptstyle \varphi_1} & & \downarrow{\scriptstyle \varphi_2} \\ \operatorname{Hom}_G(A, C_{K,S})' & \xrightarrow[\overline{\gamma}]{} & \operatorname{Hom}_G(A, C_{K,T})', \end{array}$$

where φ_1, φ_2 are duality isomorphisms and $\overline{\gamma}$ is the homomorphism dual to γ.

PROOF. One can directly verify that for $k_\alpha \subseteq k_\beta$, the diagram

$$\begin{array}{ccc} H^2(F_\alpha, A) & \longrightarrow & H^2(F_\beta, A) \\ \downarrow & & \downarrow \\ H^0(F_\alpha, \operatorname{Hom}(A, C_{k_\alpha,S}))' & \longrightarrow & H^0(F_\beta, \operatorname{Hom}(A, C_{k_\beta,T}))' \end{array}$$

is commutative. Passing to the limit, we prove the proposition. □

§4. The structure of factors of the decreasing p-central series of a free operator group

Let F be a finite group, U_d a free group with generators $u_{i,f}$, where $i = 1, \dots, d$; $f \in F$. The group U_d is an F-operator group if we set $u_{i,f}^q = u_{i,fq}$.

Consider the decreasing p-central series of the group U_d, i.e., the sequence of subgroups $U_{d,c}$ which is constructed recurrently in the following way:

$$U_{d,1} = U_d, \quad U_{d,2} = U_{d,1}^p[U_{d,1}, U_{d,1}], \quad \dots, \quad U_{d,i+1} = U_{d,i}^p[U_{d,i}, U_{d,1}],$$

where $[U_{d,i}, U_{d,1}]$ is the subgroup of the group $U_{d,i}$ generated by the mutual commutators, i.e., by the elements $[u_i, u_1]$; $u_i \in U_{d,i}$, $u_1 \in U_1$. Together with the p-central series of the group U_d, we shall also consider the central series of U_d, i.e, the sequence of subgroups $\overline{U}_{d,c}$ defined as follows:

$$\overline{U}_{d,1} = U_d, \quad \overline{U}_{d,2} = [\overline{U}_{d,1}, \overline{U}_{d,1}], \quad \dots, \quad \overline{U}_{d,i+1} = [\overline{U}_{d,i}, \overline{U}_{d,1}].$$

Let R_d be the Magnus algebra, i.e., the algebra of formal noncommutative power series with integral coefficients and variables $x_{i,f}$, where $i = 1, \dots, d$; $f \in F$. One can define the action of the group F on the elements of R_d by setting

$$x_{i,f}^g = x_{i,fg}.$$

We observe that there exists an embedding ψ of the group U_d in R_d given on the generators by the relations $\psi(u_{i,f}) = 1 + x_{i,f}$ and, by multiplicativity, extended to all elements of U_d. It is clear that

$$\psi(u_{i,f}^{-1}) = \sum_{n=0}^{\infty} (-1)^n x_{i,f}^n.$$

We define recurrently Lie monomials in the algebra R_d. The elements $x_{i,f}$ are Lie monomials l_1 of degree one. Then, $l_2 = l_1 l_1' - l_1' l_1$ are Lie monomials of degree two, where l_1 and l_1' are linear Lie monomials; $l_n = l_{n-1}l_1 - l_1 l_{n-1}$, where l_{n-1} and l_1 are Lie monomials of degree $n-1$ and 1, respectively, is a Lie monomial of degree n. An algebraic sum of Lie monomials of degree n is a Lie polynomial of degree n. It is clear that the Lie polynomials of degree n form an additive group denoted by L_n.

Let H and $\overline{H}$ be the ideals of the algebra R_d generated by the elements p, $x_{i,f}$ and $x_{i,f}$, respectively.

Recall that if $u \in \overline{U}_{d,c}$, then $\psi(u) = 1 + \bar{l}_c + \bar{h}_{c+1}$, where $\bar{l}_c \in L_c$, $\bar{h}_{c+1} = \overline{H}^{c+1}$; conversely, if $u \in U_d$ and $\psi(u) = 1 + \bar{l}_c + \bar{h}_{c+1}$, then $u \in \overline{U}_{d,c}$, i.e., there exists an isomorphism of the groups $\overline{U}_{d,c}/\overline{U}_{d,c+1}$ and L_c (the proof can be found in [**52**, Chapter 11]).

We study the structure of the group $U_{d,c}/U_{d,c+1}$. First we prove an auxiliary assertion.

LEMMA A.4. *If $u \in U_{d,c}$, then*

$$\psi(u) \equiv 1 + p^{c-1}\bar{l}_1 + p^{c-2}\bar{l}_2 + \cdots + \bar{l}_c \pmod{H^{c+1}}$$

for odd p,

$$\psi(u) \equiv 1 + 2^{c-1}\bar{l}_1 + 2^{c-2}(\bar{l}_1^2 + \bar{l}_2) + 2^{c-3}\bar{l}_3 + \cdots + \bar{l}_c \pmod{H^{c+1}}$$

for $p = 2$, where $\bar{l}_i \in L_i$.

We prove the lemma by induction on c. For $c = 1$, the lemma is trivial. Let it be proved for the elements of the group $U_{d,c-1}$. It is clear that it suffices to verify the lemma for the elements $[g_1, u]$ and g_2^p, where $g_1, g_2 \in U_{d,c-1}$; $u \in U_d$.

Consider the case $p = 2$ and let

$$\begin{aligned}
\psi(g_1) &= 1 + 2^{c-2}\bar{l}_{11} + 2^{c-3}(\bar{l}_{11}^2 + \bar{l}_{21}) + \cdots + \bar{l}_{c-1,1} + h_{c1},\\
\psi(g_2) &= 1 + 2^{c-2}\bar{l}_{12} + 2^{c-3}(\bar{l}_{12}^2 + \bar{l}_{22}) + \cdots + \bar{l}_{c-1,2} + h_{c2},\\
\psi(u) &= 1 + \bar{l}_{13} + h_{23},
\end{aligned}$$

where h_{c1}, $h_{c2} \in H^c$, $h_{23} \in H^2$.

We have

$$\begin{aligned}
\psi([g_1, u]) &= (1 + 2^{c-2}\bar{l}_{11} + \cdots + \bar{l}_{c-1,1} + h_{c1})^{-1}(1 + \bar{l}_{13} + h_{23})^{-1}\\
&\quad \times (1 + 2^{c-2}\bar{l}_{11} + \cdots + \bar{l}_{c-1,1} + h_{c1})(1 + \bar{l}_{13} + h_{23})\\
&\equiv 1 + 2^{c-2}(\bar{l}_{11}\bar{l}_{13} - \bar{l}_{13}\bar{l}_{11}) + 2^{c-3}(\bar{l}_{11}^2\bar{l}_{13} - \bar{l}_{13}\bar{l}_{11}^2 + \bar{l}_{21}\bar{l}_{13} - \bar{l}_{13}\bar{l}_{21})\\
&\quad + \cdots + (\bar{l}_{c-1,1}\bar{l}_{13} - \bar{l}_{13}\bar{l}_{c-1,1}) \pmod{H^{c+1}}.
\end{aligned}$$

We set

$$\bar{l}_3 = (\bar{l}_{11}\bar{l}_{13} - \bar{l}_{13}\bar{l}_{11})\bar{l}_{11} - \bar{l}_{11}(\bar{l}_{11}\bar{l}_{13} - \bar{l}_{13}\bar{l}_{11}).$$

Then

$$\bar{l}_3 = \bar{l}_{11}\bar{l}_{13}\bar{l}_{11} - \bar{l}_{13}\bar{l}_{11}^2 - \bar{l}_{11}^2\bar{l}_{13} + \bar{l}_{11}\bar{l}_{13}\bar{l}_{11} = \bar{l}_{11}^2\bar{l}_{13} - \bar{l}_{13}\bar{l}_{11}^2 - 2\bar{l}_{11}^2\bar{l}_{13} + 2\bar{l}_{11}\bar{l}_{13}\bar{l}_{11}.$$

It remains to note that $-2\bar{l}_{11}^2 + 2\bar{l}_{11}\bar{l}_{13}\bar{l}_{11} \in H^4$, so the element $\psi([g_1, u])$ is of the necessary form.

For g_2^2 we have

$$\begin{aligned}\psi(g_2^2) &= (1 + 2^{c-2}\bar{l}_{12} + \cdots + \bar{l}_{c-1,2} + h_{c2})^2 \\ &\equiv 1 + 2^{c-1}\bar{l}_{12} + \cdots + 2\bar{l}_{c-1,2} \pmod{H^{c+1}}.\end{aligned}$$

The case of odd p is considered similarly.

Now we define the mapping

$$\varphi\colon U_{d,c}/U_{d,c+1} \to \sum_{i=1}^{c}{}^{\oplus} L_i/pL_i.$$

Note that an arbitrary element $u \in U_{d,c}$ is of the form

$$u = u_{i_1,f_1}^{m_1 p^{c-1}} u_{i_2,f_2}^{m_2 p^{c-1}} \cdots u_{i_t,f_t}^{m_t p^{c-1}} \widetilde{u},$$

where $\widetilde{u} \in [U_d, U_d] = U_{d,2}$ and $\psi(\widetilde{u}) \equiv 1 + p^{c-2}\bar{l}_2 + \cdots + \bar{l}_c \pmod{H^{c+1}}$. We set

$$\begin{aligned}&\varphi(u \bmod U_{d,c+1}) \\ &\quad = (m_1 x_{i_1,f_1} + \cdots + m_t x_{i_t,f_t}) \bmod pL_1 + \bar{l}_2 \bmod pL_2 + \cdots + \bar{l}_c \bmod pL_c.\end{aligned}$$

THEOREM A.4. *The mapping*

$$\varphi\colon U_{d,c}/U_{d,c+1} \to \sum_{i=1}^{c}{}^{\oplus} L_i/pL_i$$

is an isomorphism.

PROOF. Clearly, φ is a homomorphism. We show that φ is surjective. Let

$$z = (m_1 x_{i_1,f_1} + \cdots + m_t x_{i_t,f_t}) \bmod pL_1 + \bar{l}_2 \bmod pL_2 + \cdots + \bar{l}_c \bmod pL_c$$

be an element of the group $\sum_{i=1}^{c}{}^{\oplus} L_i/pL_i$, where $\bar{l}_i \in L_i$ and $m_1, \ldots, m_t$ are integers. As we have noted, there exist elements $u_i \in \overline{U}_{d,i}$ such that $\psi(u_i) = 1 + \bar{l}_i \pmod{\overline{H}^{i+1}}$. We put

$$u = u_{i_1,f_1}^{m_1 p^{c-1}} \cdots u_{i_t,f_t}^{m_t p^{c-1}} u_2^{p^{c-2}} \cdots u_c.$$

Clearly, $u \in U_{d,c}$ and $\varphi(u \bmod U_{d,c+1}) = z$.

The fact that φ is injective is also checked immediately. The theorem is proved. $\square$

§5. The Frattini subgroup and its properties

The Frattini subgroup Φ of the group G is the intersection of all maximal subgroups of G. Since a subgroup conjugate to a maximal subgroup is also maximal, Φ coincides with all conjugate subgroups, hence is a normal subgroup in G.

PROPOSITION A.5.1. *If the group G is generated by a system of elements $\{b_i\}$ and the subgroup Φ, then the system $\{b_i\}$ itself generates G.*

PROOF. If $\{b_i\}$ generates a subgroup $H \neq G$, then H is contained in a maximal subgroup $A \neq G$, which contains the subgroup Φ, so that the group generated by $\{b_i\}$ and Φ is also contained in A, a contradiction. □

THEOREM A.5.1 (the Frattini argument). *Let H be a normal subgroup of a finite group G, P a Sylow subgroup in H, and let N be its normalizer in G. Then $NH = G$.*

PROOF. Let $g \in G$. Then $g^{-1}Pg \subset H$ and also is a Sylow subgroup in H. By Sylow's theorem, Sylow subgroups of H are conjugate in H, so that there exists an $h \in H$ such that $h^{-1}Ph = g^{-1}Pg$, whence $gh^{-1} \in N$, $g \in Nh$. Since this is valid for any $g \in G$, we have $NH = G$. □

COROLLARY (Frattini, [**76**]). *The Frattini subgroup of a finite group G is nilpotent.*

PROOF. Let P be a Sylow subgroup of Φ and let N be its normalizer in G. We have $N\Phi = G$, whence, by Proposition A.5.1, $N = G$. Therefore, any Sylow subgroup of Φ is normal in G, and so in Φ. Therefore, Φ is a nilpotent group. □

PROPOSITION A.5.2 (Gaschütz, [**77**]). *Let a normal subgroup H of a finite group G contain the Frattini subgroup of G. If H/Φ is nilpotent, then H is also nilpotent.*

PROOF. Let $\overline{B}$ be a Sylow p-subgroup of H/Φ, B its inverse image in H, P a Sylow p-subgroup of B. The index $(H:P) = (H:B)\cdot(B:P) = (H/\Phi:\overline{B})\cdot(B:P)$ is prime to p, therefore P is a Sylow subgroup in H containing a unique Sylow p-subgroup of Φ. This subgroup is obviously $\Phi \cap P$. Since $\overline{B} = B/\Phi$ is a p-group, then the mapping $P \to \overline{B}$ is surjective, so $B = P\Phi$. The group $\overline{B} = B/\Phi$, as a unique Sylow p-subgroup in the nilpotent group H/Φ, is characteristic in H/Φ. Hence B is normal in G. Applying the theorem, we obtain $G = NB$, where N is the normalizer of P in G, or $G = NP\Phi$. Hence, by Proposition 5.1 we have $G = NP$ or, finally, $G = N$. Thus, P is normal in G and, thus, it is a unique Sylow p-subgroup in H. This argument is valid for any p dividing the order of H. Therefore, H is nilpotent. □

THEOREM A.5.2. *For a finite solvable group G there exists a nilpotent normal subgroup H and a proper subgroup M such that $G = HM$.*

PROOF. Let Φ be the Frattini subgroup for G and $\overline{H}$ a nontrivial nilpotent normal subgroup in G/Φ. The latter subgroup exists by the solvability of G/Φ. Its inverse image H is nilpotent by the preceding proposition, and Φ is a proper subgroup of H. Among the maximal subgroups of G, there exists a subgroup M such that M does not contain H.

By the maximality of M, we have $HM = G$. The theorem is proved. □

§6. A duality theorem in cohomology of finite groups

1°. Let G be a finite group. Although in applications multiplicative G-modules are mainly considered, here we write them additively and assume that the elements of G act as right operators.

Consider the additive group of the group ring $O = \mathbb{Z}[G]$ with basis $\{e_g\}_{g\in G}$ as a G-module, where $e_g h = e_{gh}$ $(g, h \in G)$. Denote by I a submodule of O generated by the differences $e_g - e_1$ (in other words, I consists of the sums $\sum m_g e_g$ for which $\sum m_g = 0$, $m_g \in \mathbb{Z}$). Let J be the quotient module of O by the one-dimensional submodule generated by the element $\sum e_g$. The modules I and J are called the *Chevalley modules.*

Thus, we have the exact sequences of G-modules

$$0 \to I \to O \xrightarrow{\theta} \mathbb{Z} \to 0, \tag{1}$$

$$0 \to \mathbb{Z} \xrightarrow{\theta'} O \to J \to 0, \tag{1'}$$

where $\theta(\sum m_g e_g) = \sum m_g$, $\theta'(m) = \sum m e_g$ (the action of G on $\mathbb{Z}$ is trivial). Since here all the modules are lattices, the two sequences split over $\mathbb{Z}$.

Now let A be an arbitrary G-module. Then the following sequences are exact

$$0 \to A \otimes I \to A \otimes O \to A \to 0, \tag{2}$$

$$0 \to A \to A \otimes O \to A \otimes J \to 0. \tag{2'}$$

Here the tensor product is regarded over $\mathbb{Z}$ and is provided with the structure of a G-module under the action of elements from G on the two factors.

We observe that the subgroup $\widetilde{A}$ of the module $A \otimes O$ consisting of elements left fixed relative to all $g \in G$ is isomorphic to A as a $\mathbb{Z}$-module. The module $\widetilde{A} \otimes O$ (induced for the group $\widetilde{A}$) is operator isomorphic to $A \otimes O$. Thus [**1**, Chapter IV], the module $A \otimes O$ is cohomologously trivial, i.e., $\widehat{H}^n(G, A \otimes O) = 0$ for all integer n ($\widehat{H}$ is the cohomology of Tate, in the sequel we omit the sign $\widehat{\ }$). Therefore, the cohomology sequence gives us the isomorphisms

$$H^n(G, A) \to H^{n+1}(G, A \otimes I), \qquad H^{n-1}(G, A \otimes J) \to H^n(G, A).$$

Considering tensor powers of the Chevalley modules I^n and J^n, we obtain the following proposition.

PROPOSITION A.6.1. *The following isomorphisms hold:*

$$H^n(G, A) \approx H^0(G, A \otimes J^n),$$

$$H^{-n}(G, A) \approx H^0(G, A \otimes I^n) \quad (n \geq 0).$$

It is easy to check that the tensor product $I \otimes J$ is isomorphic, as a G-module, to the direct sum of $\mathbb{Z}$ and the $(|G| - 2)$-fold direct sum of the module $O = \mathbb{Z}[G]$.

2°. Let L be a lattice, i.e., an Abelian torsion free group with finitely many generators. Consider the dual lattice $L^* = \operatorname{Hom}_{\mathbb{Z}}(L, \mathbb{Z})$. If L is a G-module, then a structure of G-module is also introduced on L^* by the formula $fg(x) = f(xg^{-1})$, where $f \in L^*$, $g \in G$, $x \in L$. The dual basis $\{v_1^*, \dots, v_m^*\}$ for a basis $\{v_1, \dots, v_m\}$ of L is defined by the formulas $v_i^*(v_j) = \delta_{ij}$ (the Kronecker symbol).

Consider also the character group of an arbitrary group A. Namely, by such a group we mean $\operatorname{Hom}_{\mathbb{Z}}(A, C)$, where $C = \mathbb{Q}/\mathbb{Z}$. Denote it by A'. If A is a G-module with a right action of operators, then the same structure is also defined on A' by the formula $(\chi g)a = \chi(ag^{-1})$ $(\chi \in A', g \in G, a \in A)$.

PROPOSITION A.6.2. *Let A and L be G-modules, and let L be a lattice. Then a G-isomorphism*

$$(A \otimes L)' \approx A' \otimes L^*$$

exists.

PROOF. The mapping $\varphi_1\colon A' \otimes L^* \to \mathrm{Hom}(L, A')$ defined by the formula $(\varphi_1(\chi \otimes f))(x) = f(x)\chi$, where $\chi \in A'$, $f \in L^*$, $x \in L$, is a $\mathbb{Z}$-isomorphism; the inverse mapping is of the form $\psi \to \sum_i \psi(v_i) \otimes v_i^*$; $\psi \in \mathrm{Hom}(L, A')$, where $\{v_i\}$ is a basis of L and $\{v_i^*\}$ is the dual basis.

One can easily construct a $\mathbb{Z}$-isomorphism $\varphi\colon \mathrm{Hom}(L, A') \to (A \otimes L)'$; namely, for $\omega \in \mathrm{Hom}(L, A')$ we set $(\varphi_2\omega)(a \otimes x) = (\omega x)(a)$, where $x \in L$, $a \in A$.

Therefore, the composition $\varphi = \varphi_2\varphi_1\colon A' \otimes L^* \to (A \otimes L)'$ is also a $\mathbb{Z}$-isomorphism and is defined by the formula

$$(\varphi(\chi \otimes f))(a \otimes x) = f(x)\chi(a) \quad (a \in A,\ x \in L,\ f \in L^*,\ \chi \in A').$$

The isomorphisms φ_1 and φ_2 are not G-operator isomorphisms, but their composition φ is a G-isomorphism. Indeed, for $g \in G$ we have

$$\begin{aligned}(\varphi((\chi \otimes f)g))(a \otimes x) &= (\varphi(\chi g \otimes fg))(a \otimes x) = (fg)(x)(\chi g)(a)\\ &= f(xg^{-1})\chi(ag^{-1}),\\ ((\varphi(\chi \otimes f))g)(a \otimes x) &= (\varphi(\chi \otimes f))(ag^{-1} \otimes xg^{-1})\\ &= f(xg^{-1})\chi(ag^{-1}).\end{aligned}$$

The proposition is proved. □

LEMMA A.6.1. *If L_1 and L_2 are lattices, then $(L_1 \otimes L_2)^*$ is isomorphic to $L_1^* \otimes L_2^*$.*

Indeed, if $\{u_i\}$ and $\{v_j\}$ are bases of L_1 and L_2, respectively, then $u_i^* \otimes v_j^*$ is a basis of $(L_1 \otimes L_2)^*$.

LEMMA A.6.2. *The Chevalley modules I and J are lattices dual to each other, i.e., $J^* \approx I$ and $I^* \approx J$.*

Indeed, let $w^* \in J^*$. Consider the element $\omega = \sum_g \omega^*(\overline{e}_g)e_g$, where $\overline{e}_g$ is the image of the basis element $e_g \in O$ in J. Since $\sum_g \omega^*(\overline{e}_g) = \omega^*(\sum \overline{e}_g) = 0$, then $\omega \in I$. Conversely, for $\omega = \sum_g m_g e_g \in I$ we have $\sum_g m_g = 0$, and we set $\omega^*(e_g) = m_g$. Since $\omega^*(\sum e_g) = 0$, the map ω^* is a homomorphism of J to $\mathbb{Z}$. Thus, we have constructed a natural isomorphism $J^* \approx I$. Since the module $(J^*)^*$ is naturally isometric to J, we also have $J \approx I^*$. The lemma is proved.

COROLLARY. *The following natural isomorphisms hold:*

$$(I^n)^* \approx J^n, \qquad (J^n)^* \approx I^n.$$

3°. Let A be a G-module. Denote by A^G the submodule of A of elements invariant under $g \in G$. We also denote by IA' the submodule of A' generated by the differences $\chi g - \chi$ ($\chi \in A'$, $g \in G$).

LEMMA A.6.3. *The relation $(A/A^G)' = IA'$ is valid. (Here we regard $(A/A^G)'$ as a submodule of A'.)*

PROOF. For $a \in A^G$ we have $(\chi g)(a) = \chi(a)$, whence $\chi g - \chi \in (A/A^G)'$, i.e., $IA' \subset (A/A^G)'$. Now we prove the inclusion $(A/A^G)' \subset IA'$.

Let $\theta_g(a) = ag - a$ for $a \in A$ and $g \in G$. Denote by A_g the image of θ_g and consider the external direct sum $\sum_{g\neq 1}^{\oplus} A_g = B$. We define the homomorphism $\theta \colon A \to B$ as the homomorphism θ_g on each component A_g. It is clear that $\ker\theta = A^G$. Now we have the exact sequences

$$0 \to A^G \xrightarrow{i} A \to A/A^G \to 0, \tag{3}$$

$$0 \to A^G \xrightarrow{i} A \xrightarrow{\theta} B. \tag{4}$$

Since C is a divisible module, we have

$$0 \to (A/A^G)' \to A' \xrightarrow{i'} (A^G)' \to 0, \tag{3$'$}$$

$$B' \xrightarrow{\theta'} A' \xrightarrow{i'} (A^G)' \to 0. \tag{4$'$}$$

Therefore, we obtain $(A/A^G)' = \operatorname{Ker} i' = \operatorname{Im}\theta'$. We need to show that $\operatorname{Im}\theta' \subset IA'$.

Let $\psi \in B'$. Then $(\theta'\psi)(a) = \psi(\theta a)$, where $a \in A$. The components θa in A_g are of the form $a_g - a$, therefore

$$\psi(\theta a) = \sum_g \psi(ag - a) = \sum_g ((\psi g^{-1})(a) - \psi(a)) = \sum_g (\psi g^{-1} - \psi)(a).$$

Thus, $\theta'\psi = \sum_g (\psi g^{-1} - \psi) \in IA'$. The lemma is proved. □

4°. The purpose of this section is to prove the following duality theorem.

THEOREM A.6. *Let G be a finite group and let A be a G-module. Then for $n \geq 0$ the duality isomorphism*

$$(H^n(G,A))' \to H^{-n-1}(G,A')$$

is well defined.

PROOF. First consider the case $n = 0$. For the Tate cohomology we have $H^0(G,A) \approx A^G/A_0$, where A_0 is generated by the elements $\sum_{g\in G} ag$, $a \in A$. Consider the exact sequence of G-modules (with natural maps)

$$0 \to H^0(G,A) \to A/A_0 \to A/A^G \to 0. \tag{5}$$

With this sequence one can associate the dual sequence

$$0 \to (A/A^G)' \to (A/A_0)' \to (H^0(G,A))' \to 0. \tag{6}$$

The group $(A/A_0)'$ consists of the characters $\chi \in A'$ trivial on the subgroup A_0, and its subgroup $(A/A^G)'$, by Lemma 3, coincides with IA'. Thus, $(H^0(G,A))'$ is isomorphic to the corresponding quotient group. But the cohomology group

$H^{-1}(G, A')$ (see [**1**]) is also equal to this quotient group. Thus, for $n = 0$ the theorem is valid.

To prove the general case, we use Proposition 1 from this section. We have

$$(H^n(G, A))' \approx (H^0(G, A \otimes J^n))' \approx H^{-1}(G, (A \otimes J^n)').$$

By Proposition 2 and Lemma 2, we obtain

$$(A \otimes J^n)' \approx A' \otimes (J^n)^* \approx A' \otimes I^n.$$

Thus, we have

$$(H^n(G, A))' \approx H^{-1}(G, A' \otimes I^n) \approx H^{-1-n}(G, A')$$

(again by Proposition 1). The theorem is proved. □

Bibliography

1. J. W. S. Cassels and A. Fröhlich (eds.), *Algebraic number theory*, Academic Press, London; Thompson Book Co., Washington, DC, 1967.
2. M. I. Bashmakov, *The problem of embedding of fields*, Mat. Zametki **4** (1968), no. 2, 137–140. (Russian)
3. G. V. Belyĭ, *Galois extensions of a maximal cyclotomic field*, Izv. Akad. Nauk SSSR Ser. Mat. **43** (1979), no. 2, 267–276; English transl., Math. USSR-Izv. **14** (1980), no. 2, 247–256.
4. Z. I. Borevich and I. R. Shafarevich, *Number theory*, 3rd ed., "Nauka", Moscow, 1985; English transl. of 1st ed., Academic Press, New York–London, 1966.
5. N. Bourbaki, *Éléments de mathématique.* Vol. 14, Pt. 1, Livre 2, Ch. 7, Hermann, Paris, 1952; *Éléments de mathématique.* Vol. 23, Pt. 1, Livre 2, Ch. 8, Hermann, Paris, 1958; *Éléments de mathématique.* Vol. 24, Pt. 1, Livre 2, Ch. 9, Hermann, Paris, 1959.
6. B. L. van der Waerden, *Algebra.* Teil 1, Springer, Berlin, 1966; Teil 2, Springer, Berlin, 1967.
7. N. V. Volnina, *On fields with extended polygon groups*, Dokl. Akad. Nauk SSSR **105** (1955), no. 5, 889–892. (Russian)
8. E. Galois, *Oeuvres mathématiques*, Gauthier-Villars, Paris, 1897.
9. N. L. Gordeev, *Embedding problem with given localizations and restricted ramification*, Zap. Nauchn. Sem. Leningrad. Otdel. Mat. Inst. Steklov. (LOMI) **57** (1976), 85–99; English transl., J. Soviet Math. **11** (1979), no. 4, 585–597.
10. O. M. Grigoryan and B. B. Lur′e, *Completely solvable embedding problems with abelian kernel for local fields*, Zap. Nauchn. Sem. Leningrad. Otdel. Mat. Inst. Steklov. (LOMI) **75** (1978), 67–73; English transl., J. Soviet Math. **37** (1987), no. 2, 959–964.
11. B. N. Delone and D. K. Faddeev, *Investigations in the geometry of the Galois theory*, Mat. Sb. (N. S.) **15(57)** (1944), 243-284. (Russian; English summary)
12. S. P. Demushkin, *The group of a maximal p-extension of a local field*, Dokl. Akad. Nauk SSSR **128** (1959), no. 4, 657–660. (Russian)
13. ———, *The group of a maximal p-extension of a local field*, Izv. Akad. Nauk SSSR Ser. Mat. **25** (1961), 329–346. (Russian)
14. ———, *An embedding problem for algebraic number fields*, Dokl. Akad. Nauk SSSR **144** (1962), 27–28; English transl. in Soviet Math. Dokl. **3** (1962).
15. ———, *On 2-extensions of a local field*, Sibirsk. Mat. Zh. **4** (1963), 951–955; English transl. in Amer. Math. Soc. Transl. (2) **50** (1966).
16. ———, *Topological 2-groups with an even number of generators and one complete defining relation*, Izv. Akad. Nauk SSSR Ser. Mat. **29** (1965), no. 1, 3–10; English transl. in Amer. Math. Soc. Transl. (2) **66** (1968).
17. S. P. Demushkin and I. R. Shafarevich, *The embedding problem for local fields*, Izv. Akad. Nauk SSSR Ser. Mat. **23** (1959), 823–840; English transl. in Amer. Math. Soc. Transl. (2) **27** (1963), 267–288.
18. ———, *The second obstruction for the embedding problem of algebraic number fields*, Izv. Akad. Nauk SSSR Ser. Mat. **26** (1962), no. 6, 911–924; English transl. in Amer. Math. Soc. Transl. (2) **58** (1966).
19. I. G. Zel′venskiĭ, *The algebraic closure of a local field for* $p = 2$, Izv. Akad. Nauk SSSR Ser. Mat. **36** (1972), 933–946; English transl., Math. USSR-Izv. **6** (1972), 925–948 (1973).
20. N. P. Zyapkov and A. V. Yakovlev, *Galois universally compatible extensions*, Zap. Nauchn. Sem. Leningrad. Otdel. Mat. Inst. Steklov. (LOMI) **71** (1977), 133–152; English transl. in J. Soviet Math. **20** (1982), no. 6.
21. V. V. Ishkhanov, *The embedding problem with restricted ramification*, Izv. Akad. Nauk SSSR Ser. Mat. **36** (1972), no. 4, 742–748; English transl. in Math. USSR-Izv. **6** (1972).

22. ______, *The embedding problem with prescribed localizations*, Izv. Akad. Nauk SSSR Ser. Mat. **39** (1975), no. 3, 512–522; English transl. in Math. USSR-Izv. **9** (1975), no. 3.
23. ______, *On the semidirect embedding problem with a nilpotent kernel*, Izv. Akad. Nauk SSSR Ser. Mat. **40** (1976), no. 1, 3–25; English transl. in Math. USSR-Izv. **10** (1976), no. 1.
24. ______, *On the semidirect embedding problem with a metabelian kernel*, Zap. Nauchn. Sem. Leningrad. Otdel. Mat. Inst. Steklov. (LOMI) **100** (1980), 81–85; English transl. in J. Soviet Math. **19** (1982), no. 6.
25. ______, *A case of embedding problem with a solvable kernel*, Zap. Nauchn. Sem. Leningrad. Otdel. Mat. Inst. Steklov. (LOMI) **106** (1981), 76–81; English transl., J. Soviet Math. **23** (1983), no. 2, 2158–2162.
26. H. Cartan and S. Eilenberg, *Homological algebra*, Princeton Univ. Press, Princeton, NJ, 1956.
27. H. Koch, *Galoissche Theorie der p-Erweiterungen*, VEB Deutscher Verlag Wiss., Berlin, 1970.
28. L. V. Kuz′min, *Homologies of profinite groups, Schur's multiplicator, and class field theory*, Izv. Akad. Nauk SSSR Ser. Mat. **33** (1969), no. 6, 1220–1254; English transl. in Math. USSR-Izv. **3** (1969).
29. C. W. Curtis and I. Reiner, *Representation theory of finite groups and associative algebras*, John Wiley and Sons, New York–London, 1962.
30. S. Lang, *Algebraic numbers*, Addison-Wesley, Reading, MA–London, 1964.
31. B. B. Lur′e, *On the problem of embedding with a kernel without center*, Izv. Akad. Nauk SSSR Ser. Mat. **28** (1964), 1135–1138. (Russian)
32. ______, *On the embedding problem with noncommutative kernel of order p^3*, Trudy Mat. Inst. Steklov. **80** (1965), 98–101; English transl. in Proc. Steklov Inst. Math. **1968**.
33. ______, *Embeddability conditions when the kernel is a non-Abelian p-group*, Mat. Zametki **2** (1967), 233–238; English transl. in Math. Notes **2** (1967).
34. ______, *On the embedding problem for local fields with non-Abelian kernel*, Zap. Nauchn. Sem. Leningrad. Otdel. Mat. Inst. Steklov. (LOMI) **31** (1973), 106–114; English transl. in J. Soviet Math. **6** (1976), no. 3.
35. ______, *On the compatibility condition in the fields embedding problem*, Zap. Nauchn. Sem. Leningrad. Otdel. Mat. Inst. Steklov. (LOMI) **71** (1977), 155–162; English transl. in J. Soviet Math. **20** (1982), no. 6.
36. ______, *Completely solvable embedding problems for local fields*, Zap. Nauchn. Sem. Leningrad. Otdel. Mat. Inst. Steklov. (LOMI) **75** (1978), 121–126; English transl., J. Soviet Math. **37** (1987), no. 2, 998–1002.
37. S. Mac Lane, *Homology*, Grundlehren Math. Wiss., vol. 114, Academic Press, New York; Springer-Verlag, Berlin–Heidelberg, 1963.
38. A. S. Merkur′ev, *On the embedding problem with non-Abelian kernel*, Izv. Akad. Nauk SSSR Ser. Mat. **41** (1977), no. 5, 1043–1052; English transl., Math. USSR-Izv. **11** (1977), no. 5, 991–1000 (1978).
39. O. Neumann, *On algebraic number p-closed fields with restricted ramification*, Izv. Akad. Nauk SSSR Ser. Mat. **39** (1975), 259–271; English transl. in Math. USSR-Izv. **9** (1975).
40. L. S. Pontryagin, *Continuous groups*, 3rd ed., "Nauka", Moscow, 1973; English transl. of 2nd ed., *Topological groups*, Gordon and Breach, New York, 1966.
41. J.-P. Serre, *Cohomologie galoisienne*, Springer-Verlag, Berlin etc., 1964.
42. A. I. Skopin, *The factor groups of an upper central series of free groups*, Dokl. Akad. Nauk SSSR **74** (1950), 425–428. (Russian)
43. S. Eilenberg and N. Steenrod, *Foundations of algebraic topology*, Princeton Univ. Press, Princeton, NJ, 1952.
44. S. V. Tvaradze and A. V. Yakovlev, *On associated problems in field embedding theory*, Vestnik Leningrad. Univ. Mat. Mekh. Astronom. **1968**, vyp. 1, 67–76. (Russian)
45. D. K. Faddeev, *Construction of algebraic domains whose Galois group is the quaternion group*, Leningrad. Gos. Univ. Uchen. Zap. **3** (1937), no. 17, 17–23. (Russian)
46. ______, *Construction of algebraic fields of prime degree having solvable Galois group*, Leningrad. Gos. Ped. Inst. Uchen. Zap. **28** (1939), 127–139. (Russian)
47. ______, *Construction of algebraic number fields, whose Galois group is the group of quaternion units*, Dokl. Akad. Nauk SSSR **47** (1945), no. 6, 404–407. (Russian)
48. ______, *On a conjecture of Hasse*, Dokl. Akad. Nauk SSSR **94** (1954), 1013–1016. (Russian)

49. ———, *An interpretation of the supplementary condition in the theory of embedding of fields with an Abelian kernel*, Zap. Nauchn. Sem. Leningrad. Otdel. Mat. Inst. Steklov. (LOMI) **94** (1979), 116–118; English transl. in J. Soviet Math. **19** (1982), no. 1.
50. ———, *Galois theory (in the Mathematics Institute of the Academy of Sciences)*, Trudy Mat. Inst. Steklov. **168** (1984), 46–71; English transl. in Proc. Steklov Inst. Math. **1986**, no. 3, 47–73.
51. D. K. Faddeev and R. A. Shmidt, *Conditions for the embeddability of a field in the case of a cyclic normal subgroup of order* 8, Vestnik Leningrad. Univ. Mat. Mekh. Astronom. **1959**, vyp. 4, 36–42. (Russian)
52. M. Hall, Jr., *The theory of groups*, Macmillan, New York, 1959.
53. N. G. Chebotarev, *Foundations of Galois theory.* Vol. I, "ONTI", Moscow–Leningrad, 1934; German transl., *Grundzüge der Galois'schen Theorie*, Noordhoff, Groningen, 1950.
54. ———, *Foundations of Galois theory.* Vol. II, "ONTI", Moscow–Leningrad, 1937; German transl., *Grundzüge der Galois'schen Theorie*, Noordhoff, Groningen, 1950.
55. I. R. Shafarevich, *On p-extensions*, Mat. Sb. (N. S.) **20(62)** (1947), 351–363; English transl., Amer. Math. Soc. Transl. (2) **4** (1956), 59–72.
56. ———, *On the problem of embedding fields*, Dokl. Akad. Nauk SSSR **95** (1954), 459–461. (Russian)
57. ———, *On the construction of fields with a given Galois group of order* l^α, Izv. Akad. Nauk SSSR Ser. Mat. **18** (1954), no. 3, 261–296; English transl., Amer. Math. Soc. Transl. (2) **4** (1956), 107–142.
58. ———, *On an existence theorem in the theory of algebraic numbers*, Izv. Akad. Nauk SSSR Ser. Mat. **18** (1954), no. 4, 327–334; English transl., Amer. Math. Soc. Transl. (2) **4** (1956), 143–150.
59. ———, *On the problem of embedding fields*, Izv. Akad. Nauk SSSR Ser. Mat. **18** (1954), no. 5, 389–418; English transl., Amer. Math. Soc. Transl. (2) **4** (1956), 151–183.
60. ———, *Construction of fields of algebraic numbers with given solvable Galois group*, Izv. Akad. Nauk SSSR Ser. Mat. **18** (1954), 525–578; English transl., Amer. Math. Soc. Transl. (2) **4** (1956), 185–237.
61. ———, *The embedding problem for split extensions*, Dokl. Akad. Nauk SSSR **120** (1958), 1217–1219. (Russian)
62. ———, *Extensions with given points of ramification*, Inst. Hautes Études Sci. Publ. Math. **18** (1963), 295–316 (Russian); English transl. in Amer. Math. Soc. Transl. (2) **59** (1966).
63. R. A. Shmidt and A. V. Yakovlev, *An embeddability condition in the case of a cyclic normal divisor of order* 2^n, Vestnik Leningrad. Univ. Mat. Mekh. Astronom. **1963**, vyp. 3, 137–139. (Russian)
64. A. V. Yakovlev, *The embedding problem for fields*, Dokl. Akad. Nauk SSSR **150** (1963), 1009–1011; English transl. in Soviet Math. Dokl. **4** (1963).
65. ———, *The embedding problem for fields*, Izv. Akad. Nauk SSSR Ser. Mat. **28** (1964), 645–660. (Russian)
66. ———, *The embedding problem for number fields*, Izv. Akad. Nauk SSSR Ser. Mat. **31** (1967), 211–224; English transl. in Math. USSR-Izv. **1** (1967).
67. ———, *The Galois group of the algebraic closure of a local field*, Izv. Akad. Nauk SSSR Ser. Mat. **32** (1968), 1283–1322; English transl. in Math. USSR-Izv. **2** (1968).
68. N. Adachi, *On the relatively cyclic embedding problem with given local behavior*, J. Math. Soc. Japan **22** (1970), no. 3, 298–307.
69. A. A. Albert, *Structure of algebras*, Amer. Math. Soc., Providence, RI, 1961.
70. E. Artin, *Galois theory*, Univ. Notre Dame, Notre Dame, Indiana, 1944.
71. E. Artin and J. Tate, *Class field theory*, Harvard, 1961.
72. G. V. Belyĭ, *On extensions of the maximal cyclotomic field having a given classical Galois group*, J. Reine Angew. Math. **341** (1983), 147–156.
73. R. Brauer, *Über die Konstruktion der Schiefkörper, die von endlichem Rang in bezug auf ein gegebenes Zentrum sind*, J. Reine Angew. Math. **168** (1932), no. 1, 44–64.
74. J. Brinkhuis, *Galois modules and embedding problems*, J. Reine Angew. Math. **346** (1984), 141–165.
75. A. Brumer, *Galois groups of extensions of algebraic number fields with given ramification*, Michigan Math. J. **13** (1966), no. 1, 33–40.

76. K. Haberland, *Galois cohomology of algebraic number fields*, VEB Deutscher Verlag Wiss., Berlin, 1978.
77. H. Hasse, *Existenz und Mannigfaltigkeit abelscher Algebren mit vorgegebener Galoisgruppe über einem Teilkörper des Grundkörpers*, Math. Nachr. **1** (1948), no. 1, 40–61; ibid. **1** (1948), no. 4, 213–217.
78. K. Hoechsmann, *Über die Gruppe der maximalen l-Erweiterung eines globalen Körpers*, J. Reine Angew. Math. **222** (1966), 142–147.
79. ______, *Zum Einbettungsproblem*, J. Reine Angew. Math. **229** (1968), 81–106.
80. M. Ikeda, *Zur Existenz eigentlicher galoisscher Körper beim Einbettungsproblem für galoissche Algebren*, Abh. Math. Sem. Univ. Hamburg **24** (1960), 126–131.
81. K. Iwasawa, *On solvable extensions of algebraic number fields*, Ann. of Math. (2) **58** (1953), no. 3, 548–572.
82. H. Koch, *Über die Faktorgruppen einer absteigenden Zentralreihe*, Math. Nachr. **22** (1960), no. 3–4, 159–161.
83. R. Kochendörffer, *Zwei Reduktionssätze zum Einbettungsproblem für abelsche Algebren*, Math. Nachr. **10** (1953), no. 1–2, 75–84.
84. S. Lang, *Rapport sur la cohomologie des groupes*, W. A. Benjamin,, New York–Amsterdam, 1967.
85. W. Magnus, *Über Beziehungen zwischen höheren Kommutatoren*, J. Reine Angew. Math. **177** (1937), 105–115.
86. H. Matzat, *Zum Einbettungsproblem der algebraischen Zahlentheorie mit nicht abelschem Kern*, Invent. Math. **80** (1985), 365–374.
87. J. Neukirch, *Freie Produkte pro-endlicher Gruppen und ihre Kohomologie*, Arch. Math. **22** (1971), no. 4, 337–357.
88. ______, *Einbettungsprobleme mit lokaler Vorgabe und freie Produkte lokaler Galoisgruppen*, J. Reine Angew. Math. **259** (1973), 1–47.
89. ______, *Über das Einbettungsproblem der algebraischen Zahlentheorie*, Invent. Math. **21** (1973), no. 1–2, 59–116.
90. ______, *Über die absolute Galoisgruppe algebraischer Zahlkörper*, Jahresber. Deutsch. Math.-Verein. **76** (1974), 18–37.
91. ______, *Zur Theorie der nilpotenten algebraischen Zahlkörper*, Math. Ann. **219** (1976), no. 1, 65–83.
92. ______, *Über die absoluten Galoisgruppen algebraischer Zahlkörper*, Astérisque **41–42** (1977), 67–79.
93. ______, *On solvable number fields*, Invent. Math. **53** (1979), no. 2, 135–164.
94. O. Neumann, *Über das Einbettungsproblem für globale Körper bei beschränkter Verzweigung*, Math. Nachr. **71** (1976), 147–152.
95. ______, *Einige Klassen von endlichen Galois-Moduln mit Hasse-Prinzip*, Math. Nachr. **72** (1976), 305–320.
96. O. Ore, *Contribution to the theory of groups of finite order*, Duke Math. J. **5** (1939), no. 2, 431-460.
97. H. Reichardt, *Konstruktion von Zahlkörpern mit gegebener Galoisgruppe von Primzahlpotenzordnung*, J. Reine Angew. Math. **177** (1937), 1–5.
98. A. Scholz, *Über die Bildung algebraischer Zahlkörper mit auflösbarer Galoisscher Gruppe*, Math. Z. **30** (1929), 332–356.
99. ______, *Konstruktion algebraischer Zahlkörper mit beliebiger Gruppe von Primzahlpotenzordnung*. I, Math. Z. **42** (1937), 161–188.
100. J.-P. Serre, *Corps locaux*, Hermann, Paris, 1962.
101. J. Sonn, *On the embedding problem for nonsolvable Galois groups of algebraic number fields: reduction theorems*, Bull. Amer. Math. Soc. **78** (1972), no. 4, 553–557.
102. A. Speiser, *Zahlentheoretische Sätze aus der Gruppentheorie*, Math. Z. **5** (1919), 1–6.
103. J. Tate, *The higher dimensional cohomology groups of class field theory*, Ann. of Math. (2) **56** (1952), no. 2, 294–297.
104. ______, *Duality theorems in Galois cohomology over number fields*, Proc. Internat. Congr. Mathematicians (Stockholm, 1962), Inst. Mittag-Leffler, Djursholm, 1963, pp. 288–295.
105. K. Uchida, *On Tate's duality theorems on Galois cohomology*, Tôhoku Math. J. **21** (1969), no. 1, 92–101.

106. E. Witt, *Konstruktion von galoisschen Körpern der Charakteristik p zu vorgegebener Gruppe der Ordnung p^f*, J. Reine Angew. Math. **174** (1936), no. 4, 237–245.
107. ———, *Treue Darstellung Liescher Ringe*, J. Reine Angew. Math. **177** (1937), 152–160.
108. P. Wolf, *Algebraische Theorie der galoisschen Algebren*, VEB Deutscher Verlag Wiss., Berlin, 1956.

Subject Index

The numbers given in this index refer to the chapter followed by the section, i.e., 3.15 is Chapter 3, Section 15. Boldface is used to indicate the numbers of sections that contain definitions and statements of theorems.

Selected Titles in This Series

(Continued from the front of this publication)

127 **F. L. Zak,** Tangents and secants of algebraic varieties, 1993
126 **M. L. Agranovskiĭ,** Invariant function spaces on homogeneous manifolds of Lie groups and applications, 1993
125 **Masayoshi Nagata,** Theory of commutative fields, 1993
124 **Masahisa Adachi,** Embeddings and immersions, 1993
123 **M. A. Akivis and B. A. Rosenfeld,** Élie Cartan (1869–1951), 1993
122 **Zhang Guan-Hou,** Theory of entire and meromorphic functions: deficient and asymptotic values and singular directions, 1993
121 **I. B. Fesenko and S. V. Vostokov,** Local fields and their extensions: A constructive approach, 1993
120 **Takeyuki Hida and Masuyuki Hitsuda,** Gaussian processes, 1993
119 **M. V. Karasev and V. P. Maslov,** Nonlinear Poisson brackets. Geometry and quantization, 1993
118 **Kenkichi Iwasawa,** Algebraic functions, 1993
117 **Boris Zilber,** Uncountably categorical theories, 1993
116 **G. M. Fel′dman,** Arithmetic of probability distributions, and characterization problems on abelian groups, 1993
115 **Nikolai V. Ivanov,** Subgroups of Teichmüller modular groups, 1992
114 **Seizô Itô,** Diffusion equations, 1992
113 **Michail Zhitomirskiĭ,** Typical singularities of differential 1-forms and Pfaffian equations, 1992
112 **S. A. Lomov,** Introduction to the general theory of singular perturbations, 1992
111 **Simon Gindikin,** Tube domains and the Cauchy problem, 1992
110 **B. V. Shabat,** Introduction to complex analysis Part II. Functions of several variables, 1992
109 **Isao Miyadera,** Nonlinear semigroups, 1992
108 **Takeo Yokonuma,** Tensor spaces and exterior algebra, 1992
107 **B. M. Makarov, M. G. Goluzina, A. A. Lodkin, and A. N. Podkorytov,** Selected problems in real analysis, 1992
106 **G.-C. Wen,** Conformal mappings and boundary value problems, 1992
105 **D. R. Yafaev,** Mathematical scattering theory: General theory, 1992
104 **R. L. Dobrushin, R. Kotecký, and S. Shlosman,** Wulff construction: A global shape from local interaction, 1992
103 **A. K. Tsikh,** Multidimensional residues and their applications, 1992
102 **A. M. Il′in,** Matching of asymptotic expansions of solutions of boundary value problems, 1992
101 **Zhang Zhi-fen, Ding Tong-ren, Huang Wen-zao, and Dong Zhen-xi,** Qualitative theory of differential equations, 1992
100 **V. L. Popov,** Groups, generators, syzygies, and orbits in invariant theory, 1992
99 **Norio Shimakura,** Partial differential operators of elliptic type, 1992
98 **V. A. Vassiliev,** Complements of discriminants of smooth maps: Topology and applications, 1992 (revised edition, 1994)
97 **Itiro Tamura,** Topology of foliations: An introduction, 1992
96 **A. I. Markushevich,** Introduction to the classical theory of Abelian functions, 1992
95 **Guangchang Dong,** Nonlinear partial differential equations of second order, 1991
94 **Yu. S. Il′yashenko,** Finiteness theorems for limit cycles, 1991
93 **A. T. Fomenko and A. A. Tuzhilin,** Elements of the geometry and topology of minimal surfaces in three-dimensional space, 1991
92 **E. M. Nikishin and V. N. Sorokin,** Rational approximations and orthogonality, 1991
91 **Mamoru Mimura and Hirosi Toda,** Topology of Lie groups, I and II, 1991